Y0-AAA-386

Lecture Notes in Statistics **156**

Springer
New York
Berlin
Heidelberg
Barcelona
Hong Kong
London
Milan
Paris
Singapore
Tokyo

Gordon E. Willmot
X. Sheldon Lin

Lundberg Approximations for Compound Distributions with Insurance Applications

Springer

SEP/AE
MATH

Gordon E. Willmot
Department of Statistics and Actuarial Science
University of Waterloo
Waterloo, Ontario N2L 3G1
Canada
gewillmot@setosa.uwaterloo.ca

X. Sheldon Lin
Department of Statistics and Actuarial Science
University of Iowa
Iowa City, IA 52242-1409
USA
shlin@stat.uiowa.edu

Library of Congress Cataloging-in-Publication Data

Willmot, Gordon E., 1957–
Lundberg approximations for compound distributions with insurance applications / Gordon E. Willmot, X. Sheldon Lin.
p. cm.—(Lecture notes in statistics; 156)
Includes bibliographical references and indexes.
ISBN 0-387-95135 (softcover: alk. paper)
1. Insurance—Statistical methods. 2. Distribution (Probability theory) I. Lin, X. Sheldon. II. Title. III. Lecture notes in statistics (Springer-Verlag); v. 156.

HG8781 .W55 2000
368'.01—dc21 00-061264

Printed on acid-free paper.

Camera-ready copy provided by the authors.
Printed and bound by Sheridan Books, Ann Arbor, MI.
Printed in the United States of America.

9 8 7 6 5 4 3 2 1

ISBN 0-387-95135-0 SPIN 10780042

Springer-Verlag New York Berlin Heidelberg
A member of BertelsmannSpringer Science+Business Media GmbH

Preface

These notes represent our summary of much of the recent research that has been done in recent years on approximations and bounds that have been developed for compound distributions and related quantities which are of interest in insurance and other areas of application in applied probability. The basic technique employed in the derivation of many bounds is inductive, an approach that is motivated by arguments used by Sparre-Andersen (1957) in connection with a renewal risk model in insurance. This technique is both simple and powerful, and yields quite general results. The bounds themselves are motivated by the classical Lundberg exponential bounds which apply to ruin probabilities, and the connection to compound distributions is through the interpretation of the ruin probability as the tail probability of a compound geometric distribution. The initial exponential bounds were given in Willmot and Lin (1994), followed by the nonexponential generalization in Willmot (1994).

Other related work on approximations for compound distributions and applications to various problems in insurance in particular and applied probability in general is also discussed in subsequent chapters. The results obtained or the arguments employed in these situations are similar to those for the compound distributions, and thus we felt it useful to include them in the notes. In many cases we have included exact results, since these are useful in conjunction with the bounds and approximations developed. These exact results, which normally involve mixtures of Erlang distributions or combinations of exponentials, provide a means of comparison of the approximate results with the exact results.

The results depend quite heavily on monotonicity ideas which are summarized in notions from reliability theory. As such, chapter 2 is devoted to a discussion of those ideas which are relevant in the main part of the notes. Chapter 3 considers properties of mixed Poisson distributions, an important modeling component of many insurance models. The Lundberg bounds are the subject matter of chapters 4 through 6, and the important special cases involving compound geometric and negative binomial distributions are considered in chapter 7. An approximation due to Tijms (1986) in a ruin or queueing theoretic context is discussed and generalized in chapter 8, where a close connection to the Lundberg bounds of the earlier chapters is established. Many quantities of interest in insurance and other applied probability models are known to satisfy defective renewal equations. The close connection to compound geometric distributions allows for the application of the results from earlier chapters to these situations. Defective renewal equations are thus discussed in chapter 9, including an important general equation in insurance modeling due to Gerber and Shiu (1998). The severity of ruin is considered in chapter 10, where a mixture representation is given which allows for the derivation of various useful results. Finally, the renewal risk model of Sparre-Andersen (1957) is discussed in chapter 11.

Three other aspects of the notes deserve mention as well. First, many new results which have not been published elsewhere appear here. Second, the treatment of the topics included is not meant to be comprehensive. That is, we have concentrated on particular aspects of the models, and more complete treatments of the subject may be found in the cited references. Third, an alternative to the use of mathematical induction in the derivation of the Lundberg bounds is the use of Wald-type martingale arguments. We believe that the induction arguments are useful in their own right due to their power and simplicity, and provide much insight into the types of results which may be obtained. On the other hand, martingale arguments often provide insight in complex situations.

The notes have undergone much revision over the past two years, and have been used in graduate courses at the University of Waterloo and at the University of Western Ontario. We wish to thank many individuals at these institutions for their valuable comments and input. In particular, Jun Cai, Steve Drekic, Bruce Jones, David Stanford, Ken Seng Tan, and Cary Tsai deserve special mention. We also wish to thank Jan Grandell and two anonymous reviewers for their valuable suggestions. Special thanks goes to Lynda Clarke for her expert typing of the manuscript, and to John Kimmel of Springer-Verlag for his enthusiasm and support of the project. We wish to acknowledge the financial support of the Committee on Knowledge Extension Research of the Society of Actuaries as well, and Curtis Huntington in particular for his assistance.

Finally, we wish to thank our wives Deborah (for GW) and Feng (for XL), and our children Rachel, Lauren, and Kristen (for GW), and Jason (for XL), for their patience and tolerance of this project.

Contents

1 Introduction **1**

2 Reliability background **7**
2.1 The failure rate 8
2.2 Equilibrium distributions 14
2.3 The residual lifetime distribution and its mean 18
2.4 Other classes of distributions 24
2.5 Discrete reliability classes 28
2.6 Bounds on ratios of discrete tail probabilities 34

3 Mixed Poisson distributions **37**
3.1 Tails of mixed Poisson distributions 38
3.2 The radius of convergence 40
3.3 Bounds on ratios of tail probabilities 42
3.4 Asymptotic tail behaviour of mixed Poisson distributions . 46

4 Compound distributions **51**
4.1 Introduction and examples 52
4.2 The general upper bound 65
4.3 The general lower bound 73
4.4 A Wald-type martingale approach 78

5 Bounds based on reliability classifications **81**
5.1 First order properties 81

5.2 Bounds based on equilibrium properties 87

6 Parametric Bounds **93**
6.1 Exponential bounds 94
6.2 Pareto bounds 97
6.3 Product based bounds 100

7 Compound geometric and related distributions **107**
7.1 Compound modified geometric distributions 108
7.2 Discrete compound geometric distributions 114
7.3 Application to ruin probabilities 129
7.4 Compound negative binomial distributions 132

8 Tijms approximations **141**
8.1 The asymptotic geometric case 141
8.2 The modified geometric distribution 147
8.3 Transform derivation of the approximation 148

9 Defective renewal equations **151**
9.1 Some properties of defective renewal equations 152
9.2 The time of ruin and related quantities 159
9.3 Convolutions involving compound geometric distributions . 174

10 The severity of ruin **183**
10.1 The associated defective renewal equation 183
10.2 A mixture representation for the conditional distribution . . 186
10.3 Erlang mixtures with the same scale parameter 192
10.4 General Erlang mixtures 198
10.5 Further results 205

11 Renewal risk processes **209**
11.1 General properties of the model 210
11.2 The Coxian-2 case 216
11.3 The sum of two exponentials 224
11.4 Delayed and equilibrium renewal risk processes 226

Bibliography **235**

Symbol Index **243**

Author Index **245**

Subject Index **248**

1
Introduction

One of the standard stochastic models used in various areas of applied probability such as insurance risk theory and queueing theory is the random sum model. Random sums are defined here as a sum of independent and identically distributed (iid) random variables, the number of which is also random. The literature on such models is voluminous, both from an analytic and a numerical viewpoint. The difficulty in evaluation arises from the presence of convolutions, but such evaluation is important for many applications.

In this work we focus on the right tail of the distribution of the random sum (referred to as a compound distribution). The complement of the distribution function (df), referred to as the tail, has been the subject of much study. In particular, many asymptotic results are available. These range from the light-tailed Lundberg type asymptotics of Embrechts, Maejima, and Teugels (1985) to the heavy-tailed subexponential asymptotics of Embrechts, Goldie, and Veraverbeke (1979) and include an intermediate class discussed by Embrechts and Goldie (1982). In general, it appears that the light-tailed Lundberg asymptotics appear to be more suited for practical numerical use from the viewpoint of accuracy. We will discuss the light-tailed Lundberg results in what follows.

Related to the Lundberg asymptotics are Lundberg type bounds on the tail of the compound distribution. These are similar to, and both generalize and refine the well known Cramer-Lundberg inequality of insurance ruin theory (see Gerber, 1979). In many cases the parameter involved in the bound is the same as that in the Lundberg asymptotic formula, thus providing a link between bounds and asymptotics. The classical bound involves an exponential term, and this is both simple and convenient. In many situations of interest, however, the joint distribution of the summand terms is long-tailed, and no such exponential bound is available. In these cases it is convenient to replace the exponential term by the tail of a distribution which has a somewhat thicker tail than the exponential, such as a Pareto-type tail.

The results obtained turn out to be very much dependent on reliability based properties of the distributions. Of central importance in this regard is the failure rate (also called the hazard rate or the force of mortality) if the distribution is absolutely continuous, and this is the subject of section 2.1, along with the related concepts of logconvexity and logconcavity. Monotonicity of the failure rate is considered in some detail. Equilibrium distributions, which play a crucial role in many areas of applied probability, including the subject matter of these notes, are introduced in section 2.2. Other concepts which are also very important in what follows are the residual lifetime distribution and (even more importantly) its mean, discussed in section 2.3. The close connection between the failure rate and the mean residual lifetime is highlighted, as is monotonicity of the latter quantity. Various reliability classes of distributions are presented in section 2.3. While there are many such classes, only those which are relevant in the present application are mentioned. Discrete reliability classes also impact the results greatly, but in a different way than their continuous counterparts. Moreover, there are some basic differences between the discrete and continuous cases, which are highlighted in section 2.5. These discrete classes are of central importance in the analysis of the ratios of discrete tail probabilities (whose importance will become apparent in chapter 4), and this is the subject matter of section 2.6.

For the purposes of insurance models, mixed Poisson distributions are a natural and convenient class of counting distributions, and they are considered in chapter 3. In particular, discrete reliability properties of the mixed Poisson distribution are closely related to those of the mixing distribution, and many of these ideas are discussed. The focus in this context is on the mixed Poisson tail probabilities and the radius of convergence of the associated probability generating function, concepts which directly impact the tail behavior of the associated compound distribution. Asymptotic tail behavior of the mixed Poisson probabilities are also briefly discussed, where it is demonstrated that the tail probabilities of the mixed Poisson distribution are essentially determined by the corresponding tail probabilities of the mixing distribution in many situations.

Compound distributions are introduced in chapter 4, where basic properties of the compound tail are considered. In general, closed form expressions for the tail are not available, but a few examples are provided for which analytic results are obtainable. A particularly useful model for the incidence of claims in the context of claims inflation as well as 'incurred but not reported claims' is the mixed Poisson process, which is also briefly discussed. Lundberg asymptotics for compound tails are also presented. Also in this chapter the main upper and lower bounds on the compound tail are introduced. The derivation of these results involves mathematical induction, which is seen to be a simple yet powerful approach for this type of problem. An alternative approach which may also be used involves optional stopping with a Wald-type martingale (e.g. Lin, 1996, and Cai and Garrido, 1999). This approach is considered in section 4.4. The bounds in this chapter impose little or no restriction on the distribution of the summand terms, and the superiority of these Lundberg bounds over simpler and well known Markov-type inequalities is discussed as well.

Simpler bounds than the general results in chapter 4 may be obtained by imposition of restrictions on the distribution of the summand terms, and this is the subject matter of chapter 5. Reliability classifications are a natural and convenient approach to follow in this situation, together with some ideas from stochastic ordering. First-order properties involving reliability properties of the tail itself are followed by second-order classifications involving properties of the equilibrium tail. In the latter situation, convexity is an integral property and is discussed in some detail in relation to the problem at hand.

The bounds of chapters 4 and 5 are in terms of the tail of an arbitrary distribution (from a reliability class), and from an applications-oriented viewpoint it is useful to provide guidance on the choice of such a distribution. Exponential tails are the simplest as well as the most natural (due in part to the connection with the Lundberg asymptotics) and these are considered in some detail. In cases where the exponential bounds are inappropriate or inapplicable, moment-based Pareto tails are often useful. Finally, in some cases, particularly for the medium-tailed subexponential type models considered by Embrechts and Goldie (1982), products of exponential and other tails may be employed. These parametric choices for the bounding tail are considered in chapter 6.

The remaining chapters deal with applications of these results, as well as related concepts in various insurance and applied probability contexts. Compound geometric distributions are considered in chapter 7, and the associated tails are particularly well suited both for modeling as well as for employment of the bounds discussed earlier. In particular, these models are of central importance in areas of applied probability such as insurance ruin theory and queueing theory. In general, the maximum in a large class of random walks has a compound geometric distribution (e.g. Feller, 1971, or Prabhu, 1998). Important special cases include the ruin probability in the renewal risk model, which may be expressed as a compound geometric tail probability (e,g. Embrechts et al., 1997, Rolski et al., 1999, or Kalashnikov, 1996). Similarly, the equilibrium waiting time distribution in the so-called G/G/1 queue has a compound geometric distribution (e.g. Ross, 1996). Applications of compound geometric distributions in reliability, regenerative processes, and queueing theory may be found in many references including Brown (1990), Gertsbakh (1984), Kalashnikov (1994a, 1994b, 1997a) and Neuts (1986). See also Stoyan (1983), who emphasizes bounds and inequalities, many of which are similar in nature to those discussed here.

Discrete compound geometric distributions are also important from a modeling standpoint, and these are considered in section 7.2. This class may be viewed as a discrete reliability class, as is briefly discussed. Also, discrete analogues of the exponential bounds described earlier are given. Applications to a discrete ruin model and the equilibrium M/G/1 queue length are considered. The classical ruin model of insurance risk is reviewed in section 7.3, and this may also be viewed as the equilibrium waiting-time tail in the M/G/1 queue (e.g. Asmussen, 1987, p. 281). Compound negative binomial distributions are one of the most useful distributions from a modeling viewpoint in many types of insurance, and convolution of bounds leads to useful results in this situation.

Tijms-type approximations are considered in chapter 8. The idea here is to approximate the compound tail by the corresponding Lundberg asymptotic term plus an additive correction term designed to provide a close match for small values of the argument. These approximations are well suited for the situation where the number of terms in the compound distribution has a geometric (or more generally a compound geometric) distribution. The link between the parameter in the Lundberg bound and that in the Lundberg asymptotic is instrumental in analysis of the right-tail behavior of the Tijms approximation. These approximations, which exactly reproduce the true values of the compound tail in some cases, provide a convenient analytic tool for the analysis of more complicated quantities which are both of interest in various insurance and applied probability contexts and have the property that they may be expressed as functions of compound geometric tails. They also provide a convenient tool in situations where the compound geometric Laplace transform is available, but transform inversion is difficult or impossible. This is precisely the situation in connection with many special cases of the equilibruim waiting time distribution in the G/G/1 queue (e.g. De Smit, 1995, or Cohen, 1982).

Defective renewal equations are introduced in chapter 9. Many quantities of interest in insurance applications in particular and applied probability in general are known (or may be shown) to satisfy a defective renewal equation. A close connection to an associated compound geometric distribution is established, and this allows for application to the quantity of interest of the many compound geometric properties such as bounds, asymptotics, and approximations. An important defective renewal equation in connection with the time of and deficit at insurance ruin due to Gerber and Shiu (1998) is analyzed in some detail in section 9.2. Analytic properties of the associated compound geometric distribution are studied, and the interpretation of the solution as the Laplace transform of the time of ruin allows for derivation of the mean time to ruin itself. Convolutions of a compound geometric distribution with another random variable may be viewed as the solution to a defective renewal equation, as is shown in section 9.3. Applications to ruin perturbed by a diffusion and an approximation to the equilibrium waiting time distribution in the M/G/c queue are examples of this convolution formulation, as is discussed. In fact, many variants of the Poisson arrival queue lend themselves to such an analysis, as is discussed in detail by Neuts (1986).

The conditional distribution of the deficit at ruin, given that ruin occurs, is the subject matter of chapter 10. This quantity may be analyzed as a special case of the Gerber-Shiu defective renewal equation of section 9.2, but is treated separately due both to its importance and its many interesting and useful properties. The associated distribution may be viewed as a mixture of 'equilibruim residual lifetime' distributions, an interpretation which allows for the derivation of bounds as well as exact analytic results in the case where the claim size distribution is a mixture of Erlangs or a combination of exponentials. The mixing distribution in the limiting asymptotic case where the initial surplus tends to infinity is simpler, and the limiting distribution itself is closely connected to the general 'claim size' distribution associated with the Gerber-Shiu defective renewal equation of section 9.2.

In chapter 11, the ruin problem is considered in the situation where the number of claims process is a renewal process rather than an ordinary Poisson process. Equivalently, the subject under consideration is also the equilibruim waiting time distribution in the G/G/1 queue (e.g. Ross, 1996, section 7.4). Lundberg upper and lower bounds are derived using inductive arguments which are similar in nature to those used in earlier chapters in connection with compound distributions. Again, these are simple alternatives to Wald-type martingale arguments which have been used previously to derive these results (e.g. Ross, 1974). The compound geometric structure of the classical model is retained by the present renewal risk model, but the identification of the compound geometric components is difficult in general. In section 11.2 this identification is done in the case where the interclaim/interarrival time distribution is a Coxian-2 or Phase type-2 distribution. Again, the compound geometric 'claim size' distribution is closely connected to that associated with the Gerber-Shiu defective renewal equation of section 9.2. Further simplification results in the special case where the interclaim/interarrival distribution is the convolution of two exponential distributions, as is discussed in section 11.3. Finally, in section 11.4, the delayed and equilibruim (also referred to as stationary) renewal risk process are considered, where it is shown that many properties of the ordinary renewal risk process are carried over to these more general models.

2

Reliability background

In reliability theory, classes of distribution functions are introduced to study lifetimes of systems, devices or components. These distribution functions are often characterized in terms of failure rates, the conditional distribution of residual lifetimes, etc. Various classes of life distributions are proposed. Examples include the increasing failure rate (IFR) class, decreasing failure rate (DFR) class, new better than used (NBU) class, new worse than used (NWU) class, decreasing mean residual lifetime (DMRL) class, increasing mean residual lifetime (IMRL) class, new better than used in convex ordering (NBUC), and the new worse than used in convex ordering (NWUC) class. Other distributions according to properties of related equilibrium distributions have also been proposed. See Barlow and Proschan (1965, 1975), Block and Savits (1976), Cao and Wang (1991), and Fagiuoli and Pellerey (1993, 1994) for more details. Applications of classification of life distributions have been found in insurance and actuarial science. In particular, these distributions have been used to classify claim size distributions as well as the number of claims distribution in insurance portfolio management. See Grandell (1997), Gerber (1979), Kalashnikov (1999), Lin (1996), Willmot (1994), and Lin and Willmot (1999, 2000).

In this chapter, we recall concepts used in classifying life distributions in reliability theory. We begin with failure rates and two classes of distributions: IFR and DFR distributions. Equilibrium distributions are then introduced. One of the most important quantities related to the equilibrium distribution of a given distribution is its mean residual lifetime (MRL) which is discussed in section 2.3. Other classes of distributions related to properties of equilibrium distributions are introduced in section 2.4. The relations between these distributional classes are given. The concepts and classification of life distributions are extended to counting random variables in section 2.5. Of central interest in this regard are ratios of the tail probabilities for a given counting distribution and its connection with the radius of convergence of its probability generating function. Logconvexity and logconcavity are discussed in detail. We also present a class of counting distributions called the (a, b) class since it has been widely used in insurance applications.

2.1 The failure rate

Consider a positive random variable Y with distribution function (df) $F(y) = \Pr(Y \leq y), y \geq 0$. The random variable Y may represent the time-until-death of an individual, or in the present context as the amount of an insurance loss. It is of importance to quantify and analyze the thickness of the right tail for valuation purposes. In order to do so we use some notions from the theory of reliability. Suppose that the df $F(y)$ is absolutely continuous, i.e. has probability density function (pdf) $f(y) = F'(y)$. Then the failure rate, hazard rate, or force of mortality of Y are defined by

$$\mu(y) = \frac{f(y)}{\overline{F}(y)} = -\frac{d}{dy} \ln \overline{F}(y), \quad y \geq 0, \tag{2.1.1}$$

where $\overline{F}(y) = 1 - F(y)$ denotes the survival function (sf) of Y. We remark that hereafter we will adopt the following notational convention: for a given distribution function, say $F(y)$, the associated survival function is denoted as $\overline{F}(y)$.

Since

$$\begin{aligned}
\mu(y) &= \lim_{h \to 0} \frac{F(y+h) - F(y)}{h\overline{F}(y)} \\
&= \lim_{h \to 0} \frac{1}{h} \left\{ 1 - \frac{\overline{F}(y+h)}{\overline{F}(y)} \right\} \\
&= \lim_{h \to 0} \frac{1}{h} \left\{ 1 - Pr\left(Y > y + h | Y > y\right) \right\}, \quad y \geq 0. \qquad (2.1.2)
\end{aligned}$$

Thus, small values of $\mu(y)$ are an indication of a thick right tail whereas large values indicate a thinner right tail.

From (2.1.1),

$$\int_0^y \mu(x) dx = -\ln \overline{F}(y),$$

in other words

$$\overline{F}(y) = e^{-\int_0^y \mu(x) dx}, \quad y \geq 0, \tag{2.1.3}$$

and so $\mu(y)$ uniquely determines the distribution of Y.

Suppose now that the mean $E(Y)$ of Y exists, i.e. $E(Y) = \int_0^\infty y dF(y) < \infty$. Then integration by parts yields

$$\begin{aligned}\int_0^\infty y dF(y) &= -y\overline{F}(y)\Big|_0^\infty + \int_0^\infty \overline{F}(y)dy \\ &= -\lim_{y\to\infty} y\overline{F}(y) + \int_0^\infty \overline{F}(y)dy.\end{aligned}$$

But,

$$0 \le y\overline{F}(y) = y\int_y^\infty dF(x) \le \int_y^\infty x dF(x),$$

and since $E(Y) < \infty$, it follows that

$$0 \le \lim_{y\to\infty} y\overline{F}(y) \le \lim_{y\to\infty} \int_y^\infty x dF(x) = 0,$$

i.e. $\lim_{y\to\infty} y\overline{F}(y) = 0$. Thus,

$$E(Y) = \int_0^\infty \overline{F}(y)dy. \tag{2.1.4}$$

From (2.1.1) and (2.1.4), therefore, we have

$$E(Y) = \int_0^\infty \overline{F}(y)dy = \int_0^\infty \frac{f(y)}{\mu(y)}dy,$$

that is,

$$E(Y) = E\left\{1/\mu(Y)\right\}. \tag{2.1.5}$$

Equation (2.1.5) is in agreement with our intuition that small values of $\mu(y)$ are associated with large values of Y. Furthermore, (2.1.5) generalizes the well known exponential case when $\mu(y)$ is constant.

In many situations of practical interest, the failure rate $\mu(y)$ is strictly monotone nonincreasing (nondecreasing) in y, and this is associated with the situation where the distribution has a thick (thin) right tail.

The df $F(y)$ is said to be decreasing (increasing) failure rate or DFR (IFR) if $\overline{F}(x+y)/\overline{F}(y)$ is nondecreasing (nonincreasing) in y for fixed $x \ge 0$, i.e. if $\overline{F}(y)$ is log-convex (log-concave). It is evident from (2.1.2) that if $F(y)$ is absolutely continuous, then DFR (IFR) is equivalent to $\mu(y)$ nonincreasing (nondecreasing) in y.

The exponential df $F(y) = 1 - e^{-\mu y}$, $y \ge 0$ is both DFR and IFR since $\mu(y) = \mu$ for all y. In other situations, it may be difficult to analyze the behavior of $\mu(y)$, often because $\overline{F}(y)$ or even $f(y)$ is of a complicated form. Various approaches often yield useful results, however, as in the following example.

Example 2.1.1 Mixture of exponentials Suppose that $F(y)$ is the df of a mixture of exponentials, i.e., it may be expressed as

$$F(y) = \int_0^\infty \left(1 - e^{-\theta y}\right) dH(\theta), \quad y \geq 0, \tag{2.1.6}$$

where $H(\theta)$ is itself the df of a positive random variable. The failure rate is given by

$$\mu(y) = -\frac{d}{dy} \ln \overline{F}(y) = \frac{\int_0^\infty \theta e^{-\theta y} dH(\theta)}{\int_0^\infty e^{-\theta y} dH(\theta)}. \tag{2.1.7}$$

In order to analyze $\mu(y)$, consider the df $H_y(\theta) = Pr(\Theta_y \leq \theta)$ satisfying

$$dH_y(\theta) = \frac{e^{-\theta y} dH(\theta)}{\int_0^\infty e^{-\theta y} dH(\theta)}.$$

Clearly,

$$E(\Theta_y^k) = \frac{\int_0^\infty \theta^k e^{-\theta y} dH(\theta)}{\int_0^\infty e^{-\theta y} dH(\theta)}.$$

Differentiation of (2.1.7) yields

$$\begin{aligned} \mu'(y) &= -\frac{\int_0^\infty \theta^2 e^{-\theta y} dH(\theta)}{\int_0^\infty e^{-\theta y} dH(\theta)} + \left\{ \frac{\int_0^\infty \theta e^{-\theta y} dH(\theta)}{\int_0^\infty e^{-\theta y} dH(\theta)} \right\}^2 \\ &= -E(\Theta_y^2) + \{E(\Theta_y)\}^2 = -Var(\Theta_y). \end{aligned}$$

Thus, $\mu'(y) \leq 0$ and so the mixed exponential df (2.1.6) is DFR. More and stronger properties of mixed exponential df's may be found in Steutel (1970) and Feller (1971). □

The result in the above example is a special case of a much more general result. Recall that the exponential df $F(y|\theta) = 1 - e^{-\theta y}$ is DFR. It can be shown using the Cauchy-Schwarz inequality that if

$$F(y) = \int_{-\infty}^\infty F(y|\theta) \, dH(\theta) \tag{2.1.8}$$

where $F(y|\theta)$ is a DFR df and $H(\theta)$ is a df, then $F(y)$ is DFR. That is, mixing preserves the DFR property, and so mixing may be regarded as a process which generates thick tailed distributions. Conversely, if X and Y are independent with IFR df's, then $X + Y$ has an IFR df, and so convolution preserves the IFR property and may thus be viewed as a process which generates thinner tails. See Barlow and Proschan (1975), for example.

The following approach is also useful in analyzing $\mu(y)$. It follows from (2.1.1) that

$$\frac{1}{\mu(y)} = \frac{\overline{F}(y)}{f(y)} = \int_y^\infty \frac{f(x)}{f(y)}dx = \int_0^\infty \frac{f(x+y)}{f(y)}dx, \tag{2.1.9}$$

and the behavior of $\mu(y)$ may often be ascertained through examination of the integrand $f(x+y)/f(y)$. In particular, the pdf $f(y)$ is log-convex (log-concave) if $f(x+y)/f(y)$ is a nondecreasing (nonincreasing) function of y for fixed $x \geq 0$. From (2.1.9), it is clear that if $f(y)$ is log-convex (log-concave), then $\mu(y)$ is nonincreasing (nondecreasing) in y, i.e., $F(y)$ is DFR (IFR), and from the above discussion, this is associated with a thick (thin) right tail.

The limiting behavior of $\mu(y)$ is somewhat simpler. By L'Hopital's rule and (2.1.1)

$$\mu(\infty) = \lim_{y\to\infty} \mu(y) = \lim_{y\to\infty} \frac{f(y)}{\overline{F}(y)} = -\lim_{y\to\infty} \frac{f'(y)}{f(y)}, \tag{2.1.10}$$

as long as the limits are well defined.

In general, DFR and IFR df's are absolutely continuous, except possibly for a mass point at the origin in the DFR case and at the right end of the interval of support in the IFR case. See Barlow and Proschan (1965, pp. 25-26), for details. If $\mu(y) \leq (\geq)\ \mu$ where $0 < \mu < \infty$, then $\int_0^y \mu(x)dx \leq (\geq) \int_0^y \mu dx = \mu y$, and from (2.1.3)

$$\overline{F}(y) \geq (\leq)\ e^{-\mu y}, \quad y \geq 0, \tag{2.1.11}$$

and so Y is stochastically larger (smaller) than an exponential random variable with mean $1/\mu$.

We now present two examples illustrating these ideas.

Example 2.1.2 The gamma distribution

Suppose that

$$f(y) = \frac{\beta\,(\beta y)^{\alpha-1}\, e^{-\beta y}}{\Gamma(\alpha)}, \quad y > 0, \tag{2.1.12}$$

where $\alpha > 0$ and $\beta > 0$ are parameters. In general, $F(y)$ is not easily expressed in simple analytic form, but when α is a positive integer the Erlang distribution results, and one obtains by repeated integration by parts that for $y \geq 0$

$$F(y) = 1 - e^{-\beta y} \sum_{j=0}^{\alpha-1} (\beta y)^j / j!; \quad \alpha = 1, 2, 3, \cdots. \tag{2.1.13}$$

Thus $\mu(y)$ is also not easily expressed in simple analytic form unless $\alpha = 1, 2, 3, \cdots$, but

$$f'(y) = \frac{\beta^\alpha}{\Gamma(\alpha)} \left\{(\alpha-1)y^{\alpha-2} - \beta y^{\alpha-1}\right\} e^{-\beta y}.$$

Hence, from (2.1.10),

$$\mu(\infty) = \lim_{y\to\infty} \frac{\beta y^{\alpha-1} - (\alpha-1)y^{\alpha-2}}{y^{\alpha-1}} = \beta - \lim_{y\to\infty} \frac{\alpha-1}{y} = \beta.$$

Also,

$$\frac{f(x+y)}{f(y)} = \left(\frac{x+y}{y}\right)^{\alpha-1} e^{-\beta x}$$

which is nonincreasing (nondecreasing) in y for fixed $x \geq 0$ if $\alpha \geq (\leq)\ 1$. Thus, $f(y)$ is log-convex (log-concave) if $\alpha \leq (\geq)\ 1$, implying that $F(y)$ is DFR (IFR) if $\alpha \leq (\geq)\ 1$. When $\alpha = 1$ the exponential distribution results which is both DFR and IFR, and $\mu(y) = \beta$ in this case. When $\alpha < 1 (> 1)$, $\mu(y)$ is decreasing (increasing) to its limit $\mu(\infty) = \beta$. □

Example 2.1.3 Mixture of Erlangs
Suppose that

$$f(y) = \sum_{k=1}^{r} q_k \frac{\beta(\beta y)^{k-1} e^{-\beta y}}{(k-1)!}, \quad y > 0, \tag{2.1.14}$$

where $\{q_1, q_2, \cdots, q_r\}$ is a probability distribution. Then, using (2.1.13),

$$\begin{aligned}\overline{F}(y) &= \sum_{k=1}^{r} q_k \int_y^\infty \frac{\beta\,(\beta x)^{k-1}\, e^{-\beta x}}{(k-1)!}\, dx \\ &= e^{-\beta y} \sum_{k=1}^{r} q_k \sum_{j=0}^{k-1} \frac{(\beta y)^j}{j!}.\end{aligned}$$

Interchanging the order of summation yields

$$\overline{F}(y) = e^{-\beta y} \sum_{j=0}^{r-1} \frac{(\beta y)^j}{j!} \sum_{k=j+1}^{r} q_k, \quad y \geq 0. \tag{2.1.15}$$

or

$$\overline{F}(y) = e^{-\beta y} \sum_{j=0}^{r-1} \overline{Q}_j \frac{(\beta y)^j}{j!},$$

where $\overline{Q}_j = \sum_{k=j+1}^{r} q_k$, for $j \leq r-1$, and $\overline{Q}_j = 0$ for $j \geq r$. The failure rate is then

$$\mu(y) = \frac{f(y)}{\overline{F}(y)} = \frac{\beta \sum_{j=0}^{r-1} q_{j+1} \frac{(\beta y)^j}{j!}}{\sum_{j=0}^{r-1} \overline{Q}_j \frac{(\beta y)^j}{j!}}.$$

It is clear from this representation that $\mu(0) = \beta q_1$ and $\mu(\infty) = \beta$ since $\overline{Q}_{r-1} = q_r$. This implies that $F(y)$ is DFR only if $q_1 = 1$. In this case $F(y)$ is exponential and thus DFR.

Since $q_{j+1} = \overline{Q}_j - \overline{Q}_{j+1}$, one has

$$\mu(y) = \beta \left\{ 1 - \frac{\sum_{j=0}^{r-2} \overline{Q}_{j+1} \frac{(\beta y)^j}{j!}}{\sum_{j=0}^{r-1} \overline{Q}_j \frac{(\beta y)^j}{j!}} \right\},$$

from which it is clear that $\mu(y) \le \beta = \mu(\infty)$ for all y.

Conditions under which $F(y)$ is IFR are discussed by Esary, Marshall, and Proschan (1973). In particular, a sufficient condition is that $\overline{Q}_{k+1}\overline{Q}_{k+j} \ge \overline{Q}_k\overline{Q}_{k+j+1}$ for $k = 0, 1, 2, \cdots$, and $j = 1, 2, 3, \cdots$. This condition is implied if $\overline{Q}_{k+1}/\overline{Q}_k$ is nonincreasing in k. In particular, with $k = 0$ and $j = 1$ one obtains $\overline{Q}_1^2 \ge \overline{Q}_0\overline{Q}_2$.

Denote

$$\gamma(y) = \frac{\sum_{j=0}^{r-2} \overline{Q}_{j+1} \frac{(\beta y)^j}{j!}}{\sum_{j=0}^{r-1} \overline{Q}_j \frac{(\beta y)^j}{j!}}. \tag{2.1.16}$$

Then $F(y)$ is IFR if and only if for all y, $\gamma'(y) \le 0$. From (2.1.16),

$$\gamma'(y) \left\{ \sum_{j=0}^{r-1} \overline{Q}_j \frac{(\beta y)^j}{j!} \right\}^2$$
$$= \beta \left\{ \left[\sum_{j=0}^{r-3} \overline{Q}_{j+2} \frac{(\beta y)^j}{j!} \right] \left[\sum_{j=0}^{r-1} \overline{Q}_j \frac{(\beta y)^j}{j!} \right] - \left[\sum_{j=0}^{r-2} \overline{Q}_{j+1} \frac{(\beta y)^j}{j!} \right]^2 \right\}. \tag{2.1.17}$$

In particular,

$$\gamma'(0) = \beta[\overline{Q}_2\overline{Q}_0 - \overline{Q}_1^2] = \beta[q_1(1 - q_1) - q_2].$$

Thus, $\overline{Q}_1^2 \ge \overline{Q}_0\overline{Q}_2$, or equivalently $q_2 \ge q_1(1 - q_1)$, is also a necessary condition for $F(y)$ to be IFR.

We now consider some special cases. Let $E_{r_1, r_2, \cdots}$ denote a mixture of Erlangs such that $q_j = 0$ for all j except $q_{r_1}, q_{r_2}, \cdots$. From example 2.1.2, E_r is IFR. Then, $E_{1,r}$ is neither IFR nor DFR for $r \ge 3$ since $q_2 = 0$ and it is not exponential. If $r = 2$, it follows from (2.1.16) that $\gamma(y)$ is always decreasing. Hence, $E_{1,2}$ is IFR.

Let us consider $E_{1,2,3}$. With a little algebra, the right-hand side of (2.1.17) may be expressed as

$$\beta\left[\overline{Q}_2\overline{Q}_0 - \overline{Q}_1^2 - \overline{Q}_2\overline{Q}_1(\beta y) - \overline{Q}_2^2\frac{(\beta y)^2}{2}\right].$$

Thus, a mixture of Erlangs with $r = 3$ is IFR if and only if $\overline{Q}_1^2 \geq \overline{Q}_0\overline{Q}_2$, or equivalently, $q_2 \geq q_1(1-q_1)$.

For more discussion on $E_{1,r}$ and $E_{r-1,r}$ distributions and their applications, see Tijms(1994, p.358-9). □

Mixtures of Erlangs are an important distributional class since any absolutely continuous distribution on $(0,\infty)$ may be approximated arbitrarily accurately by a distribution of this type. Tijms (1994, p.162-164) demonstrates that for a given df $F(x)$, $F(0) = 0$ and arbitrary $\Delta > 0$, the df $F_\Delta(x)$ defined such that

$$F'_\Delta(x) = \sum_{k=1}^{\infty} p_k(\Delta)\frac{(1/\Delta)^k x^{k-1} e^{-x/\Delta}}{(k-1)!},$$

where $p_k(\Delta) = F(k\Delta) - F((k-1)\Delta)$, satisfies $\lim_{\Delta\to 0} F_\Delta(x) = F(x)$ for any continuity point of $F(x)$. Tijms (1994) suggests that for computational purposes, it is convenient to use an Erlang mixture with similar moments to a given distribution. In what follows we shall see that Erlang mixtures are very tractable in a wide variety of situations.

2.2 Equilibrium distributions

Equation (2.1.4) may be divided by $E(Y)$ to give $\int_0^\infty \{\overline{F}(y)/E(Y)\}\,dy = 1$, which implies that $f_1(y) = \overline{F}(y)/E(Y)$ is a pdf (even if $F(y)$ is not absolutely continuous). The corresponding df

$$F_1(y) = 1 - \overline{F}_1(y) = \int_0^y \{\overline{F}(x)/E(Y)\}\,dx, \quad y \geq 0, \tag{2.2.1}$$

is called the equilibrium df of $F(y)$. The n-th moment is, by integration by parts,

$$\begin{aligned}\int_0^\infty y^n\frac{\overline{F}(y)}{E(Y)}dy &= \left.\frac{y^{n+1}\overline{F}(y)}{(n+1)E(Y)}\right|_0^\infty + \int_0^\infty \frac{y^{n+1}dF(y)}{(n+1)E(Y)} \\ &= \lim_{y\to\infty}\frac{y^{n+1}\overline{F}(y)}{(n+1)E(Y)} + \int_0^\infty \frac{y^{n+1}dF(y)}{(n+1)E(Y)}.\end{aligned}$$

Now,

$$0 \leq y^{n+1}\overline{F}(y) = y^{n+1}\int_y^{\infty} dF(x) \leq \int_y^{\infty} x^{n+1}dF(x).$$

Thus, if $E\left(Y^{n+1}\right) = \int_0^{\infty} x^{n+1}dF(x) < \infty$,

$$0 \leq \lim_{y\to\infty} y^{n+1}\overline{F}(y) \leq \lim_{y\to\infty}\int_y^{\infty} x^{n+1}dF(x) = 0,$$

implying that $\lim_{y\to\infty} y^{n+1}\overline{F}(y) = 0$, and so for $n \geq 0$

$$\int_0^{\infty} y^n dF_1(y) = \frac{E\left(Y^{n+1}\right)}{(n+1)E(Y)}. \tag{2.2.2}$$

For $n = 1$, we have the equilibrium mean

$$\int_0^{\infty} y dF_1(y) = \frac{E\left(Y^2\right)}{2E(Y)}. \tag{2.2.3}$$

There is a useful identity involving $F(y)$ and $F_1(y)$. Integration by parts yields, for $y \geq 0$,

$$\int_y^{\infty} x dF(x) = -x\overline{F}(x)\Big|_y^{\infty} + \int_y^{\infty} \overline{F}(x)dx.$$

As shown in section 2.1, $E(Y) < \infty$ implies that $\lim_{x\to\infty} x\overline{F}(x) = 0$. Thus,

$$\int_y^{\infty} x dF(x) = y\overline{F}(y) + E(Y)\overline{F}_1(y), \quad y \geq 0. \tag{2.2.4}$$

It is sometimes convenient to solve (2.2.4) for $\overline{F}_1(y)$, yielding

$$\overline{F}_1(y) = \frac{\int_y^{\infty}(x-y)dF(x)}{E(Y)}, \quad y \geq 0. \tag{2.2.5}$$

Example 2.2.1 Mixture of Erlangs

In this example, we derive the equilibrium distribution of a mixture of Erlangs. As in example 2.1.3, suppose that

$$f(y) = \sum_{k=1}^{r} q_k \frac{\beta(\beta y)^{k-1}e^{-\beta y}}{(k-1)!}, \quad y > 0, \tag{2.2.6}$$

where $\{q_1, q_2, \cdots, q_r\}$ is a probability distribution. Then

$$E(Y) = \int_0^{\infty} y f(y)dy$$

$$= \sum_{k=1}^{r} q_k \left(\frac{k}{\beta}\right) \int_0^\infty \frac{\beta(\beta y)^k e^{-\beta y}}{k!} \, dy$$

$$= \left(\sum_{k=1}^{r} k q_k\right) / \beta.$$

Hence from (2.1.15)

$$f_1(y) = \frac{\overline{F}(y)}{E(Y)} = \sum_{k=1}^{r} q_k^* \frac{\beta(\beta y)^{k-1} e^{-\beta y}}{(k-1)!}, \quad y > 0, \tag{2.2.7}$$

where

$$q_k^* = \frac{\sum_{j=k}^{r} q_j}{\sum_{j=1}^{r} j q_j}, \quad k = 1, 2, 3, \cdots, r. \tag{2.2.8}$$

Clearly,

$$\sum_{k=1}^{r} q_k^* = \frac{\sum_{k=1}^{r}\sum_{j=k}^{r} q_j}{\sum_{j=1}^{r} j q_j} = \frac{\sum_{j=1}^{r}\sum_{k=1}^{j} q_j}{\sum_{j=1}^{r} j q_j} = \frac{\sum_{j=1}^{r} j q_j}{\sum_{j=1}^{r} j q_j} = 1.$$

Thus, $\{q_1^*, q_2^*, \cdots, q_r^*\}$ is a probability distribution, and comparing (2.2.7) with (2.2.6), the equilibrium distribution is also a mixture of (the same) Erlangs, but with different weights. Similarly,

$$\overline{F}_1(y) = e^{-\beta y} \sum_{j=0}^{r-1} \frac{(\beta y)^j}{j!} \sum_{k=j+1}^{r} q_k^*, \quad y \geq 0, \tag{2.2.9}$$

as is clear from (2.1.15) and the above discussion. □

As a second example, consider the equilibrium distribution of a gamma distribution.

Example 2.2.2 The gamma distribution
Suppose that

$$f(y) = \frac{\beta(\beta y)^{\alpha-1} e^{-\beta y}}{\Gamma(\alpha)}, \quad y > 0. \tag{2.2.10}$$

Then if α is a positive integer, we may use the previous example with $r = \alpha$ and $q_\alpha = 1$ to obtain $q_k^* = 1/\alpha$ for $k = 1, 2, \cdots$, to obtain from (2.2.7)

$$f_1(y) = \frac{\beta e^{-\beta y}}{\alpha} \sum_{k=0}^{\alpha-1} \frac{(\beta y)^k}{k!}, \tag{2.2.11}$$

and from (2.2.9)

$$\overline{F}_1(y) = e^{-\beta y} \sum_{j=0}^{\alpha-1} \left(1 - \frac{j}{\alpha}\right) \frac{(\beta y)^j}{j!}, \quad y \geq 0. \tag{2.2.12}$$

When α is not a positive integer, the situation is more complicated. From (2.2.5) with $E(Y) = \alpha/\beta$,

$$\begin{aligned} \overline{F}_1(y) &= \int_y^\infty \frac{(x-y)\beta(\beta x)^{\alpha-1}e^{-\beta x}}{\alpha\Gamma(\alpha)/\beta} dx \\ &= \frac{\beta^{\alpha+1}}{\Gamma(\alpha+1)} \int_y^\infty (x-y)x^{\alpha-1}e^{-\beta x} dx. \end{aligned}$$

A change of variables from x to $t = -1 + x/y$ results in

$$\overline{F}_1(y) = \frac{(\beta y)^{\alpha+1}e^{-\beta y}}{\Gamma(\alpha+1)} \int_0^\infty t(1+t)^{\alpha-1}e^{-\beta y t} dt.$$

In terms of Kummer's confluent hypergeometric function $U(a,b,x)$, (e.g. Abramowitz and Stegun, 1965, p. 505),

$$\overline{F}_1(y) = \frac{(\beta y)^{\alpha+1}e^{-\beta y}}{\Gamma(\alpha+1)} U(2, \alpha+2, \beta y), \quad y \geq 0, \tag{2.2.13}$$

valid for general $\alpha > 0$. □

Example 2.2.1 is actually a special case of a more general result concerning equilibrium mixtures (Hesselager et al., 1998), since from (2.1.8) with $E(Y|\theta) = \int_0^\infty y dF(y|\theta)$ and $f_1(y|\theta) = \overline{F}(y|\theta)/E(Y|\theta)$, one obtains

$$f_1(y) = \frac{\overline{F}(y)}{E(Y)} = \int_{-\infty}^\infty \frac{f_1(y|\theta)E(Y|\theta)dH(\theta)}{E(Y)}. \tag{2.2.14}$$

Define a new mixing density

$$dH_*(\theta) = \frac{E(Y|\theta)dH(\theta)}{\int_{-\infty}^\infty E(Y|\theta)dH(\theta)}. \tag{2.2.15}$$

Since $E(Y) = \int_{-\infty}^\infty E(Y|\theta)dH(\theta)$, it follows from (2.2.14) that the equilibrium df of (2.1.8) is

$$F_1(y) = \int_{-\infty}^\infty F_1(y|\theta)\, dH_*(\theta), \quad y \geq 0, \tag{2.2.16}$$

where $dH_*(\theta)$ is given by (2.2.15). We now discuss an important application of these mixture results.

Example 2.2.3 Mixture of exponentials
Suppose $F(y|\theta) = 1 - e^{-\theta y}, y \geq 0$. Then $E(Y|\theta) = 1/\theta$, and $F_1(y|\theta) = F(y|\theta) = 1 - e^{-\theta y},\ y > 0$. Then, as in example 2.1.1,

$$F(y) = \int_0^\infty \left(1 - e^{-\theta y}\right) dH(\theta) \tag{2.2.17}$$

and the equilibrium df (2.2.16) becomes

$$F_1(y) = \int_0^\infty \left(1 - e^{-\theta y}\right) dH_*(\theta) \tag{2.2.18}$$

where

$$dH_*(\theta) = \frac{\theta^{-1} dH(\theta)}{\displaystyle\int_0^\infty \theta^{-1} dH(\theta)}. \tag{2.2.19}$$

Thus the equilibrium df is a different mixture of the same exponentials. In the discrete case, if the pdf is

$$f(y) = \sum_{j=1}^r q_j \theta_j e^{-\theta_j y}, \quad y > 0, \tag{2.2.20}$$

then the equilibrium pdf is

$$f_1(y) = \sum_{j=1}^r q_j^* \theta_j e^{-\theta_j y} \tag{2.2.21}$$

where

$$q_j^* = \frac{q_j/\theta_j}{\displaystyle\sum_{k=1}^r q_k/\theta_k}, \quad j = 1, 2, \cdots, r, \tag{2.2.22}$$

as is clear from the above. □

2.3 The residual lifetime distribution and its mean

Consider the residual lifetime random variable $T_y = Y - y|Y > y$ for $Y > y$, and T_y is undefined otherwise. Then $\Pr(T_y > t) = \Pr(Y - y > t|Y > y) = \overline{F}(y+t)/\overline{F}(y)$ for $t \geq 0$, and so T_y has df

$$Pr(T_y \leq t) = 1 - \frac{\overline{F}(y+t)}{\overline{F}(y)}, \quad t \geq 0. \tag{2.3.1}$$

The expected value of T_y, termed the mean residual lifetime (MRL), is given by

$$r(y) = E(T_y) = \frac{\int_y^\infty (t-y)dF(t)}{\overline{F}(y)}, \quad y \geq 0. \tag{2.3.2}$$

Either from integration by parts or from (2.1.4),

$$r(y) = \int_0^\infty \Pr(T_y > t)\, dt = \int_0^\infty \frac{\overline{F}(y+t)}{\overline{F}(y)} dt. \tag{2.3.3}$$

Equations (2.2.5) and (2.3.2) yield

$$r(y) = \frac{\int_y^\infty \overline{F}(x)dx}{\overline{F}(y)} = \frac{E(Y)\overline{F}_1(y)}{\overline{F}(y)} \tag{2.3.4}$$

using (2.2.1). Obviously, $r(0) = E(Y)$. The mean residual lifetime is closely related to the failure rate $\mu(y)$ where the latter exists, but does not require absolute continuity for its existence. It is very useful for analysis of tail thickness, and large values of $r(y)$ are associated with a thick tail. It is also very closely connected with the equilibrium distribution $F_1(y)$ discussed in the previous section. We have from (2.2.1) that

$$-\frac{d}{dy}\ln \overline{F}_1(y) = \frac{\overline{F}(y)/E(Y)}{\overline{F}_1(y)} = \frac{1}{r(y)}, \tag{2.3.5}$$

which implies that the reciprocal $1/r(y)$ of the mean residual lifetime $r(y)$ is the failure rate associated with the equilibrium df $F_1(y)$, and from (2.1.3)

$$\overline{F}_1(y) = e^{-\int_0^y \{r(x)\}^{-1}dx}, \quad y \geq 0. \tag{2.3.6}$$

Equation (2.3.6), together with $F_1'(y) = \overline{F}(y)/r(0)$, shows that $F(y)$ is uniquely determined by $r(y)$.

The df $F(y)$ is said to be increasing (decreasing) mean residual lifetime or IMRL (DMRL) if $r(y)$ is nondecreasing (nonincreasing) in y. Recalling that $F(y)$ is DFR (IFR) is equivalent to $\overline{F}(y+t)/\overline{F}(y)$ nondecreasing (nonincreasing) in y for fixed $t \geq 0$, it follows easily from (2.3.3) that DFR (IFR) implies IMRL (DMRL). That is, the DFR (IFR) class is contained in the IMRL (DMRL) class.

We now show that the IMRL class, like the DFR class, is preserved under mixing, a result obtained by Bondesson (1983). Equation (2.2.16) shows that if $F(y)$ is a mixture then its equilibrium distribution is also a mixture. Furthermore, the mixed df $F(y|\theta)$ being IMRL is equivalent to the equilibrium df $F_1(y|\theta)$ being DFR, as follows from (2.3.5). Thus, (2.2.16) expresses $F_1(y)$ as a mixture of DFR df's which is also DFR. Again, from (2.3.5), $F_1(y)$ is DFR is equivalent to $F(y)$ is IMRL. To summarize, if (2.1.8) holds and $F(y|\theta)$ is IMRL for each θ, then $F(y)$ is IMRL. On the other hand, unlike the IFR class, the DMRL class is not preserved under convolution (i.e., Bondesson, 1983).

Example 2.3.1 Mixture of Erlangs
As in examples 2.1.3 and 2.2.1, suppose that

$$f(y) = \sum_{k=1}^{r} q_k \frac{\beta(\beta y)^{k-1} e^{-\beta y}}{(k-1)!}, \quad y > 0, \tag{2.3.7}$$

where $\{q_1, q_2, \cdots, q_r\}$ is a probability distribution. Then from (2.3.1), T_y has density

$$f_{T_y}(t) = \frac{d}{dt} Pr(T_y \leq t) = \frac{f(y+t)}{\overline{F}(y)}, \quad t > 0. \tag{2.3.8}$$

Substitution of (2.1.15) and (2.3.7) into (2.3.8) yields

$$\begin{aligned} f_{T_y}(t) &= \frac{\sum_{k=1}^{r} q_k \frac{\beta^k (y+t)^{k-1} e^{-\beta(y+t)}}{(k-1)!}}{\sum_{m=0}^{r-1} \overline{Q}_m \frac{(\beta y)^m e^{-\beta y}}{m!}} \\ &= e^{-\beta t} \frac{\sum_{k=1}^{r} q_k \frac{\beta^k}{(k-1)!} \sum_{j=0}^{k-1} \binom{k-1}{j} y^{k-1-j} t^j}{\sum_{m=0}^{r-1} \overline{Q}_m \frac{(\beta y)^m}{m!}}, \end{aligned}$$

after a binomial expansion. In the numerator, change the index of summation from j to $i = j + 1$. This results in

$$\begin{aligned} f_{T_y}(t) &= e^{-\beta t} \frac{\sum_{k=1}^{r} q_k \frac{\beta^k}{(k-1)!} \sum_{i=1}^{k} \binom{k-1}{i-1} y^{k-i} t^{i-1}}{\sum_{m=0}^{r-1} \overline{Q}_m \frac{(\beta y)^m}{m!}} \\ &= e^{-\beta t} \frac{\sum_{i=1}^{r} \sum_{k=i}^{r} q_k \frac{\beta^k y^{k-i} t^{i-1}}{(k-i)!(i-1)!}}{\sum_{m=0}^{r-1} \overline{Q}_m \frac{(\beta y)^m}{m!}}, \end{aligned}$$

where the last line follows by interchanging the order of summation in the numerator. That is,

$$f_{T_y}(t) = \sum_{i=1}^{r} q_i(y) \frac{\beta(\beta t)^{i-1} e^{-\beta t}}{(i-1)!} \tag{2.3.9}$$

where

$$q_i(y) = \frac{\sum_{k=i}^{r} q_k \frac{(\beta y)^{k-i}}{(k-i)!}}{\sum_{m=0}^{r-1} \overline{Q}_m \frac{(\beta y)^m}{m!}}, \quad i = 1, 2, \cdots, r. \tag{2.3.10}$$

It follows from (2.3.8) and (2.3.9) that

$$1 = \int_0^\infty f_{T_y}(t)dt = \sum_{i=1}^{r} q_i(y) \int_0^\infty \frac{\beta(\beta t)^{i-1}e^{-\beta t}}{(i-1)!}dt = \sum_{i=1}^{r} q_i(y),$$

and from (2.3.10) that $q_i(y) \geq 0$. Thus, $\{q_1(y), q_2(y), \cdots, q_r(y)\}$ is a probability distribution, and the residual lifetime distribution is a (different) mixture of the same Erlangs. Alternatively, one may sum (2.3.10) directly.

The mean residual lifetime $r(y)$ may be obtained from (2.3.9), or directly as follows. Using examples 2.1.3 and 2.2.1, the failure rate of the equilibrium distribution $F_1(y)$ is

$$\frac{\beta \sum_{j=0}^{r-1} q_{j+1}^* \frac{(\beta y)^j}{j!}}{\sum_{j=0}^{r-1} \overline{Q}_j^* \frac{(\beta y)^j}{j!}}$$

where $\overline{Q}_j^* = \sum_{k=j+1}^{r} q_k^*$. But from (2.3.5), this is the reciprocal of $r(y)$, i.e.

$$r(y) = \frac{\sum_{j=0}^{r-1} \overline{Q}_j^* \frac{(\beta y)^j}{j!}}{\beta \sum_{j=0}^{r-1} q_{j+1}^* \frac{(\beta y)^j}{j!}}.$$

In the notation of example 2.1.3, consider $E_{1,3}$. That is, $r = 3$, $q_2 = 0$, and $q_3 = 1 - q_1$. As mentioned, $E_{1,r}$ is not IFR. From example 2.2.1, the equilibrium distribution of $E_{1,3}$ is an Erlang mixture with $r = 3$, and weights $q_1^* = 1/(3-2q_1)$, $q_2^* = q_3^* = (1-q_1)/(3-2q_1)$. Thus, from example 2.1.3, the equilibrium distribution is IFR if and only if $q_2^* \geq q_1^*(1-q_1^*)$ which is a restatement of $q_1 \leq 1/2$. But from (2.3.5), the equilibrium distribution is IFR if and only if $E_{1,3}$ is DMRL. To summarize, $E_{1,3}$ is not IFR, but $E_{1,3}$ is DMRL if and only if $q_1 \leq 1/2$. □

If $F(y)$ is absolutely continuous with failure rate $\mu(y)$, then from (2.3.4) and L'Hopital's rule,

$$r(\infty) = \lim_{y\to\infty} \frac{E(Y)\overline{F}_1(y)}{\overline{F}(y)} = \lim_{y\to\infty} \frac{-\overline{F}(y)}{-f(y)} = \lim_{y\to\infty} \frac{1}{\mu(y)} = \frac{1}{\mu(\infty)}, \tag{2.3.11}$$

as long as $\mu(\infty)$ is well defined.

Example 2.3.2 The gamma distribution
In this example, we analyze the mean residual lifetime $r(y)$ of a gamma distribution. Recall that the gamma distribution has pdf

$$f(y) = \frac{\beta(\beta y)^{\alpha-1}e^{-\beta y}}{\Gamma(\alpha)}, \quad y > 0.$$

One has $E(Y) = \alpha/\beta$. The mean residual lifetime $r(y)$ is complicated in general. When its shape parameter α is a positive integer, one has from (2.2.12), (2.1.13), and (2.3.4), for $y \geq 0$,

$$r(y) = \frac{\sum_{j=0}^{\alpha-1}(\alpha - j)\frac{(\beta y)^j}{j!}}{\beta\sum_{j=0}^{\alpha-1}\frac{(\beta y)^j}{j!}}. \tag{2.3.12}$$

For general $\alpha > 0$ no simple form exists. However, from example 2.1.2 if $\alpha = 1$ then $r(y) = 1/\beta$, $y \geq 0$, whereas if $\alpha < 1$ then $F(y)$ is DFR and hence $r(y)$ is increasing in y, whereas if $\alpha > 1$ then $F(y)$ is IFR and hence $r(y)$ is decreasing in y. Also $r(\infty) = 1/\mu(\infty) = 1/\beta$.

To summarize, if $\alpha = 1$ then $r(y) = 1/\beta$, whereas if $\alpha < (>)\ 1$, then $r(y)$ is increasing (decreasing) from $r(0) = \alpha/\beta$ to its limit $r(\infty) = 1/\beta$. □

The mean residual lifetime is also instrumental in the following example.

Example 2.3.3 The equilibrium residual lifetime distribution
Since T_y has mean $r(y)$, it follows that the equilibrium df of $Pr(T_y \leq t)$ is, from (2.3.1) and (2.3.4),

$$\begin{aligned} 1 - \frac{\int_t^\infty Pr(T_y > x)dx}{r(y)} &= 1 - \frac{\int_t^\infty \overline{F}(x+y)dx}{r(y)\overline{F}(y)} \\ &= 1 - \frac{\int_{y+t}^\infty \overline{F}(x)dx}{E(Y)\overline{F}_1(y)} \\ &= 1 - \frac{\overline{F}_1(y+t)}{\overline{F}_1(y)}. \end{aligned}$$

Thus we have the intriguing result that the equilibrium df of the residual lifetime df $1 - \overline{F}(y+t)/\overline{F}(y)$ is $1 - \overline{F}_1(y+t)/\overline{F}_1(y)$, the residual lifetime df of the equilibrium df $F_1(y)$. □

Some other relationships are worth mentioning as well. It follows from (2.2.4) and (2.3.4) that

$$\int_y^\infty x dF(x) = \overline{F}(y)\{y + r(y)\}$$

which implies that

$$\overline{F}(y) = \frac{\int_y^\infty x dF(x)}{y + r(y)}, \tag{2.3.13}$$

and replacing y by 0 in the numerator results in

$$\overline{F}(y) \leq \frac{E(Y)}{y + r(y)}, \quad y \geq 0, \tag{2.3.14}$$

a refinement of Markov's inequality. Again from (2.2.4) and (2.3.4),

$$\int_y^\infty x dF(x) = E(Y)\overline{F}_1(y)\left\{1 + \frac{y}{r(y)}\right\},$$

implying that

$$\overline{F}_1(y) = \frac{r(y)\int_y^\infty x dF(x)}{E(Y)\{y + r(y)\}}, \quad y \geq 0. \tag{2.3.15}$$

Again replacing y by 0 in the numerator results in

$$\overline{F}_1(y) \leq \frac{r(y)}{y + r(y)}, \quad y \geq 0. \tag{2.3.16}$$

Also, if $F(y)$ is absolutely continuous with failure rate $\mu(y)$, then from (2.1.3) and (2.3.3) we have

$$r(y) = \int_0^\infty e^{-\int_y^{y+t} \mu(x)dx} dt, \quad y \geq 0, \tag{2.3.17}$$

i.e. $r(y)$ is expressed as a function of $\mu(y)$. Thus, if $\mu(y) \leq (\geq)\ \mu$ where $0 < \mu < \infty$, then from (2.3.17),

$$r(y) \geq (\leq) \int_0^\infty e^{-\int_y^{y+t} \mu \cdot dx} dt = \int_0^\infty e^{-\mu t} dt = \frac{1}{\mu}. \tag{2.3.18}$$

As in (2.1.11), it follows from (2.3.6) that $r(y) \geq (\leq)\ r$ implies that

$$\overline{F}_1(y) \geq (\leq)\ e^{-y/r}, \quad y \geq 0. \tag{2.3.19}$$

Combining the above, $\mu(y) \leq (\geq)\ \mu$ where $0 < \mu < \infty$ implies that $\overline{F}(y) \geq (\leq)\ e^{-\mu y}, \ \ y \geq 0$, and $\overline{F}_1(y) \geq (\leq)\ e^{-\mu y}, y \geq 0$. Similarly, if $F(y)$ is DFR (IFR) with failure rate $\mu(y)$, then from (2.3.17),

$$r(y) \geq (\leq) \int_0^\infty e^{-\int_y^{y+t} \mu(y)dx} dt = \int_0^\infty e^{-\mu(y)t} dt = \frac{1}{\mu(y)}, \tag{2.3.20}$$

i.e. $\mu(y) \geq (\leq) \, 1/r(y)$, which is a failure rate ordering between $F(y)$ and $F_1(y)$.

It is worth noting that if $r(y) \leq r$, then (2.3.16) yields $\overline{F}_1(y) \leq r/(y+r)$, but $e^{-y/r} \leq r/(y+r)$ may be restated as $e^{y/r} \geq 1 + y/r$, which is clearly true. That is, (2.3.19) is a tighter inequality than (2.3.16) with $r(y)$ replaced by r. Wolff (1989) discusses the condition $r(y) \geq (\leq) \, r$ as well.

2.4 Other classes of distributions

In the last two sections, we have introduced the notions of failure rate and mean residual lifetime, and classifications based on these notions. There are many other classes of distributions, some of which are of interest for the present application. These distributions are classified in terms of their survival function or the survival function of their equilibrium distributions. See Fagiuoli and Pellerey (1993, 1994) and references therein.

The DFR (IFR) class is defined by $\overline{F}(x+y)/\overline{F}(y)$ nondecreasing (nonincreasing) in y for fixed $x \geq 0$. Thus, $\overline{F}(x+y)/\overline{F}(y)$ is no smaller (larger) than when $y = 0$, i.e. $\overline{F}(x+y) \geq (\leq) \, \overline{F}(x)\overline{F}(y), \quad x \geq 0, y \geq 0$. The df $F(x)$ is said to be new worse (better) than used or NWU (NBU) if $\overline{F}(x+y) \geq (\leq) \, \overline{F}(x)\overline{F}(y)$ for all $x \geq 0$ and $y \geq 0$. The name has its origin in the fact that the inequality is a restatement of $Pr\,(T_y > x) \geq (\leq) \, Pr(Y > x)$, i.e. the residual lifetime is stochastically larger (smaller) than the original or new lifetime Y. Clearly, the DFR (IFR) class of distributions is a subclass of the NWU (NBU) class.

Similarly, the df $F(y)$ is said to be 2-NWU (2-NBU) if its equilibrium df $F_1(y)$ is NWU (NBU), i.e. if $\overline{F}_1(x+y) \geq (\leq) \, \overline{F}_1(x)\overline{F}_1(y)$ for all $x \geq 0$ and $y \geq 0$. As discussed in the previous section, $F(y)$ is IMRL (DMRL) is equivalent to $F_1(y)$ is DFR(IFR) which implies that $F_1(y)$ is NWU (NBU), i.e. $F(y)$ is 2-NWU (2-NBU). In other words, the IMRL (DMRL) class is a subclass of the 2-NWU (2-NBU) class.

Another larger class than the IMRL (DMRL) class is the 3-DFR (3-IFR) class for which $F_1(y)$ is IMRL (DMRL). This follows from the fact that if $F(y)$ is IMRL (DMRL) then $F_1(y)$ is DFR (IFR) and hence $F(y)$ is 3-DFR (3-IFR).

We now introduce the new worse (better) than used in expectation or NWUE (NBUE) class. The df $F(y)$ is NWUE (NBUE) if the mean residual lifetime $r(y)$ satisfies $r(y) \geq (\leq) \, r(0)$, i.e. if $E(T_y) \geq (\leq) \, E(Y)$, and the residual lifetime has a larger (smaller) expectation than the original or new lifetime Y. Clearly, from equation (2.3.4), NWUE (NBUE) is equivalent to $\overline{F}_1(y) \geq (\leq) \, \overline{F}(y)$. If $F(y)$ is NWU (NBU), then from (2.3.3) it follows that

$$r(y) = \int_0^\infty \frac{\overline{F}(y+t)}{\overline{F}(y)} dt \geq (\leq) \int_0^\infty \overline{F}(t) dt = E(Y) = r(0),$$

i.e. the NWU (NBU) class is a subclass of the NWUE (NBUE) class.

Before proceeding, we present the following simple result, which is closely related to failure rate ordering (e.g. Shaked and Shanthikumar, 1994, p. 12), and is needed in what follows.

Proposition 2.4.1 Suppose $A(y)$ and $H(y)$ are absolutely continuous dfs for $y \geq 0$ with failure rates $\mu_A(y) = -\frac{d}{dy}\ln \overline{A}(y)$ and $\mu_H(y) = -\frac{d}{dy}\ln \overline{H}(y)$ respectively, where $\overline{A}(y) = 1 - A(y)$ and $\overline{H}(y) = 1 - H(y)$. If

$$\frac{\overline{H}(x+y)}{\overline{H}(y)} \geq (\leq)\ \overline{A}(x), \quad x \geq 0,\ y \geq 0,$$

then

$$\mu_H(y) \leq (\geq)\ \mu_A(0), \quad x \geq 0.$$

Proof: One has

$$e^{-\int_0^x \mu_H(y+t)dt} \geq (\leq)\ e^{-\int_0^x \mu_A(t)dt}$$

which may be restated as

$$\frac{\int_0^x \mu_H(y+t)dt}{x} \leq (\geq)\ \frac{\int_0^x \mu_A(t)dt}{x}.$$

Taking the limit as $x \to 0$ and using L'Hopital's rule yields

$$\lim_{x\to 0} \mu_H(y+x) \leq (\geq)\ \lim_{x\to 0} \mu_A(x)$$

and the result follows. □

The next implication result follows directly from the above proposition.

Theorem 2.4.1
If $F(y)$ is 2-NWU (2-NBU) then $F(y)$ is NWUE (NBUE).

Proof: It follows from proposition 2.4.1 with $A(y) = H(y) = F_1(y)$ and the failure rate property (2.3.5) that $1/r(y) \leq (\geq)\ 1/r(0)$, i.e. $r(y) \geq (\leq)\ r(0)$. □

Another class is the new worse (better) than used in convex ordering or NWUC (NBUC) class. The df $F(y)$ is NWUC (NBUC) if

$$\overline{F}_1(x+y) \geq (\leq)\ \overline{F}_1(y)\overline{F}(x)$$

for all $x \geq 0, y \geq 0$. That is, the residual lifetime of the equilibrium df $F_1(y)$ is stochastically larger (smaller) than Y. Substitution of $y = 0$ yields

$\overline{F}_1(x) \geq (\leq) \overline{F}(x)$, i.e. the NWUC (NBUC) class is a subclass of the NWUE (NBUE) class. Also, from theorem 2.4.1 it follows that if $F(y)$ is 2-NWU (2-NBU) then $\overline{F}_1(x+y) \geq (\leq) \overline{F}_1(y)\overline{F}_1(x) \geq (\leq) \overline{F}_1(y)\overline{F}(x)$. That is, 2-NWU (2-NBU) implies NWUC (NBUC) and the 2-NWU (2-NBU) class is a subclass of the NWUC (NBUC) class. Similarly, $F(y)$ is NWU (NBU) implies that $\overline{F}(x+t) \geq (\leq) \overline{F}(x)\overline{F}(t)$. Thus $\int_y^\infty \overline{F}(x+t)dt \geq (\leq) \overline{F}(x)\int_y^\infty \overline{F}(t)dt$. Since $\overline{F}_1(x+y) = \int_{x+y}^\infty \overline{F}(t)dt/E(Y) = \int_y^\infty \overline{F}(x+t)dt/E(Y)$, it follows that $\overline{F}_1(x+y) \geq (\leq) \overline{F}(x)\int_y^\infty \overline{F}(t)dt/E(Y) = \overline{F}(x)\overline{F}_1(y)$, i.e. $F(y)$ is NWUC (NBUC). Thus, the NWU (NBU) class is also a subclass of the NWUC (NBUC) class.

The following diagram lists the relations between the classes of the distributions discussed.

$$\begin{array}{ccc} \text{DFR(IFR)} & \Rightarrow & \text{NWU(NBU)} \\ \Downarrow & & \Downarrow \\ \text{IMRL(DMRL)} & \Rightarrow \text{2-NWU(2-NBU)} \Rightarrow & \text{NWUC(NBUC)} \\ \Downarrow & & \Downarrow \\ \text{3-DFR(3-IFR)} & & \text{NWUE(NBUE)} \end{array}$$

We now give several properties of NWUE (NBUE) distributions, which also hold for all of its subclasses as well. Since $r(y) \geq (\leq)\ r(0) = E(Y)$, it follows immediately from (2.3.19) that NWUE (NBUE) implies that the equilibrium df satisfies

$$\overline{F}_1(y) \geq (\leq)\ e^{-y/E(Y)}, \quad y \geq 0. \tag{2.4.1}$$

Thus, the equilibrium df is stochastically larger (smaller) than an exponential df with the same mean.

Integration of both sides of (2.4.1) together with (2.2.3) yields

$$\frac{E(Y^2)}{2E(Y)} = \int_0^\infty \overline{F}_1(y)dy \geq (\leq) \int_0^\infty e^{-y/E(Y)}dy = E(Y).$$

In other words, NWUE (NBUE) implies that

$$\mathrm{Var}(Y) = E(Y^2) - \{E(Y)\}^2 \geq (\leq)\ \{E(Y)\}^2, \tag{2.4.2}$$

a result which agrees with intuition about tail thickness. A measure of tail thickness often used is the coefficient of variation C_Y of Y, defined by $C_Y = \sqrt{\mathrm{Var}(Y)}/E(Y)$. Clearly, NWUE (NBUE) implies that $C_Y^2 \geq (\leq)\ 1$. In particular, a DFR (IFR) df must have $C_Y^2 \geq (\leq)\ 1$.

Since NWUE implies that $r(y) \geq E(Y)$, it follows from (2.3.14) that if $F(y)$ is NWUE then

$$\overline{F}(y) \leq \frac{E(Y)}{y + E(Y)}, \quad y \geq 0. \tag{2.4.3}$$

If m is any median of Y, then $\overline{F}(m) = 1/2$, and from (2.4.3), $1/2 \leq E(Y)/\{m + E(Y)\}$, i.e. $m \leq E(Y)$. That is, if $F(y)$ is NWUE then the expected value of Y is at least as large as any median of Y.

The ideas can also be extended to nonnegative rather than strictly positive random variables, as for example in the NWU case. In the NBU case however, $\overline{B}(0+x) \leq \overline{B}(0)\overline{B}(x)$ implies that $\overline{B}(0) \geq 1$, i.e. $\overline{B}(0) = 1$. That is, the random variable is strictly positive in this case.

A related class of distributions (Alzaid, 1994) is the used worse (better) than aged or UWA (UBA) class. We shall say (slightly generalizing Alzaid, 1994) that a df $F(y)$ is UWA (UBA) if the mean residual lifetime $r(y) = \int_0^\infty \{\overline{F}(y+t)/\overline{F}(y)\}dt$ satisfies $0 < r(\infty) = \lim_{y\to\infty} r(y) < \infty$ and $\overline{F}(x + y) \leq (\geq)\ \overline{F}(y)e^{-x/r(\infty)}$ for all $x \geq 0$ and $y \geq 0$.

If $F(y)$ is absolutely continuous with failure rate $\mu(y) = -\frac{d}{dy}\ln \overline{F}(y)$ then using (2.3.11) it is easy to see that $F(y)$ is UWA (UBA) if $\mu(y) \geq (\leq)\ \mu(\infty)$ where $0 < \mu(\infty) < \infty$. Clearly, from example 2.1.3, a mixture of Erlangs is UBA, and if $F(y)$ is an absolutely continuous DFR (IFR) df with $\mu(\infty) > 0$ ($\mu(\infty) < \infty$) then $F(y)$ is UWA (UBA).

The above result may be improved upon (Willmot and Cai, 2000).

Theorem 2.4.2 If $F(y)$ is IMRL (DMRL) with $r(\infty) < \infty$ ($r(\infty) > 0$) then $F(y)$ is UWA (UBA).

Proof: Since $F_1(y)$ has failure rate $1/r(y)$ from (2.3.5), it is clear from the above discussion that $0 < r(\infty) < \infty$ and $F_1(y)$ is DFR (IFR) and thus UWA (UBA). Thus, $\overline{F}_1(x+y) \leq (\geq)\ \overline{F}_1(y)e^{-x/r(\infty)}$. But from (2.3.4),

$$\overline{F}(x+y) = \frac{r(0)\overline{F}_1(x+y)}{r(x+y)} \leq (\geq)\ \frac{r(0)\overline{F}_1(y)e^{-x/r(\infty)}}{r(y)} = \overline{F}(y)e^{-x/r(\infty)}$$

since $r(y)$ is nondecreasing (nonincreasing). Therefore, $F(y)$ is UWA (UBA). □

A larger class than the UWA (UBA) class is the used worse (better) than aged in expectation or UWAE (UBAE) class. The df $F(y)$ is UWAE (UBAE) if the mean residual lifetime $r(y)$ satisfies $r(y) \leq (\geq)\ r(\infty)$ where $0 < r(\infty) < \infty$. If $F(y)$ is UWA (UBA), then from (2.3.3),

$$r(y) = \int_0^\infty \{\overline{F}(x+y)/\overline{F}(y)\}dx \leq (\geq) \int_0^\infty e^{-x/r(\infty)}dx = r(\infty),$$

i.e. the UWA (UBA) class is a subclass of the UWAE (UBAE) class.

The next result is actually a special case of the failure rate ordering criterion (1.B.5) of Shaked and Shanthikumar (1994, p. 13).

Theorem 2.4.3 The mean residual lifetime $r(y) = \int_0^\infty \{\overline{F}(y+t)/\overline{F}(y)\}dt$

satisfies $r(y) \geq (\leq)\ r$ where $0 < r < \infty$ if and only if

$$\frac{\overline{F}_1(x+y)}{\overline{F}_1(y)} \geq (\leq)\ e^{-x/r}, \quad x \geq 0,\ y \geq 0.$$

Proof: If $r(y) \geq (\leq)\ r$ then by (2.3.5)

$$\frac{\overline{F}_1(x+y)}{\overline{F}_1(y)} = e^{-\int_0^x \frac{1}{r(y+t)}dt} \geq (\leq)\ e^{-\int_0^x \frac{1}{r}dt} = e^{-\frac{x}{r}}.$$

The converse follows from proposition (2.4.1) with $H(y) = F_1(y)$ and $A(y) = 1 - e^{-y/r}$. □

In the special case with $r = r(\infty)$ it is clear that if $r(\infty) < \infty$, then $F(y)$ is UWAE (UBAE) if and only if $F_1(y)$ is UWA (UBA). Therefore, the class terminology 2-UWA (2-UBA) is redundant and will not be used. We now have the following corollary.

Corollary 2.4.1 If $F(y)$ is 3-DFR (3-IFR) with $0 < r(\infty) < \infty$ then $F(y)$ is UWAE (UBAE).

Proof: By theorem 2.4.2, $F_1(y)$ is IMRL (DMRL) and hence UWA (UBA). Thus, $F(y)$ is UWAE (UBAE) by the comment following theorem 2.4.3. □

Wolff (1989) discusses applications of bounded mean residual lifetimes in a queueing context.

2.5 Discrete reliability classes

Many of the concepts discussed previously extend naturally to the discrete case, although there are some differences.

Let N be a counting random variable with probabilities

$$p_n = Pr(N = n); \quad n = 0, 1, 2, \cdots, \tag{2.5.1}$$

and tail probabilities

$$a_n = Pr(N > n) = \sum_{k=n+1}^{\infty} p_k; \quad n = 0, 1, 2, \cdots. \tag{2.5.2}$$

In an insurance context, N often represents the number of losses or claims.

A discrete failure rate may be defined as

$$h_n = Pr(N = n | N \geq n) = \frac{p_n}{p_n + a_n}; \quad n = 0, 1, 2, \cdots. \tag{2.5.3}$$

It follows easily if $a_n > 0$ that

$$\frac{a_{n+1}}{a_n} = 1 - h_{n+1}; \quad n = 0, 1, 2, \cdots. \tag{2.5.4}$$

Clearly, the sequence $\{a_{n+1}/a_n;\ n = 0, 1, 2, \cdots\}$ is nondecreasing (nonincreasing) in n iff the sequence $\{h_n;\ n = 1, 2, 3, \cdots\}$ is nonincreasing (nondecreasing) in n. For this reason, and by analogy with the continuous case, we say that the distribution $\{p_n;\ n = 0, 1, 2, \cdots\}$ is discrete decreasing (increasing) failure rate or D-DFR (D-IFR) if $\{a_{n+1}/a_n;\ n = 0, 1, 2, 3, \cdots\}$ is nondecreasing (nonincreasing) in n. Evidently, D-DFR (D-IFR) implies that a_{n+1}/a_n is nondecreasing (nonincreasing) in n, as long as $a_n > 0$. To accommodate the possibility that a_n could be zero (which can not happen if a_{n+1}/a_n is nondecreasing and implies that $a_{n+k} = 0$ for $k \geq 0$), we consider instead

$$a_{n+1}^2 \leq (\geq)\ a_n a_{n+2}; \quad n = 0, 1, 2, \cdots. \tag{2.5.5}$$

Evidently, if $a_n > 0$ and $a_{n+1} > 0$, (2.5.5) is equivalent to $a_{n+1}/a_n \leq (\geq)\ a_{n+2}/a_{n+1}$, i.e. a_{n+1}/a_n is nondecreasing (nonincreasing). However, if $a_n = 0$, then $a_{n+1} = 0$ and $a_{n+2} = 0$ and (2.5.5) is still satisfied (and is an equality).

We say that $\{a_n, n = 0, 1, 2, \cdots\}$ is log-convex (log-concave) if (2.5.5) holds, and this is implied if $\{p_n; n = 0, 1, 2, \cdots\}$ is D-DFR (D-IFR).

A difficulty with the D-DFR (D-IFR) definition is that h_0 (and hence p_0) is completely arbitrary. For this reason, and because of the direct physical interpretation, alternative definitions have been used by various authors (e.g. Rolski et al., 1999, pp. 44-45) in terms of the sequence $\{h_n;\ n = 0, 1, 2, \cdots\}$. In order to avoid confusion, we shall say that the distribution $\{p_n;\ n = 0, 1, 2, \cdots\}$ is discrete strongly decreasing (increasing) failure rate or DS-DFR (DS-IFR) if $\{h_n;\ n = 0, 1, 2, \cdots\}$ is nonincreasing (nondecreasing) in n. Clearly, this is a stronger property than the D-DFR (D-IFR) property, i.e. the DS-DFR (DS-IFR) class is a subclass of the D-DFR (D-IFR) class. Obviously, a D-DFR (D-IFR) distribution is DS-DFR (DS-IFR) if $h_0 \geq (\leq)\ h_1$. It is not hard to see using (2.5.3) that $h_0 \geq (\leq)\ h_1$ is equivalent to each of $p_1 \leq (\geq)\ p_0(1 - p_0)$ and $a_1 \geq (\leq)\ a_0^2$.

A further difficulty from a practical standpoint with the above definitions is that $\{a_n;\ n = 0, 1, 2, \cdots\}$ is involved, normally not convenient since closed form expressions are usually not available. To this end, we say that the distribution $\{p_n;\ n = 0, 1, 2, \cdots\}$ is log-convex (log-concave) if

$$p_{n+1}^2 \leq (\geq)\ p_n p_{n+2}; \quad n = 0, 1, 2, \cdots. \tag{2.5.6}$$

As discussed above, $\{p_n;\ n = 0, 1, 2, \cdots\}$ is log-convex (log-concave) if the sequence $\{p_{n+1}/p_n;\ n = 0, 1, 2, \cdots\}$ is nondecreasing (nonincreasing) in n.

We now demonstrate that the log-convex (log-concave) class is a subclass of the DS-DFR (DS-IFR) class. As in the continuous case, it follows from

(2.5.3) that

$$\frac{1}{h_n} = 1+\frac{a_n}{p_n} = 1+\sum_{k=1}^{\infty}\frac{p_{n+k}}{p_n} = 1+\sum_{k=1}^{\infty}\prod_{j=1}^{k}\frac{p_{n+j}}{p_{n+j-1}}; \quad n = 0,1,2,\cdots. \tag{2.5.7}$$

Clearly, if $\{p_n;\ n = 0,1,2,\cdots\}$ is log-convex (log-concave), then $1/h_n$ is nondecreasing (nonincreasing) in n, i.e. $\{p_n;\ n = 0,1,2,\cdots\}$ is DS-DFR (DS-IFR).

It is instructive to note that the D-DFR (D-IFR) property does not involve h_0 and hence p_0, and thus (2.5.7) need not be considered when $n = 0$. In particular, if (2.5.6) holds except for $n = 0$, then $\{p_n;\ n = 0,1,2,\cdots\}$ is D-DFR (D-IFR).

We now consider an important class of discrete distributions in insurance modelling. See Panjer and Willmot (1992) or Klugman et al. (1998) for details.

Example 2.5.1 The (a, b) class

We now suppose that the probability function of N satisfies

$$p_{n+1} = \left(a + \frac{b}{n+1}\right) p_n; \quad n = 1,2,3,\cdots, \tag{2.5.8}$$

where a and b are constant. Distributions which satisfy (2.5.8) include the following.

Poisson

$p_n = \frac{\lambda^n e^{-\lambda}}{n!}; \quad n = 0,1,2,\cdots$
$\lambda > 0.$
$a = 0; b = \lambda$

Binomial

$p_n = \binom{M}{n} q^n (1-q)^{M-n}; \quad n = 0,1,2,\cdots,M.$
$0 < q < 1,\ M = 1,2,\cdots.$
$a = -\frac{q}{1-q},\ b = \frac{(M+1)q}{1-q}$

Negative binomial

$p_n = \frac{\alpha(\alpha+1)\cdots(\alpha+n-1)}{n!}(1-\phi)^\alpha \phi^n; \quad n = 0,1,2,\cdots$
$\alpha > 0, 0 < \phi < 1.$
$a = \phi,\ b = (\alpha-1)\phi$

Geometric

$p_n = (1-\phi)\phi^n; \quad n = 0,1,2,\cdots$
$0 < \phi < 1.$
$a = \phi,\ b = 0$

Logarithmic series

$$p_n = \frac{\phi^n}{-n\ln(1-\phi)};\quad n = 1,2,\cdots$$
$$0 < \phi < 1.$$
$$a = \phi,\ b = -\phi$$

Extended truncated negative binomial (ETNB)

$$p_1 = \frac{-\alpha\phi}{1-(1-\phi)^{-\alpha}}$$
$$p_n = \frac{-\alpha}{1-(1-\phi)^{-\alpha}}\frac{(\alpha+1)(\alpha+2)\cdots(\alpha+n-1)}{n!}\phi^n;\quad n = 2,3,4\cdots$$
$$-1 < \alpha < 0,\ 0 < \phi \leq 1.$$
$$a = \phi,\ b = (\alpha-1)\phi.$$

For the Poisson, binomial, negative binomial, and geometric distributions, (2.5.8) holds for $n = 0$ also. In this case, if $b \leq (\geq)\ 0$ it follows that $\{p_n;\ n = 0,1,2,\cdots\}$ is log-convex (log-concave) and thus DS-DFR (DS-IFR). In particular, the Poisson, binomial, negative binomial ($\alpha \geq 1$), and geometric distributions are log-concave, whereas the negative binomial ($\alpha \leq 1$) and geometric distributions are log-convex. For the logarithmic series and ETNB distributions, $p_0 = 0$ which implies that they are not DS-DFR. However, since $b < 0$ in both cases, (2.5.6) holds except for $n = 0$. Thus, the logarithmic series and ETNB distributions are both D-DFR.

It can be shown that, other than the above distributions with an arbitrary p_0 (i.e. (zero) modified), no other distributions satisfy (2.5.8). See Willmot (1988a), for details. □

The following example illustrates the relationship between the log-convex (log-concave), DS-DFR (DS-IFR), and D-DFR (D-IFR) distributions.

Example 2.5.2 Suppose that $\{f_n;\ n = 1,2,3,\cdots\}$ is a discrete probability distribution with $f_n > 0$ for $n = 1,2,3,\cdots$, and obviously $f_0 = 0$. Also, let $0 < p_0 < 1$ and $p_n = (1-p_0)f_n$ for $n = 1,2,3,\cdots$. Thus, $\{p_n;\ n = 0,1,2,\cdots\}$ is a modified discrete probability distribution. Clearly, $\{f_n;\ n = 1,2,3,\cdots\}$ is not DS-DFR since $f_1 > f_0(1-f_0) = 0$, as discussed following (2.5.5). But $a_n = \sum_{k=n+1}^{\infty} p_k = (1-p_0)\sum_{k=n+1}^{\infty} f_k = (1-p_0)\overline{F}_n$, which implies that $\{f_n;\ n = 1,2,3,\cdots\}$ is D-DFR if and only if $\{p_n;\ n = 0,1,2,\cdots\}$ is D-DFR. This is implied if, in addition,

$$\frac{f_{n+1}}{f_n} < \frac{f_{n+2}}{f_{n+1}};\quad n = 1,2,3,\cdots,$$

since (2.5.7) yields for $n = 1,2,3,\cdots$

$$\frac{1}{h_n} = 1 + \sum_{k=1}^{\infty}\prod_{j=1}^{k}\frac{p_{n+j}}{p_{n+j-1}} = 1 + \sum_{k=1}^{\infty}\prod_{j=1}^{k}\frac{f_{n+j}}{f_{n+j-1}},$$

i.e. h_n is strictly decreasing for $n \geq 1$. When $n = 1$ it follows from (2.5.3) that

$$f_1 = \frac{f_1}{f_1 + \overline{F}_1} = \frac{p_1}{p_1 + a_1} = h_1 > h_2 = \frac{p_2}{p_2 + a_2} = \frac{f_2}{\overline{F}_1} = \frac{f_2}{1 - f_1},$$

i.e. $f_2 < f_1(1 - f_1)$, or $f_1 + (f_2/f_1) < 1$. The distribution $\{p_n;\ n = 0, 1, 2, \cdots\}$ is DS-DFR if, in addition, $p_1 \leq p_0(1 - p_0)$ or equivalently $f_1 \leq p_0$, as is clear from the discussion following (2.5.5). However, $\{p_n;\ n = 0, 1, 2, \cdots\}$ is not log-convex if $p_2/p_1 < p_1/p_0$, which is a restatement of $f_2/f_1 < f_1(1 - p_0)/p_0$, i.e. $p_0 < f_1/\{f_1 + (f_2/f_1)\}$. To summarize, $\{p_n;\ n = 0, 1, 2, \cdots\}$ is DS-DFR but not log-convex if $f_1 \leq p_0 < f_1/\{f_1 + (f_2/f_1)\}$. Similarly, if

$$\frac{f_{n+1}}{f_n} > \frac{f_{n+2}}{f_{n+1}}; \quad n = 1, 2, 3, \cdots,$$

then $\{p_n;\ n = 0, 1, 2, \cdots\}$ is DS-IFR but not log-concave if $f_1/\{f_1 + (f_2/f_1)\} < p_0 \leq f_1$. □

As in the continuous case, we consider classifications involving the mean residual lifetime

$$r_n = E(N - n|N > n) = \frac{\sum_{k=n+1}^{\infty} (k - n)p_k}{a_n} = \frac{\sum_{k=n}^{\infty} a_k}{a_n}; \quad n = 0, 1, 2, \cdots, \tag{2.5.9}$$

as follows from summation by parts. The distribution $\{p_n;\ n = 0, 1, 2, \cdots\}$ is said to be discrete increasing (decreasing) mean residual lifetime or D-IMRL (D-DMRL) if $\{r_n;\ n = 0, 1, 2, \cdots\}$ is nondecreasing (nonincreasing) in n. The D-DFR (D-IFR) class is a subclass of the D-IMRL (D-DMRL) class, as is clear from the representation

$$r_n = \sum_{k=0}^{\infty} \frac{a_{n+k}}{a_n} = \sum_{k=0}^{\infty} \prod_{j=1}^{k} \frac{a_{n+j}}{a_{n+j-1}}.$$

Examples of distributions which are D-IMRL (D-DMRL) but not D-DFR (D-IFR) are not difficult to construct. See Block and Savits (1980) and Ebrahimi (1986), for example.

An alternative characterization of the D-IMRL (D-DMRL) class in terms of discrete equilibrium distributions is also useful. If $E(N) = \sum_{k=1}^{\infty} kp_k = \sum_{k=0}^{\infty} a_k < \infty$, let N^* be the discrete equilibrium random variable with probability function

$$p_n^* = Pr(N^* = n) = \frac{a_n}{E(N)}; \quad n = 0, 1, 2, \cdots, \tag{2.5.10}$$

and tail

$$a_n^* = \sum_{k=n+1}^{\infty} p_k^* = \frac{\sum_{k=n+1}^{\infty} a_k}{E(N)}; \quad n = 0, 1, 2, \cdots. \tag{2.5.11}$$

It is not hard to see that the discrete equilibrium failure rate is

$$h_n^* = Pr(N^* = n | N^* \geq n) = \frac{p_n^*}{p_n^* + a_n^*} = \frac{a_n}{a_n + \sum_{k=n+1}^{\infty} a_k} = \frac{1}{r_n}, \tag{2.5.12}$$

which implies that $\{p_n;\ n = 0, 1, 2, \cdots\}$ is D-IMRL (D-DMRL) is equivalent to $\{p_n^*;\ n = 0, 1, 2, \cdots\}$ is DS-DFR (DS-IFR).

Cai and Kalashnikov (2000) introduced the discrete strongly new worse than used or DS-NWU class of probability distributions for which

$$a_{n+m+1} \geq a_n a_m; \quad n, m = 0, 1, 2, \cdots. \tag{2.5.13}$$

The definition actually corresponds to the discrete new worse than used definition given by Fagiuoli and Pellerey (1994) and others. That is, replacement of n by $n-1$ and m by $m-1$ in (2.5.13) together with (2.5.2) yields $Pr(N \geq n+m) \geq Pr(N \geq n) Pr(N \geq m)$, which is obviously true for $n = 0$ or $m = 0$ as well. The DS-NWU class contains the DS-DFR class as a subclass. To see this, note that $h_0 \geq h_1$ is equivalent to $a_1 \geq a_0^2$, as mentioned earlier. Also, DS-DFR implies D-DFR. Thus,

$$\frac{a_{n+m+1}}{a_n} = \frac{a_{n+m+1}}{a_{n+m}} \cdot \frac{a_{n+m}}{a_n} \geq \frac{a_1}{a_0} \cdot \frac{a_m}{a_0} \geq a_m,$$

since $a_{k+1}/a_k \geq a_1/a_0$ for $k \geq 0$, and a_{n+m}/a_n is nondecreasing in n for $n \geq 0$. That is, DS-DFR implies DS-NWU.

The distribution $\{p_n;\ n = 0, 1, 2, \cdots\}$ is said to be discrete new worse (better) than used or D-NWU (D-NBU) if $a_{n+m} \geq (\leq)\ a_n a_m$ for all $n, m = 0, 1, 2, \cdots$. Clearly, DS-NWU implies D-NWU since $a_{n+m} \geq a_{n+m+1} \geq a_n a_m$. Also, D-DFR implies D-NWU since

$$\frac{a_{n+m}}{a_n} \geq \frac{a_m}{a_0} \geq a_m.$$

On the other hand, D-NBU implies that $a_{n+0} \leq a_n a_0$, i.e. $a_0 \geq 1$. Thus $p_0 = 0$, and D-IFR only implies D-NBU if $p_0 = 0$.

To summarize, we may conclude that

$$\text{DS-IFR} \Rightarrow \text{D-IFR} \Rightarrow \text{D-DMRL}$$

as well as

$$\begin{array}{ccccc} \text{DS-DFR} & \Rightarrow & \text{D-DFR} & \Rightarrow & \text{D-IMRL} \\ \Downarrow & & \Downarrow & & \\ \text{DS-NWU} & \Rightarrow & \text{D-NWU} & & \end{array}$$

Other discrete distributional classes may be (and have been) defined. See Esary et al. (1973), Klefsjo (1981), Fagiuoli and Pellerey (1994), Willmot and Cai (1999), and references therein for further discussion.

2.6 Bounds on ratios of discrete tail probabilities

Of central importance in what follows is the existence of $\phi \in (0,1)$ satisfying

$$a_{n+1} \leq (\geq)\ \phi a_n; \quad n = 0, 1, 2, \cdots. \tag{2.6.1}$$

Intuitively, (2.6.1) indicates that the counting distribution under consideration is stochastically dominated by (stochastically dominates) a geometric distribution, as is not hard to see from example 2.5.1 and $a_n \leq (\geq)\ \phi^n a_0$. For this reason, we call (2.6.1) the dominance assumption. As we shall see, identification of ϕ is possible for some of the discrete reliability classes discussed in the previous section. As long as $a_n > 0$, (2.6.1) may be rewritten as

$$\frac{a_{n+1}}{a_n} \leq (\geq)\ \phi; \quad n = 0, 1, 2, \cdots. \tag{2.6.2}$$

If $\{p_n;\ n = 0, 1, 2, \cdots\}$ is D-DFR (D-IFR), then the sequence $\{a_{n+1}/a_n;\ n = 0, 1, 2, \cdots\}$ is nondecreasing (nonincreasing) and thus has a minimum (maximum) at $n = 0$. Therefore, (2.6.2) and (2.6.1) hold with $\phi = a_1/a_0$, i.e., if $\{p_n;\ n = 0, 1, 2, \cdots\}$ is D-DFR (D-IFR) then

$$a_{n+1} \geq (\leq) \left(\frac{a_1}{a_0}\right) a_n; \quad n = 0, 1, 2, \cdots. \tag{2.6.3}$$

To find an upper (lower) bound in the D-DFR (D-IFR) case, we introduce the probability generating function (pgf) of $\{p_n; n = 0, 1, 2, \cdots\}$, namely

$$P(z) = \sum_{n=0}^{\infty} p_n z^n, \quad |z| < z_0 \tag{2.6.4}$$

where $z_0 \geq 1$ is the radius of convergence of $P(z)$. We have, for the ordinary generating function of $A(z)$,

$$A(z) = \sum_{n=0}^{\infty} a_n z^n = \sum_{n=0}^{\infty} \left(\sum_{k=n+1}^{\infty} p_k \right) z^n = \sum_{k=1}^{\infty} \sum_{n=0}^{k-1} p_k z^n$$

as long as both sides are finite, i.e. $|z| < z_0$. Thus

$$A(z) = \sum_{k=1}^{\infty} p_k \left(\sum_{n=0}^{k-1} z^n \right) = \sum_{k=1}^{\infty} \frac{p_k \left(1 - z^k\right)}{1 - z} = \sum_{k=0}^{\infty} \frac{p_k \left(1 - z^k\right)}{1 - z}.$$

In other words,

$$A(z) = \sum_{n=0}^{\infty} a_n z^n = \frac{1 - P(z)}{1 - z}, \quad |z| < z_0, \tag{2.6.5}$$

and $P(z)$ and $A(z)$ have the same radius of convergence z_0.

If $\{p_n; n = 0, 1, 2, \cdots\}$ is D-DFR (D-IFR), then the sequence $\{a_{n+1}/a_n;$ $n = 0, 1, 2, \cdots\}$ is nondecreasing (nonincreasing) and bounded from above by 1 (below by 0) and hence has a finite limit. By the ratio test for convergence, this limit is $1/z_0$, i.e.

$$\lim_{n \to \infty} \frac{a_{n+1}}{a_n} = \frac{1}{z_0}. \tag{2.6.6}$$

Furthermore, since $\{p_n; n = 0, 1, 2, \cdots\}$ is D-DFR (D-IFR), the sequence $\{a_{n+1}/a_n;\ n = 0, 1, 2, \cdots\}$ is nondecreasing (nonincreasing) to its limit, implying that

$$a_{n+1} \leq (\geq)\ a_n/z_0; \quad n = 0, 1, 2, \cdots, \tag{2.6.7}$$

and (2.6.1) holds with $\phi = 1/z_0$. The above analysis assumes that $z_0 < \infty$, which in turn requires that $a_n > 0$ for $n > 0$ (otherwise the support of N is finite and $z_0 = \infty$). It is interesting to note that if $1 < z_0 < \infty$ the inequality (2.6.7) is essentially a discrete version of the UWA (UBA) property discussed in section 2.4 (Willmot and Cai, 2000). Also, from (2.5.4) and (2.6.6), $\lim_{n \to \infty} h_n = 1 - 1/z_0$. Finally a similar argument yields

$$\lim_{n \to \infty} \frac{p_{n+1}}{p_n} = \frac{1}{z_0} \tag{2.6.8}$$

if the indicated limit exists.

In what follows, the inequality (2.6.1) where $\phi \in (0, 1)$ is of much interest and the subject of further study. The special case $\phi = 1/z_0$ given by (2.6.7) is the optimal choice of ϕ in the sense that it can be made no smaller (larger). To see this, note that (2.6.1) implies that $a_n \leq (\geq)\ \phi^n a_0$ for $n \geq 0$. Thus $A(|z|) \leq (\geq)\ a_0 \sum_{n=0}^{\infty} (\phi|z|)^n$, and so $A(|z|) < (=)\ \infty$ if $|z| <$ $(\geq)\ \phi^{-1}$. But this in turn implies that $\phi^{-1} \leq (\geq)\ z_0$ since $A(|z|) < (=)\ \infty$ if $|z| < (>)\ z_0$ and $A(z_0) \leq \infty$. That is, $\phi \geq (\leq)\ 1/z_0$, and (2.6.7) is the best possible inequality.

Hence, special attention will be paid to (2.6.7), i.e. (2.6.1) with $\phi = 1/z_0$.

We shall now demonstrate that the optimal bound (2.6.7) actually holds for the larger D-IMRL (D-DMRL) class. The following result, of interest in its own right, demonstrates that in the DS-DFR (DS-IFR) case the probabilities themselves satisfy a certain smoothness condition.

Theorem 2.6.1 If $\{p_n;\ n = 0, 1, 2, \cdots\}$ is DS-DFR (DS-IFR) then

$$p_{n+1} \leq (\geq) \left(\frac{1}{z_0}\right) p_n; \quad n = 0, 1, 2, \cdots. \tag{2.6.9}$$

Proof. It follows from (2.5.7) that

$$p_n = \frac{a_n}{\frac{1}{h_n} - 1}; \quad n = 0, 1, 2, \cdots. \tag{2.6.10}$$

Now DS-DFR (DS-IFR) implies that $\{h_n;\ n = 0, 1, 2, \cdots\}$ is nonincreasing (nondecreasing) and D-DFR (D-IFR) so that (2.6.7) holds. Therefore, from (2.6.10),

$$p_{n+1} = \frac{a_{n+1}}{\frac{1}{h_{n+1}} - 1} \leq (\geq) \frac{(\frac{1}{z_0})a_n}{\frac{1}{h_n} - 1} = (\frac{1}{z_0})p_n,$$

and the result follows. □

In light of (2.6.8), if $\{p_n;\ n = 0, 1, 2, \cdots\}$ is DS-DFR (DS-IFR) then the sequence $\{p_{n+1}/p_n;\ n = 0, 1, 2, \cdots\}$ is bounded from above (below) by its limit. It is not hard to see that if $\{p_n;\ n = 0, 1, 2, \cdots\}$ is D-DFR (D-IFR) then (2.6.9) holds except when $n = 0$.

The following important corollary is similar to theorem 2.4.2.

Corollary 2.6.1 If $\{p_n;\ n = 0, 1, 2, \cdots\}$ is D-IMRL (D-DMRL) then

$$a_{n+1} \leq (\geq) \left(\frac{1}{z_0}\right) a_n; \quad n = 0, 1, 2, \cdots. \tag{2.6.11}$$

Proof. Since $\{p_n^*;\ n = 0, 1, 2, \cdots\}$ is DS-DFR (DS-IFR) and from (2.5.10) it follows that $\{p_n;\ n = 0, 1, 2, \cdots\}$ and $\{p_n^*;\ n = 0, 1, 2, \cdots\}$ have the same radius of convergence z_0, theorem 2.6.1 yields $p_{n+1}^* \leq (\geq) \left(\frac{1}{z_0}\right) p_n^*$. Since (2.5.10) holds, the result follows. □

Finally, we remark that if (2.5.8) holds and $p_n > 0$ for $n > 0$, then it is not hard to see using (2.6.8) that $z_0 = 1/a$. In the next chapter, we consider a detailed analysis of (2.6.1) in the practically important case when $\{p_n;\ n = 0, 1, 2, \cdots\}$ is a mixed Poisson distribution.

3

Mixed Poisson distributions

One of the most important counting distributional classes in insurance modelling is the class of mixed Poisson distributions. A mixed Poisson distribution is often used to model the number of losses or claims arising from a group of risks where the risk level among the group retains heterogeneity which can not be classified by underwriting criteria. However, it may be reasonable to assume that the risk level follows a probability distribution, and given the risk level the number of losses follows a Poisson distribution. Thus, the number of losses follows a mixed Poisson distribution. Examples can be found in Klugman, Panjer and Willmot (1998). An excellent reference for mixed Poisson distributions is the book by Grandell (1997). In this chapter, we focus on relations between the tail of a mixed Poisson distribution and its mixing distribution. We begin with a representation of tail probabilities of the mixed Poisson distribution, and discuss ratios of these tail probabilities and their connection to the radius of convergence of its probability generating function. Special attention is given to the mixing distribution and its reliability classification. We derive reliability based bounds for the ratios of tail probabilities. In the final section, we present some asymptotic results for mixed Poisson distributions.

Although this chapter deals primarily with right tail probabilities of mixed Possion distributions, recursive numerical evaluation of the probability mass function is often possible. See Grandell (1997, chapter 2) and references therein for details.

3.1 Tails of mixed Poisson distributions

Suppose that $K(x)$ is a df on $(0, \infty)$. For a fixed $\lambda > 0$, define

$$p_n = \int_0^\infty \frac{(\lambda x)^n e^{-\lambda x}}{n!} dK(x); \quad n = 0, 1, 2, \cdots. \tag{3.1.1}$$

Then $\{p_n;\ n = 0, 1, 2, \cdots\}$ form a discrete mixed Poisson probability distribution. The distribution with df $K(x)$ is called the associated mixing distribution. Of special interest in the present context are the tail probabilities

$$a_n = \sum_{k=n+1}^{\infty} p_k;\ n = 0, 1, 2, \cdots,$$

where p_k is given in (3.1.1). The following is a well-known result (e.g. Grandell, 1997, p. 14).

Lemma 3.1.1 The tail probabilities satisfy

$$a_n = \sum_{k=n+1}^{\infty} p_k = \lambda \int_0^\infty \frac{(\lambda x)^n e^{-\lambda x}}{n!} \overline{K}(x) dx; \quad n = 0, 1, 2, \cdots, \tag{3.1.2}$$

where $\overline{K}(x) = 1 - K(x)$ is the survival function.

Proof: From (2.1.13), it follows that

$$\begin{aligned} a_n &= \sum_{k=n+1}^{\infty} \int_0^\infty \frac{(\lambda y)^k e^{-\lambda y}}{k!} dK(y) \\ &= \int_0^\infty \left\{ e^{-\lambda y} \sum_{k=n+1}^{\infty} \frac{(\lambda y)^k}{k!} \right\} dK(y) \\ &= \int_0^\infty \int_0^y \frac{\lambda(\lambda x)^n e^{-\lambda x}}{n!} dx dK(y) \end{aligned}$$

and interchanging the order of integration yields

$$a_n = \int_0^\infty \left\{ \int_x^\infty dK(y) \right\} \frac{\lambda(\lambda x)^n e^{-\lambda x}}{n!} dx$$

from which (3.1.2) follows. □

In what follows we focus on bounds on the ratio a_{n+1}/a_n. In other words, we are looking for the best value of ϕ in the inequality (2.6.2) (since (3.1.2) implies that $a_n > 0$ for $n > 0$). The results are very much dependent on the distributional classes of $K(x)$. Although there is a very close connection between reliability properties of $\{p_n;\ n = 0, 1, 2, \cdots\}$ and those of $K(x)$ as discussed in detail by Vinogradov (1973) and Block and Savits (1980), here we are primarily interested in (2.6.2). See also Grandell (1997, chapter 7).

In order to proceed, we derive an alternative representation for a_n. We assume without loss of generality in what follows that $\overline{K}(x) > 0$, since otherwise the range of integration becomes finite and $\overline{K}(x)$ is positive in the integrand anyway. We have the following alternative representation for the tail probabilities (Willmot and Lin, 1994).

Theorem 3.1.1 The tail probabilities satisfy

$$a_{n+1} = \lambda \int_0^\infty \phi(x) \frac{(\lambda x)^n e^{-\lambda x}}{n!} \overline{K}(x)dx; \quad n = 0, 1, 2, \cdots, \tag{3.1.3}$$

where

$$\phi(x) = \lambda \frac{\int_x^\infty e^{-\lambda y}\overline{K}(y)dy}{e^{-\lambda x}\overline{K}(x)}, \quad x \geq 0. \tag{3.1.4}$$

Proof: Replace n by $n+1$ in (3.1.2) and use integration by parts with $u = \lambda(\lambda x)^{n+1}/(n+1)!$ and $dv = e^{-\lambda x}\overline{K}(x)dx$ to obtain

$$\begin{aligned} a_{n+1} &= -\frac{\lambda(\lambda x)^{n+1}}{(n+1)!} \int_x^\infty e^{-\lambda y}\overline{K}(y)dy \Bigg|_{x=0}^{\infty} \\ &+ \lambda^2 \int_0^\infty \frac{(\lambda x)^n}{n!} \int_x^\infty e^{-\lambda y}\overline{K}(y)dydx. \end{aligned}$$

Now,

$$\begin{aligned} 0 &\leq \frac{\lambda(\lambda x)^{n+1}}{(n+1)!} \int_x^\infty e^{-\lambda y}\overline{K}(y)dy \\ &\leq \lambda \int_x^\infty \frac{(\lambda y)^{n+1} e^{-\lambda y}}{(n+1)!} \overline{K}(y)dy \leq \lambda \int_x^\infty \frac{(\lambda y)^{n+1} e^{-\lambda y}}{(n+1)!} dy \end{aligned}$$

and since the right term is the tail of a gamma distribution, it follows that

$$\lim_{x\to\infty} \frac{\lambda(\lambda x)^{n+1}}{(n+1)!} \int_x^\infty e^{-\lambda y}\overline{K}(y)dy = 0.$$

Similarly, when $x = 0, (\lambda x)^{n+1} = 0$ and

$$\int_0^\infty e^{-\lambda y}\overline{K}(y)dy \leq \int_0^\infty e^{-\lambda y}dy = \lambda^{-1} < \infty.$$

Therefore,

$$a_{n+1} = \lambda^2 \int_0^\infty \frac{(\lambda x)^n}{n!} \int_x^\infty e^{-\lambda y}\overline{K}(y)dydx$$

which is (3.1.3) since (3.1.4) holds. □

The above theorem is very useful for the present purposes. First, if $\phi(x) \leq (\geq)\ \phi$, then $a_{n+1} \leq (\geq)\ \phi a_n$. Second there is a close connection between $\phi(x)$ and the radius of convergence z_0 of the generating function of $\{p_n;\ n = 0, 1, 2, \cdots\}$ and $\{a_n;\ n = 0, 1, 2, \cdots\}$, as shown in the following sections.

3.2 The radius of convergence

As shown in section 2.6, the radius of convergence of the pgf of the counting random variable N plays an important role in identifying ϕ in (2.6.2) for certain classes of distributions. In this section, we investigate in detail the relation between the radius of convergence and the mixing distribution $K(x)$. The pgf of $\{p_n;\ n = 0, 1, 2, \cdots\}$ is, in this case

$$P(z) = \sum_{n=0}^{\infty} p_n z^n = \int_0^{\infty} e^{\lambda x(z-1)} dK(x), \quad |z| < z_0, \tag{3.2.1}$$

where z_0 is the radius of convergence. Similarly, from (3.1.2),

$$A(z) = \sum_{n=0}^{\infty} a_n z^n = \lambda \int_0^{\infty} e^{\lambda x(z-1)} \overline{K}(x) dx, \quad |z| < z_0. \tag{3.2.2}$$

In what follows, we utilize this special structure of the mixed Poisson distribution to determine z_0. First, note that if $\overline{K}(m) = 0$ for $m < \infty$, then $z_0 = \infty$. To see this, note that if $|z| < \infty$

$$|P(z)| = \left| \int_0^{m} e^{\lambda x(z-1)} dK(x) \right| \leq m \sup_{0 \leq x \leq m} \left| e^{\lambda x(z-1)} \right| < \infty.$$

Henceforth, we shall assume that $\overline{K}(x) > 0$. We have the following result.

Theorem 3.2.1 If $\phi(\infty) = \lim_{x\to\infty} \phi(x) < \infty$ where $\phi(x)$ is given by (3.1.4), then $z_0 = 1/\phi(\infty)$.

Proof: By the ratio test for convergence, it suffices to show that

$$\lim_{n\to\infty} a_{n+1}/a_n = \phi(\infty).$$

To show this, we may show that given any $\epsilon > 0$, it follows that

$$|(a_{n+1}/a_n) - \phi(\infty)| < \epsilon$$

for n sufficiently large. But from (3.1.3) and (3.1.2),

$$\left| \frac{a_{n+1}}{a_n} - \phi(\infty) \right| = \frac{\lambda}{a_n} \left| \int_0^{\infty} \frac{(\lambda x)^n}{n!} e^{-\lambda x} \overline{K}(x) \left\{ \phi(x) - \phi(\infty) \right\} dx \right|.$$

Choose $x_0 > 0$ such that $|\phi(x) - \phi(\infty)| \le \epsilon/2$ for all $x > x_0$. Then

$$\begin{aligned}\left|\frac{a_{n+1}}{a_n} - \phi(\infty)\right| &\le \frac{\lambda}{a_n}\int_0^{x_0} \frac{(\lambda x)^n}{n!} e^{-\lambda x}\overline{K}(x)\left|\phi(x) - \phi(\infty)\right| dx \\ &+ \frac{\epsilon\lambda}{2a_n}\int_{x_0}^{\infty} \frac{(\lambda x)^n}{n!} e^{-\lambda x}\overline{K}(x)dx \\ &\le \frac{\lambda}{a_n}\int_0^{x_0} \frac{(\lambda x)^n}{n!} e^{-\lambda x}\overline{K}(x)\left|\phi(x) - \phi(\infty)\right| dx + \frac{\epsilon}{2}.\end{aligned}$$

Now, from (3.1.4), since $\overline{K}(y) \le \overline{K}(x)$ for $y \ge x$,

$$\phi(x) \le \lambda \int_x^\infty e^{-\lambda y}dy/e^{-\lambda x} = 1$$

and so $\phi(\infty) \le 1$. Thus $|\phi(x) - \phi(\infty)| \le 2$ and

$$\begin{aligned}\left|\frac{a_{n+1}}{a_n} - \phi(\infty)\right| &\le \frac{\epsilon}{2} + \frac{2\lambda}{a_n}\int_0^{x_0} \frac{(\lambda x)^n}{n!} e^{-\lambda x}dx \\ \le \frac{\epsilon}{2} + \frac{2\int_0^{x_0} x^n e^{-\lambda x}dx}{\int_{2x_0}^\infty x^n e^{-\lambda x}\overline{K}(x)dx} &\le \frac{\epsilon}{2} + \left(\frac{1}{2}\right)^n \frac{2\int_0^{x_0} e^{-\lambda x}dx}{\int_{2x_0}^\infty e^{-\lambda x}\overline{K}(x)dx}.\end{aligned}$$

The second term is less than $\epsilon/2$ for n is sufficiently large, and the result follows. □

It is useful to relate $\phi(x)$ to more well known functions associated with the df $K(x)$. To this end, define the mean residual lifetime to be

$$r_K(x) = \frac{\int_x^\infty \overline{K}(y)dy}{\overline{K}(x)}, \quad x \ge 0. \tag{3.2.3}$$

We have the following result.

Theorem 3.2.2 If $r_K(\infty) = \lim_{x\to\infty} r_K(x) < \infty$ where $r_K(x)$ is given by (3.2.3), then $z_0 = 1 + 1/\{\lambda r_K(\infty)\}$.

Proof: By L'Hopital' rule,

$$\begin{aligned}\lim_{x\to\infty}\frac{\int_x^\infty e^{-\lambda y}\overline{K}(y)dy}{e^{-\lambda x}\int_x^\infty \overline{K}(y)dy} &= \lim_{x\to\infty}\frac{-e^{-\lambda x}\overline{K}(x)}{-\lambda e^{-\lambda x}\int_x^\infty \overline{K}(y)dy - e^{-\lambda x}\overline{K}(x)} \\ &= \lim_{x\to\infty}\frac{1}{\lambda r_K(x) + 1} = \frac{1}{\lambda r_K(\infty) + 1}.\end{aligned}$$

Using (3.1.4) and (3.2.3),

$$\phi(x) = \lambda r_K(x)\frac{\int_x^\infty e^{-\lambda y}\overline{K}(y)dy}{e^{-\lambda x}\int_x^\infty \overline{K}(y)dy},$$

implying that

$$\lim_{x \to \infty} \phi(x) = \frac{\lambda r_K(\infty)}{\lambda r_K(\infty) + 1},$$

and the theorem follows from theorem 3.2.1.

□

If the failure rate of the mixing distribution $K(x)$ exists, a relation between the radius of convergence and the failure rate can be obtained.

Corollary 3.2.1 If $K(x)$ is absolutely continuous with failure rate $\mu_K(x) = -\frac{d}{dx} \ln \overline{K}(x)$ which satisfies $\mu_K(\infty) = \lim_{x \to \infty} \mu_K(x) > 0$, then $z_0 = 1 + \mu_K(\infty)/\lambda$.

Proof: From (2.3.11), $\mu_K(\infty) = 1/r_K(\infty)$ and the result follows from theorem 3.2.2.

□

We remark that the results of corollary 3.2.1 and theorem 3.2.2 follow from (and may be generalized by) properties of the Laplace transform (e.g. Widder, 1946, section 2.2). For our purposes, however, they are sufficient as stated, and it is of interest to incorporate the function $\phi(x)$ into the analysis. This approach is followed in the next section.

3.3 Bounds on ratios of tail probabilities

We now consider bounds on the ratio a_{n+1}/a_n based on $\phi(x)$ given in (3.1.4). The basic idea is to obtain bounds on $\phi(x)$ and to replace $\phi(x)$ in (3.1.3) by its bounds. In order to do so, it is convenient to rewrite (3.1.4) as

$$\phi(x) = \lambda \int_0^\infty e^{-\lambda y} \frac{\overline{K}(x+y)}{\overline{K}(x)} dy, \quad x \geq 0. \tag{3.3.1}$$

Various results now follow directly from (3.3.1).

Theorem 3.3.1 If the mixing df $K(x)$ is UWA (UBA) then

$$\frac{a_{n+1}}{a_n} \leq (\geq) \frac{1}{z_0}; \quad n = 0, 1, 2, \cdots. \tag{3.3.2}$$

Proof: It follows from (3.3.1) and theorem 3.2.2 that

$$\phi(x) \leq (\geq) \lambda \int_0^\infty e^{-\lambda y - y/r_K(\infty)} dy = \lambda/\{\lambda + 1/r_K(\infty)\} = 1/z_0.$$

Then (3.3.2) follows from (3.1.3).

□

Theorem 3.3.1 yields an optimal bound (as discussed previously), but does not require monotonicity of the failure rate or the mean residual lifetime (the special case when $K(x)$ is IMRL (DMRL) may be proved using corollary 2.6.1 together with the results of Block and Savits, 1980), and the boundedness of the UWA (UBA) class is sufficient. For example, when the mixing distribution $K(x)$ is a mixture of Erlangs as in example 2.1.3, the condition in theorem 3.3.1 is satisfied. Thus, an optimal bound can be obtained for mixtures of Erlangs.

Monotonicity of the failure rate is certainly sufficient for an optimal bound however, and in this case neither absolute continuity nor a finite limit on $r_K(y)$ is needed.

Theorem 3.3.2 If $K(x)$ is DFR (IFR), then

$$\frac{a_{n+1}}{a_n} \leq (\geq) \frac{1}{z_0}; \quad n = 0, 1, 2, \cdots. \tag{3.3.3}$$

Proof: Clearly, $\overline{K}(x+y)/\overline{K}(x)$ is nondecreasing (nonincreasing) in x for fixed $y \geq 0$. Thus, from (3.3.1), $\phi(x)$ is nondecreasing (nonincreasing) in x. Since $\overline{K}(x+y)/\overline{K}(x) \leq 1$, it follows from (3.3.1) that $0 \leq \phi(x) \leq \lambda \int_0^\infty e^{-\lambda y} dy = 1$. Since $\phi(x)$ is monotone in x and bounded, $\phi(\infty) = \lim_{x\to\infty} \phi(x)$ exists. Thus, using theorem 3.2.1, $\phi(x) \leq (\geq)\ \phi(\infty) = 1/z_0$ and (3.3.3) follows from (3.1.3). □

It can be shown (e.g. Grandell, 1997, pp. 135-6) that if $K(x)$ is DFR (IFR) then $\{p_n; n = 0, 1, 2, \cdots\}$ is DS-DFR (DS-IFR). It follows from this observation and the discussion in sections 2.5 and 2.6 that

$$\frac{a_1}{a_0} \leq (\geq) \frac{a_{n+1}}{a_n} \leq (\geq) \frac{1}{z_0}; \quad n = 0, 1, 2, \cdots. \tag{3.3.4}$$

Clearly, theorem 3.3.2 follows directly from this result, which actually yields a two-sided bound. One reason for adopting the present line of reasoning is that here we are interested primarily in constructing (optimal) bounds on the ratio a_{n+1}/a_n, and not in reliability classifications of mixed Poisson distributions, which are discussed in detail by Block and Savits (1980). Theorem 3.3.1 follows from the arguments employed here, for example. We remark that Block and Savits (1980) also demonstrate that if $K(x)$ is NWU then $\{p_n; n = 0, 1, 2, \cdots\}$ is DS-NWU (see also Cai and Kalashnikov, 2000), as well as the analogous result in the case when $K(x)$ is NBU. The following bound follows from this result, or directly as we now show.

Theorem 3.3.3 If $K(x)$ is NWU (NBU), then

$$\frac{a_{n+1}}{a_n} \geq (\leq)\ 1 - p_0; \quad n = 0, 1, 2, \cdots. \tag{3.3.5}$$

Proof: Since $\overline{K}(x+y) \geq (\leq)\ \overline{K}(x)\overline{K}(y)$, (3.3.1) yields

$$\phi(x) \geq (\leq)\ \lambda \int_0^\infty e^{-\lambda y}\overline{K}(y)dy = a_0 = 1 - p_0,$$

and (3.3.5) follows from (3.1.3). □

In the special case where $K(x)$ is DFR (IFR), it follows from (3.3.5) with $n = 0$ that $a_1/a_0 \geq (\leq)\ 1 - p_0$, implying that (3.3.4) gives a tighter bound than (3.3.5) in this case.

The following general bound is based on bounds of the failure rate of $K(x)$.

Theorem 3.3.4 If $K(x)$ is absolutely continuous with failure rate $\mu_K(x) = -\frac{d}{dx}\ln \overline{K}(x)$ which satisfies $\mu_K(x) \geq (\leq)\ \mu$ where $0 \leq \mu < \infty$, then

$$\frac{a_{n+1}}{a_n} \leq (\geq)\ \frac{\lambda}{\lambda+\mu}; \quad n = 0, 1, 2, \cdots. \tag{3.3.6}$$

Proof: One has

$$\overline{K}(x+y)/\overline{K}(x) = e^{-\int_0^y \mu_K(x+t)dt} \leq (\geq)\ e^{-\mu y}$$

and from (3.3.1),

$$\phi(x) \leq (\geq)\ \lambda \int_0^\infty e^{-\lambda y - \mu y}dy = \frac{\lambda}{\lambda+\mu},$$

and (3.3.6) follows from (3.1.3). □

The following alternative representation for $\phi(x)$ is also of use.

Lemma 3.3.1 The function $\phi(x)$ given by (3.1.4) may be expressed as

$$\phi(x) = 1 - \int_0^\infty e^{-\lambda y}dK_x(y) \tag{3.3.7}$$

where

$$K_x(y) = 1 - \frac{\overline{K}(x+y)}{\overline{K}(x)}; \quad y \geq 0, \tag{3.3.8}$$

represents the df of the residual lifetime associated with $K(y)$.

Proof: Integration by parts yields

$$\lambda \int_x^\infty e^{-\lambda y}\overline{K}(y)dy = e^{-\lambda x}\overline{K}(x) - \int_x^\infty e^{-\lambda y}dK(y),$$

and from (3.1.4) it follows that

$$\phi(x) = 1 - \frac{\int_x^\infty e^{-\lambda y} dK(y)}{e^{-\lambda x}\overline{K}(x)},$$

from which the result follows. □

The above result expresses $\phi(x)$ in terms of the Laplace-Stieltjes transform of a random variable with df $K_x(y)$. Bounds on $\phi(x)$ based on stochastic ordering arguments are easily employed with the above representation. The following bounds may be obtained in this regard.

Theorem 3.3.5 If $r_K(x) \leq r < \infty$, then

$$\frac{a_{n+1}}{a_n} \leq 1 - e^{-\lambda r}; \quad n = 0, 1, 2, \cdots. \tag{3.3.9}$$

Proof: It follows from section 2.3 that $r_K(x) = \int_0^\infty y dK_x(y)$. Since $e^{-\lambda y}$ is a convex function of y, Jensen's inequality yields from (3.3.7) that

$$\phi(x) \leq 1 - e^{-\lambda r_K(x)}, \quad x \geq 0. \tag{3.3.10}$$

Thus, since $r_K(x) \leq r$, $\phi(x) \leq 1 - e^{-\lambda r}$ from (3.3.10). The result follows from (3.1.3). □

Corollary 3.3.1 If $K(y)$ is NBUE then

$$\frac{a_{n+1}}{a_n} \leq 1 - e^{-\lambda \int_0^\infty y dK(y)}; \quad n = 0, 1, 2, \cdots. \tag{3.3.11}$$

Proof: Theorem 3.3.5 applies with $r = \int_0^\infty y dK(y)$.

□

If $K(x)$ is NBU then (3.3.5) holds as well as (3.3.11). But $1 - p_0 = a_0 = \phi(0) \leq 1 - e^{-\lambda r_K(0)}$ from (3.3.10), and so (3.3.5) is tighter than (3.3.11) in this case.

Similarly, if $\mu_K(x) \geq \mu > 0$ then (3.3.6) applies, and from (2.3.18) it follows that $r_K(x) \leq 1/\mu$, implying that (3.3.9) holds with $r = 1/\mu$, but $\lambda/(\lambda + \mu) \leq 1 - e^{-\lambda/\mu}$ is obviously true since it may be restated as $e^{\lambda/\mu} \geq 1 + \lambda/\mu$. Thus, (3.3.6) is superior to (3.3.9) in this case.

Corollary 3.3.2 If $K(y)$ is UWAE then

$$\frac{a_{n+1}}{a_n} \leq 1 - e^{-\lambda r_K(\infty)}; \quad n = 0, 1, 2, \cdots. \tag{3.3.12}$$

Proof: Theorem 3.3.5 applies with $r = r_K(\infty)$. □

3.4 Asymptotic tail behaviour of mixed Poisson distributions

We now return to the pgf of Poisson mixtures given in (3.1.1), and examine the right tail behaviour of Poisson mixtures. Using the notation $a(x) \sim b(x), x \to \infty$, to mean $\lim_{x\to\infty} a(x)/b(x) = 1$, we have the following result.

Theorem 3.4.1 Suppose that

$$p_n = \int_0^\infty \frac{(\lambda x)^n e^{-\lambda x}}{n!} k(x)dx; \quad n = 0, 1, 2, \cdots \tag{3.4.1}$$

where the density function $k(x)$ satisfies

$$k(x) \sim Cx^\alpha e^{-\beta x}, \quad x \to \infty, \tag{3.4.2}$$

with $C > 0, \beta \geq 0$, and $-\infty < \alpha < \infty$. Then

$$p_n \sim \frac{C}{(\lambda+\beta)^{\alpha+1}} n^\alpha \left(\frac{\lambda}{\lambda+\beta}\right)^n, \quad n \to \infty. \tag{3.4.3}$$

Proof: Consider

$$\begin{aligned} S_n &= \left| \frac{\int_0^\infty x^n e^{-\lambda x} k(x)dx}{C\int_0^\infty x^{n+\alpha} e^{-(\lambda+\beta)x}dx} - 1 \right| \\ &= \frac{\left| \int_0^\infty x^{n+\alpha} e^{-(\lambda+\beta)x} \left\{ \frac{k(x)e^{\beta x}}{Cx^\alpha} - 1 \right\} dx \right|}{\int_0^\infty x^{n+\alpha} e^{-(\lambda+\beta)x}dx}. \end{aligned}$$

Since (3.4.2) holds, there exists $x_0 \in (0, \infty)$ such that $\left| \frac{k(x)e^{\beta x}}{Cx^\alpha} - 1 \right| < \epsilon/2$ for all $x > x_0$ and $\epsilon > 0$. Thus

$$\begin{aligned} S_n &< \frac{\left| \int_0^{x_0} x^{n+\alpha} e^{-(\lambda+\beta)x} \left\{ \frac{k(x)e^{\beta x}}{Cx^\alpha} - 1 \right\} dx \right|}{\int_0^\infty x^{n+\alpha} e^{-(\lambda+\beta)x}dx} + \frac{\epsilon}{2} \\ &< \frac{\epsilon}{2} + \frac{\frac{1}{C}\int_0^{x_0} x^n e^{-\lambda x} k(x)dx + \int_0^{x_0} x^{n+\alpha} e^{-(\lambda+\beta)x}dx}{\int_0^\infty x^{n+\alpha} e^{-(\lambda+\beta)x}dx}. \end{aligned}$$

For $n+\alpha > 0$,

$$
\begin{aligned}
S_n &< \frac{\epsilon}{2} + \frac{x_0^n \left\{ \frac{1}{C} \int_0^{x_0} e^{-\lambda x} k(x) dx + x_0^\alpha \int_0^{x_0} e^{-(\lambda+\beta)x} dx \right\}}{\int_{2x_0}^\infty x^{n+\alpha} e^{-(\lambda+\beta)x} dx} \\
&< \frac{\epsilon}{2} + \left(\frac{x_0}{2x_0}\right)^n \frac{\frac{1}{C} \int_0^\infty k(x) dx + \frac{x_0^\alpha}{\lambda+\beta} \left\{1 - e^{-(\lambda+\beta)x_0}\right\}}{x_0^\alpha \int_{2x_0}^\infty e^{-(\lambda+\beta)x} dx} \\
&= \frac{\epsilon}{2} + \left(\frac{1}{2}\right)^n \frac{\frac{1}{C} + \frac{x_0^\alpha}{\lambda+\beta} \left\{1 - e^{-(\lambda+\beta)x_0}\right\}}{\frac{x_0^\alpha}{\lambda+\beta} e^{-2(\lambda+\beta)x_0}}.
\end{aligned}
$$

For n sufficiently large, $\left(\frac{1}{2}\right)^n$ may be made smaller than $\epsilon/2$ divided by the term it is multiplied by, i.e. $S_n < \frac{\epsilon}{2} + \frac{\epsilon}{2} = \epsilon$. Therefore,

$$
\int_0^\infty x^n e^{-\lambda x} k(x) dx \sim C \int_0^\infty x^{n+\alpha} e^{-(\lambda+\beta)x} dx, \quad n \to \infty. \tag{3.4.4}
$$

For $n+\alpha > 1$,

$$
\int_0^\infty x^{n+\alpha} e^{-(\lambda+\beta)x} dx = \frac{\Gamma(n+\alpha+1)}{(\lambda+\beta)^{n+\alpha+1}},
$$

and from (3.4.1)

$$
p_n \sim \frac{C}{(\lambda+\beta)^{\alpha+1}} \frac{\Gamma(n+\alpha+1)}{n!} \left(\frac{\lambda}{\lambda+\beta}\right)^n, \quad n \to \infty. \tag{3.4.5}
$$

Now Stirling's formula $\Gamma(x) \sim \sqrt{2\pi} e^{-x} x^{x-\frac{1}{2}}, \ \ x \to \infty$ results in

$$
\begin{aligned}
\frac{\Gamma(n+\alpha+1)}{n!} &= \frac{\Gamma(n+\alpha+1)}{\Gamma(n+1)} \sim e^{-\alpha} \left(\frac{n+\alpha+1}{n+1}\right)^{n+1} \frac{(n+\alpha+1)^{\alpha-\frac{1}{2}}}{(n+1)^{-\frac{1}{2}}} \\
&= \frac{e^{-\alpha}}{n^{-\alpha}} \left(1 + \frac{\alpha}{n+1}\right)^{n+1} \left(\frac{n+1+\alpha}{n+1}\right)^{-\frac{1}{2}} \left(\frac{n+\alpha+1}{n}\right)^\alpha \\
&\sim e^{-\alpha} e^\alpha (1)(1) n^\alpha, \quad n \to \infty.
\end{aligned}
$$

That is, $\Gamma(n+\alpha+1)/n! \sim n^\alpha, \ \ n \to \infty$, and the result follows from (3.4.5).

□

Theorem 3.4.1 relates the tail behaviour of p_n to that of the mixing density $k(x)$. A similar result holds if the mixing distribution is discrete and/or C is replaced by a slowly varying function (e.g. Willmot, 1990). In the following example, we apply theorem 3.4.1 to the negative binomial distribution.

Example 3.4.1 The negative binomial distribution
Suppose that $k(x)$ is the gamma pdf given by (2.1.12), i.e.

$$k(x) = \frac{\beta^\alpha x^{\alpha-1} e^{-\beta x}}{\Gamma(\alpha)}, \quad x \geq 0,$$

and

$$\begin{aligned} p_n &= \int_0^\infty \frac{(\lambda x)^n e^{-\lambda x}}{n!} \frac{\beta^\alpha}{\Gamma(\alpha)} x^{\alpha-1} e^{-\beta x} dx \\ &= \frac{\lambda^n \beta^\alpha}{n!\Gamma(\alpha)} \int_0^\infty x^{n+\alpha-1} e^{-(\lambda+\beta)x} dx \\ &= \frac{\lambda^n \beta^\alpha}{n!\Gamma(\alpha)} (\lambda+\beta)^{-n-\alpha} \Gamma(n+\alpha) \\ &= \frac{\Gamma(n+\alpha)}{\Gamma(\alpha) n!} \left(\frac{\beta}{\lambda+\beta}\right)^\alpha \left(\frac{\lambda}{\lambda+\beta}\right)^n \end{aligned} \tag{3.4.6}$$

which is the negative binomial distribution of example 2.5.1 with $\phi = \lambda/(\lambda+\beta)$. In this case (3.4.2) is an equality for all $x \geq 0$ with α replaced by $\alpha - 1$ and $C = \beta^\alpha/\Gamma(\alpha)$. Then from (3.4.3)

$$p_n \sim \left(\frac{\beta}{\lambda+\beta}\right)^\alpha \frac{1}{\Gamma(\alpha)} n^{\alpha-1} \left(\frac{\lambda}{\lambda+\beta}\right)^n, \quad n \to \infty, \tag{3.4.7}$$

which may also be verified directly from (3.4.6) using Stirling's formula.

Using the above, it follows that if the mixing density $k(x)$ is asymptotically gamma, then the mixed Poisson probability p_n is asymptotically negative binomial.

Also, from (3.4.3),

$$\lim_{n\to\infty} \frac{p_{n+1}}{p_n} = \frac{\lambda}{\lambda+\beta}, \tag{3.4.8}$$

and from (2.6.8) it follows that the radius of convergence is $z_0 = (\lambda+\beta)/\lambda$. □

The results easily extend to the situation where the mixing distribution is shifted, as in the following example.

Example 3.4.2 Shifted mixing distributions
Suppose that the mixing distributionis is shifted μ units to the right, so that the mixed Poisson probabilities may be expressed as

$$p_n = \int_\mu^\infty \frac{(\lambda x)^n e^{-\lambda x}}{n!} k(x-\mu) dx; \; n = 0, 1, 2, \cdots. \tag{3.4.9}$$

The probability generating function is thus

$$\begin{aligned} P(z) &= \sum_{n=0}^{\infty} p_n z^n \\ &= \int_{\mu}^{\infty} e^{\lambda x(z-1)} k(x-\mu)dx \\ &= e^{\lambda\mu(z-1)} \int_{0}^{\infty} e^{\lambda x(z-1)} k(x)dx, \end{aligned}$$

from which it follows that the distribution is the convolution of a Poisson distribution with mean $\lambda\mu$ and the mixed Poisson distribution with unshifted mixing distribution. Ruohonen (1988) has discussed the special case with $k(x)$ a gamma density, and it is clear that the resulting distribution is the convolution of a Poisson and a negative binomial distribution.

If the asymptotic result (3.4.2) holds, then

$$k(x-\mu) \sim Ce^{\beta\mu} x^{\alpha} e^{-\beta x}, \quad x \to \infty,$$

and from theorem 3.4.1 it follows that

$$p_n \sim \frac{Ce^{\beta\mu}}{(\lambda+\beta)^{\alpha+1}} n^{\alpha} \left(\frac{\lambda}{\lambda+\beta}\right)^n, \quad n \to \infty, \tag{3.4.10}$$

which generalizes (3.4.3), recovered with $\mu = 0$. □

Further discussion of these and related topics may be found in Willmot (1989a, 1990) and Grandell (1997).

4
Compound distributions

Compound distributions are widely used in modeling the aggregate claims in an insurance portfolio. The counting random variable component in a compound distribution represents the number of claims arising from the insurance portfolio while the associated sequence of iid random variables represents consecutive individual claim amounts. Various quantities of interest such as stop-loss premiums and risk premiums which arise in insurance are closely connected to the tail probabilities of compound distributions. For example, a thick tail of an aggregate claims distribution will result in a high risk premium. In addition to these insurance situations, compound distributions arise in connection with various queueing theoretic models as well as the theory of dams. Asmussen (1987) describes the common probabilistic structure underlying these models. Also see Neuts (1986).

In this chapter quite general classes of upper and lower bounds are presented. These include as special cases bounds which involve the tail of a distribution from particular reliability classes described in chapter 2. The research to date has generally concentrated on finding the best possible exponential bounds. An obvious drawback to the use of exponential bounds is that they are not applicable to long tail distributions such as subexponential distributions or distributions which have only a finite number of moments. However this drawback can be overcome by considering a larger class of distributions and expressing bounds in terms of the tail of a distribution from the class. In this chapter we show that for upper bounds a convenient choice of distributional class is the NWU class and for lower bounds the NBU class.

4.1 Introduction and examples

As in sections 2.5 and 2.6, let N be a counting random variable with probability function $\{p_n;\ n = 0, 1, 2, \cdots,\}$ given by (2.5.1), tail probabilities given by (2.5.2), and respective generating functions given by (2.6.4) and (2.6.5).

Next, let $\{Y_1, Y_2, \cdots\}$ be an independent and identically distributed sequence of positive random variables, independent of N, with common df $F(y) = Pr\{Y \leq y\}$, $y \geq 0$, where Y is an arbitrary Y_i. As usual, let $\overline{F}(y) = 1 - F(y)$. For convenience, we call $\{p_n;\ n = 0, 1, 2, \cdots,\}$ the number of claims distribution and $F(y)$, $y \geq 0$ the individual claim amount distribution.

Also, let $F^{*n}(y) = Pr\,(Y_1 + Y_2 + \cdots + Y_n \leq y)$ for $n = 1, 2, \cdots$ be the n-fold convolution and $\overline{F}^{*n}(y) = Pr\,(Y_1 + Y_2 + \cdots + Y_n > y)$ be the tail.

Of interest here is the random sum, termed the aggregate claims amount, $X = Y_1 + Y_2 + \cdots + Y_N$ (with $X = 0$ if $N = 0$). The df of X, $G(x) = Pr\,(X \leq x)$, is termed as a compound distribution. It is easy to see that

$$G(x) = \sum_{n=0}^{\infty} p_n F^{*n}(x), \quad x \geq 0, \tag{4.1.1}$$

where $F^{*0}(x) = 1$, and therefore

$$\overline{G}(x) = \sum_{n=1}^{\infty} p_n \overline{F}^{*n}(x), \quad x \geq 0. \tag{4.1.2}$$

In general, evaluation of the tail of the aggregate claims $\overline{G}(x)$ is difficult due to the presence of the convolutions. One approach is to identify the Laplace transform of X utilizing the identity

$$E\left(e^{-sX}\right) = P\left\{E\left(e^{-sY}\right)\right\}, \tag{4.1.3}$$

where $P(z)$ is the pgf of N. If the individual claim amount distribution $F(y)$ is closed under convolution, then simplification occurs. See Klugman, Panjer and Willmot (1998, chapter 4), for example. The idea extends easily to mixtures, as in the following example.

Example 4.1.1 Mixture of Erlangs

Recall that a mixture of Erlangs has pdf

$$f(y) = \sum_{k=1}^{r} q_k \frac{\beta(\beta y)^{k-1} e^{-\beta y}}{(k-1)!}, \quad y > 0, \tag{4.1.4}$$

where $\{q_1, q_2, \cdots, q_r\}$ is itself a probability distribution. Therefore, define $Q(z) = \sum_{k=1}^{r} q_k z^k$, and from (4.1.4)

$$E\left(e^{-sY}\right) = Q\left(\frac{\beta}{\beta + s}\right), \tag{4.1.5}$$

which is itself of the form (4.1.3). Then

$$E\left(e^{-sX}\right) = C\left(\frac{\beta}{\beta+s}\right) = \sum_{n=0}^{\infty} c_n \left(\frac{\beta}{\beta+s}\right)^n \tag{4.1.6}$$

where $C(z) = \sum_{n=0}^{\infty} c_n z^n = P\{Q(z)\}$. Clearly, (4.1.6) implies that X is of the random sum form with p_n replaced by c_n and $F(y)$ by $1 - e^{-\beta y},\ y \geq 0$. Then from (2.1.13) and (4.1.2)

$$\overline{G}(x) = \sum_{n=1}^{\infty} c_n \left\{ e^{-\beta x} \sum_{j=0}^{n-1} \frac{(\beta x)^j}{j!} \right\},$$

and interchanging the order of summation,

$$\overline{G}(x) = e^{-\beta x} \sum_{j=0}^{\infty} \overline{C}_j \frac{(\beta x)^j}{j!}, \quad x \geq 0 \tag{4.1.7}$$

where $\overline{C}_j = \sum_{n=j+1}^{\infty} c_n;\ j = 0, 1, 2, \cdots$.

The coefficients $\{c_n;\ n = 0, 1, 2, \cdots\}$ may sometimes be evaluated recursively, as in the case when (2.5.8) holds, for example. See Panjer and Willmot (1992; chapter 6) for details. Furthermore, in the modified geometric case, the coefficients $\{\overline{C}_0, \overline{C}_1, \cdots\}$ may be evaluated recursively. See example 7.1.1 for details. □

Recursive numerical evaluation of compound distributions has been the subject of much attention. In particular, for many choices of $\{p_n; n = 0, 1, 2, \cdots\}$, this is quite feasible. Klugman, Panjer, and Willmot (1998) consider this issue using the models of example 2.5.1, and Grandell (1997, section 8.4) the mixed Poisson class, for example.

There is also some asymptotic help available. Embrechts, Maejima, and Teugels (1985) show that if

$$p_n \sim C n^{\alpha} \tau^n, \quad n \to \infty, \tag{4.1.8}$$

where $C > 0, -\infty < \alpha < \infty$, and $0 < \tau < 1$, it follows that

$$\overline{G}(x) \sim \frac{C x^{\alpha} e^{-\kappa x}}{\kappa \left\{\tau E\left(Y e^{\kappa Y}\right)\right\}^{\alpha+1}}, \quad x \to \infty \tag{4.1.9}$$

where $\kappa > 0$ satisfies

$$E\left(e^{\kappa Y}\right) = \int_0^\infty e^{\kappa y} dF(y) = \frac{1}{\tau}, \tag{4.1.10}$$

if Y has a non-arithmetic (and thus not a discrete counting) distribution. Equation (4.1.10) is quite important for asymptotic analysis as well as bounds, and is referred to as the Lundberg adjustment equation.

It is worth noting that if (4.1.8) holds, then

$$\lim_{n\to\infty} \frac{p_{n+1}}{p_n} = \lim_{n\to\infty} \left(1 + \frac{1}{n}\right)^{\alpha} \tau = \tau,$$

and so $z_0 = 1/\tau$ where z_0 is the radius of convergence of the pgf $P(z)$, as follows from the ratio test for convergence. Also, there is a similar formula to (4.1.9) if Y has a discrete counting distribution, (e.g. Willmot, 1989a) and other asymptotic formulas hold for subexponential distributions where no $\kappa > 0$ satisfies (4.1.10) (e.g. Embrechts and Goldie, 1982, and references therein).

Example 4.1.2 Compound Pascal-exponential

Suppose that $F(y) = 1 - e^{-\beta y}, y > 0$ and N has a Pascal (negative binomial with α an integer) distribution with

$$p_n = \frac{\alpha(\alpha+1)\cdots(\alpha+n-1)}{n!}(1-\phi)^{\alpha}\phi^n; \quad n = 0, 1, 2, \cdots,$$

as in example 2.5.1. Then

$$P(z) = \left(\frac{1-\phi}{1-\phi z}\right)^{\alpha}, \quad |z| < 1/\phi,$$

and from (4.1.3),

$$\begin{aligned}
E\left(e^{-sX}\right) &= \left(\frac{1-\phi}{1-\phi\frac{\beta}{\beta+s}}\right)^{\alpha} \\
&= \left\{\frac{(1-\phi)(\beta+s)}{s+\beta(1-\phi)}\right\}^{\alpha} \\
&= \left\{1-\phi\left(\frac{s}{s+\beta(1-\phi)}\right)\right\}^{\alpha} \\
&= \left\{1-\phi+\phi\frac{\beta(1-\phi)}{s+\beta(1-\phi)}\right\}^{\alpha} \\
&= \sum_{j=0}^{\alpha}\binom{\alpha}{j}(1-\phi)^{\alpha-j}\phi^j\left\{\frac{\beta(1-\phi)}{s+\beta(1-\phi)}\right\}^j.
\end{aligned}$$

Since

$$\left\{\frac{\beta(1-\phi)}{s+\beta(1-\phi)}\right\}^j = \int_0^\infty e^{-sy}\frac{\{\beta(1-\phi)\}^j y^{j-1}e^{-\beta(1-\phi)y}}{(j-1)!}dy,$$

it follows that the Pascal-exponential is also a binomial-exponential and from (2.1.13) and (4.1.2)

$$\begin{aligned}\overline{G}(x) &= \sum_{j=1}^{\alpha}\binom{\alpha}{j}(1-\phi)^{\alpha-j}\phi^j\int_x^\infty \frac{\{\beta(1-\phi)\}^j y^{j-1}e^{-\beta(1-\phi)y}}{(j-1)!}dy \\ &= \sum_{j=1}^{\alpha}\binom{\alpha}{j}(1-\phi)^{\alpha-j}\phi^j\sum_{i=0}^{j-1}\frac{\{\beta(1-\phi)x\}^i}{i!}e^{-\beta(1-\phi)x},\end{aligned}$$

and interchanging the order of summation,

$$\overline{G}(x) = e^{-\beta(1-\phi)x}\sum_{i=0}^{\alpha-1}\frac{\{\beta(1-\phi)x\}^i}{i!}\left\{\sum_{j=i+1}^{\alpha}\binom{\alpha}{j}(1-\phi)^{\alpha-j}\ \phi^j\right\},\ x\geq 0, \tag{4.1.11}$$

a special case of (4.1.7). In the compound geometric-exponential case with $\alpha = 1$, (4.1.11) reduces to

$$\overline{G}(x) = \phi e^{-\beta(1-\phi)x},\quad x\geq 0. \tag{4.1.12}$$

It is not difficult to verify using Stirling's formula that

$$p_n \sim \frac{(1-\phi)^\alpha}{\Gamma(\alpha)}n^{\alpha-1}\phi^n, n\to\infty,$$

i.e. (4.1.8) is satisfied. Then (4.1.9) becomes

$$\overline{G}(x) \sim \frac{\phi^\alpha}{(\alpha-1)!}\{\beta(1-\phi)x\}^{\alpha-1}e^{-\beta(1-\phi)x},\quad x\to\infty, \tag{4.1.13}$$

which is also easily verified directly from (4.1.11). Evidently (4.1.13) is an equality for all $x \geq 0$ when $\alpha = 1$, as is clear from (4.1.12). □

Example 4.1.3 Modified geometric-combinations of exponentials

Suppose that $0 < \phi < 1$ and

$$p_n = (1-p_0)(1-\phi)\phi^{n-1};\quad n = 1,2,3,\cdots. \tag{4.1.14}$$

Then

$$P(z) = \sum_{n=0}^\infty p_n z^n = p_0 + (1-p_0)\frac{(1-\phi)z}{1-\phi z}. \tag{4.1.15}$$

Integration by parts yields the Laplace transform of $\overline{G}(x)$ (e.g. Feller, 1971, p. 435), namely

$$\int_0^\infty e^{-sx}\overline{G}(x)dx = \frac{1}{s}\left\{1 - E\left(e^{-sX}\right)\right\}. \tag{4.1.16}$$

Using (4.1.3), one obtains

$$\int_0^\infty e^{-sx}\overline{G}(x)dx = \frac{1}{s}\left\{1 - p_0 - (1-p_0)\frac{(1-\phi)E(e^{-sY})}{1-\phi E(e^{-sY})}\right\}$$

which may be expressed as

$$\int_0^\infty e^{-sx}\overline{G}(x)dx = \frac{1-p_0}{s}\left\{\frac{1-E(e^{-sY})}{1-\phi E(e^{-sY})}\right\}. \tag{4.1.17}$$

Now let

$$F(y) = 1 - qe^{-\alpha y} - (1-q)e^{-\beta y}, \quad y > 0, \tag{4.1.18}$$

where we assume without loss of generality that $\alpha < \beta$. If $0 < q < 1$ then $F(y)$ is the df of the mixture of two exponential distributions whereas if $q = \beta/(\beta - \alpha) > 1$ then $F(y)$ is the df of the sum of two independent exponentially distributed random variables with parameters α and β. Then

$$E(e^{-sY}) = \int_0^\infty e^{-sy}dF(y) = q\left(\frac{\alpha}{\alpha+s}\right) + (1-q)\left(\frac{\beta}{\beta+s}\right), \tag{4.1.19}$$

and substitution into (4.1.17) yields

$$\begin{aligned}
\int_0^\infty e^{-sx}\overline{G}(x)dx &= \frac{1-p_0}{s}\left\{\frac{q\left(\frac{s}{\alpha+s}\right) + (1-q)\left(\frac{s}{\beta+s}\right)}{1-\phi\left\{q\left(\frac{\alpha}{\alpha+s}\right) + (1-q)\left(\frac{\beta}{\beta+s}\right)\right\}}\right\} \\
&= (1-p_0)\left\{\frac{q(\beta+s) + (1-q)(\alpha+s)}{(\alpha+s)(\beta+s) - \phi\left\{q\alpha(\beta+s) + (1-q)\beta(\alpha+s)\right\}}\right\} \\
&= (1-p_0)\left\{\frac{s + \alpha(1-q) + \beta q}{s^2 + \{(1-q\phi)\alpha + (1-(1-q)\phi)\beta\}s + (1-\phi)\alpha\beta}\right\}.
\end{aligned}$$

The quadratic in the denominator has distinct, real roots since the discriminant may be expressed as

$$\begin{aligned}
&\left\{\alpha(1-q\phi) + \beta\left\{1-(1-q)\phi\right\}\right\}^2 - 4\alpha\beta(1-\phi) \\
= &\left\{\alpha(1-q\phi) + \beta q\phi - \beta(1-\phi)\right\}^2 + 4\beta q\phi(1-\phi)(\beta-\alpha).
\end{aligned}$$

Thus let r_1 and r_2 defined by

$$s^2 + \{\alpha(1-q\phi) + \beta(1-(1-q)\phi)\}s + (1-\phi)\alpha\beta = (s+r_1)(s+r_2)$$

and let $\psi = \alpha(1-q) + \beta q$ for notational convenience. Then

$$\int_0^\infty e^{-sx}\overline{G}(x)dx = (1-p_0)\frac{s+\psi}{(s+r_1)(s+r_2)}$$

and a partial fraction expansion yields

$$\int_0^\infty e^{-sx}\overline{G}(x)dx = \frac{1-p_0}{r_2-r_1}\left\{\frac{\psi-r_1}{s+r_1}+\frac{r_2-\psi}{s+r_2}\right\}.$$

Thus, by the uniqueness of the Laplace transform, it follows that

$$\overline{G}(x) = \frac{1-p_0}{r_2-r_1}\left\{(\psi-r_1)e^{-r_1x}+(r_2-\psi)e^{-r_2x}\right\}, \quad x \geq 0. \qquad (4.1.20)$$

□

The following example gives the general asymptotic formula for a large class of distributions $\{p_n; n = 0, 1, 2, \cdots\}$.

Example 4.1.4 Compound mixed Poisson distributions
Suppose that

$$p_n = \int_0^\infty \frac{(\lambda x)^n e^{-\lambda x}}{n!} k(x)dx; \quad n = 0, 1, 2, \cdots$$

where

$$k(x) \sim Cx^\alpha e^{-\beta x}, \quad x \to \infty,$$

with $C > 0, \beta > 0$ and $-\infty < \alpha < \infty$. Then (3.4.3) holds and is of the form (4.1.8), and so if $\kappa > 0$ satisfies

$$E\left(e^{\kappa Y}\right) = \int_0^\infty e^{\kappa y}dF(y) = \frac{\lambda+\beta}{\lambda},$$

then from (4.1.9), if $F(y)$ is non-arithmetic

$$\overline{G}(x) \sim \frac{C(\lambda+\beta)^{-\alpha-1}x^\alpha e^{-\kappa x}}{\kappa\left\{\frac{\lambda}{\lambda+\beta}E\left(Ye^{\kappa Y}\right)\right\}^{\alpha+1}}, \quad x \to \infty,$$

i.e.

$$\overline{G}(x) \sim \frac{Cx^\alpha e^{-\kappa x}}{\kappa\left\{\lambda E\left(Ye^{\kappa Y}\right)\right\}^{\alpha+1}}, \quad x \to \infty.$$

□

We now show that in some situations the distribution of a random sum of independent but not identically distributed random variables is the same as that of a random sum of independent and identically distributed random variables. As we shall demonstrate, this model has applications in insurance in terms of modeling claims inflation as well as incurred but not reported (IBNR) claims.

In what follows we shall assume that the value of a claim depends on when the claim occurs, and we are interested in the cost associated with all claims which are incurred in the time period $(0,\ t)$. It is convenient if the distribution of the value of the claim is allowed to depend on the time of incurral (say x, where $0 < x < t$) as well as t. Thus, let this df be $F_{t,x}(y),\ y > 0$, and the Laplace-Stieltjes transform

$$\tilde{f}_{t,x}(s) = \int_0^\infty e^{-sy} dF_{t,x}(y). \tag{4.1.21}$$

Furthermore, we shall assume that the value of different claims are independent.

Next, we assume that the number of claims incurred process $\{N_t,\ t \geq 0\}$ is a mixed Poisson process. That is, given that $\Theta = \theta$, $\{N_t,\ t \geq 0\}$ is a Poisson process with parameter θ, and Θ has df $K(\theta)$. See Grandell (1997, section 9.1) for a thorough description of this process, who shows that

$$p_n(t) = Pr\{N_t = n\} = \int_0^\infty \frac{(\theta t)^n e^{-\theta t}}{n!} dK(\theta),\ n = 0, 1, 2, \cdots. \tag{4.1.22}$$

Also, the n $(n \geq 1)$ incurred times in $(0,\ t)$, given that n claims were incurred in $(0,\ t)$, are independent and uniformly distributed over $(0,\ t)$, with constant density $1/t$. In the ordinary Poisson process case with Θ degenerate, this property is discussed in detail by Ross (1996, section 2.3).

Let X_t be the sum of the values of all claims which are incurred in $(0,\ t)$ with $X_t = 0$ if $N_t = 0$, and we shall compute its Laplace-Stieltjes transform. By conditioning on the number and time of the claims incurred, we find that

$$E(e^{-sX_t}) = p_0(t) + \sum_{n=1}^\infty p_n(t) \int_0^t \int_0^t \cdots \int_0^t \left\{ \prod_{i=1}^n \frac{1}{t} \tilde{f}_{t,x_i}(s) dx_i \right\}.$$

The n-fold integral is actually the same integral repeated n times, and thus

$$E(e^{-sX_t}) = \sum_{n=0}^\infty p_n(t) \left\{ \frac{1}{t} \int_0^t \tilde{f}_{t,x}(s) dx \right\}^n.$$

In terms of the pgf of N_t, namely,

$$P_t(z) = \sum_{n=0}^\infty p_n(t) z^n = \int_0^\infty e^{\theta t(z-1)} dK(\theta), \tag{4.1.23}$$

it follows that

$$E(e^{-sX_t}) = P_t\left\{ \tilde{f}_t(s) \right\} \tag{4.1.24}$$

where

$$\tilde{f}_t(s) = \frac{1}{t}\int_0^t \tilde{f}_{t,x}(s)dx. \tag{4.1.25}$$

The representation (4.1.24) of the Laplace-Stieltjes transform is actually quite remarkable. Since (4.1.25) is the Laplace-Stieltjes transform of a distribution (the average of $\tilde{f}_{t,x}(s)$ over $x \in (0,\ t)$), it follows that (4.1.24) is of the form (4.1.3) in the random sum situation. That is, X_t has the same distribution as the simpler random sum described earlier in this section, where the 'individual claim amount distribution' has transform (4.1.25). Consequently, the results of this section may be applied to the analysis of the distribution of X_t, such as the asymptotic result in example 4.1.4. Of course, if $F_{t,x}(y)$ and hence $\tilde{f}_{t,x}(s)$ do not depend on x, then the present model reduces to the simpler model.

By suitable choice of $F_{t,x}(y)$, X_t may be chosen to represent useful quantities in insurance modeling, as in the following example.

Example 4.1.5 An inflation model

We are interested in the distribution of the (discounted) total claims over $(0,\ t)$ in the presence of inflation and interest. The present model is taken from Willmot (1989b) and Grandell (1997, p. 213). Let δ be the force of net interest, i.e. actual interest less claims inflation. Let Y represent the amount payable on a claim occuring at time 0, with df $F(y)$ and Laplace-Stieltjes transform $\tilde{f}(s) = E(e^{-sY})$. The value at time 0 of a claim occuring at time x is then $Ye^{-\delta x}$. Therefore, one has

$$F_{t,x}(y) = Pr\{Ye^{-\delta x} \le y\} = Pr\{Y \le e^{\delta x}y\} = F(e^{\delta x}y),$$

and from (4.1.21),

$$\tilde{f}_{t,x}(s) = \tilde{f}(e^{-\delta x}s). \tag{4.1.26}$$

Then from (4.1.25)

$$F_t(y) = \frac{1}{t}\int_0^t F(e^{\delta x}y)dx,$$

where $F_t(y)$ has Laplace-Stieltjes transform $\tilde{f}_t(s) = \int_0^\infty e^{-sy}dF_t(y)$. Following Grandell (1997, p. 213), a change of variables from x to $r = e^{\delta x}y$ results in

$$F_t(y) = \int_y^{e^{\delta t}y} \frac{F(r)}{\delta t r}dr,$$

which demonstrates upon differentiation that the 'individual claim amount' df $F_t(y)$ is absolutely continuous with density

$$f_t(y) = \frac{F(e^{\delta t}y) - F(y)}{\delta t y},\ y > 0. \tag{4.1.27}$$

The moments associated with (4.1.27) are also easy to obtain. For $k \neq 0$,

$$\begin{aligned}\int_0^\infty y^k f_t(y)dy &= \frac{1}{\delta t}\int_0^\infty y^{k-1}\int_y^{e^{\delta t}y} dF(x)dy \\ &= \frac{1}{\delta t}\int_0^\infty \int_{e^{-\delta t}x}^{x} y^{k-1}dydF(x) \\ &= \frac{1}{\delta t}\int_0^\infty \frac{x^k}{k}\{1-e^{-\delta tk}\}dF(x).\end{aligned}$$

That is, for $k \neq 0$,

$$\int_0^\infty y^k f_t(y)dy = \frac{1-e^{-\delta tk}}{\delta tk}\int_0^\infty y^k dF(y). \tag{4.1.28}$$

If $F(y) = 1 - e^{-\beta y}$, $y > 0$, then (4.1.27) becomes

$$f_t(y) = \frac{e^{-\beta y} - e^{-\beta e^{\delta t}y}}{\delta t y}, \quad y > 0. \tag{4.1.29}$$

To express (4.1.25) in a different manner, let $\beta_t = \beta e^{\delta t}$ and $\phi_t = 1 - e^{-\delta t}$ if $\delta > 0$, whereas if $\delta < 0$ let $\beta_t = \beta$ and $\phi_t = 1 - e^{\delta t}$. Then (4.1.29) may be expressed as

$$\begin{aligned}f_t(y) &= \frac{e^{-\beta_t(1-\phi_t)y} - e^{-\beta_t y}}{-y\ln(1-\phi_t)} \\ &= \sum_{n=1}^\infty \frac{\phi_t^n}{-n\ln(1-\phi_t)} \cdot \frac{\beta_t(\beta_t y)^{n-1}e^{-\beta_t y}}{(n-1)!}.\end{aligned}$$

The Laplace transform is thus

$$\begin{aligned}\int_0^\infty e^{-sy}f_t(y)dy &= \sum_{n=1}^\infty \frac{\phi_t^n}{-n\ln(1-\phi_t)}\left(\frac{\beta_t}{\beta_t+s}\right)^n \\ &= \frac{\ln\left(1-\phi_t\frac{\beta_t}{\beta_t+s}\right)}{\ln(1-\phi_t)}.\end{aligned} \tag{4.1.30}$$

It is not hard to see from example 2.5.1 and (4.1.3) that the 'individual claim amount' density $f_t(y)$ is a compound logarithmic series-exponential pdf. If, in addition, $\{N_t,\ t \geq 0\}$ is an ordinary Poisson process with rate λ, i.e. $Pr(\Theta = \lambda) = 1$, then from (4.1.24) and (4.1.30) one has

$$\begin{aligned}E(e^{-sX_t}) &= \exp\left\{\lambda t\left\{\frac{\ln\left(1-\phi_t\frac{\beta_t}{\beta_t+s}\right)}{\ln(1-\phi_t)} - 1\right\}\right\} \\ &= \exp\left\{\frac{\lambda t}{\ln(1-\phi_t)}\ln\left\{\frac{1-\phi_t\frac{\beta_t}{\beta_t+s}}{1-\phi_t}\right\}\right\}.\end{aligned}$$

In other words,

$$E(e^{-sX_t}) = \left\{ \frac{1-\phi_t}{1-\phi_t \frac{\beta_t}{\beta_t+s}} \right\}^{\alpha}, \tag{4.1.31}$$

where $\alpha = \lambda t/\{-\ln(1-\phi_t)\} = \lambda/|\delta|$. Comparison of (4.1.31) with (4.1.3) reveals that X_t has a compound negative binomial-exponential distribution. It is not difficult to show using (4.1.9) and (4.1.31) that (as in example 4.1.2)

$$Pr(X_t > x) \sim \frac{\phi_t^{\alpha}}{\Gamma(\alpha)} \{\beta_t(1-\phi_t)x\}^{\alpha-1} e^{-\beta_t(1-\phi_t)x}, \quad x \to \infty. \tag{4.1.32}$$

Also, if $\delta > 0$ then it is not difficult to show that $(1-\phi_t)/\{1-\phi_t\beta_t/(\beta_t + s)\} = (\beta + se^{-\delta t})/(\beta + s)$ and thus from (4.1.31)

$$\lim_{t\to\infty} E(e^{-sX_t}) = \left(\frac{\beta}{\beta+s}\right)^{\alpha}, \tag{4.1.33}$$

i.e. X_∞ has a gamma distribution, as shown by Gerber (1979, p. 136), and Ross (1996, p.395). □

The results are easily adapted to the situation where the 'value of the claim' is a discrete counting random variable. In this case, we use the pgf rather than the Laplace-Stieltjes transform, and define $Q_{t,x}(z) = \tilde{f}_{t,x}(-\ln z)$ where $\tilde{f}_{t,x}(s)$ is given in (4.1.21). Then $Q_{t,x}(z)$ is the pgf of the 'value of the claim' at time x. Substitution of $-\ln z$ for s in (4.1.24) and (4.1.25) yields the pgf

$$E(z^{X_t}) = P_t\{Q_t(z)\}, \tag{4.1.34}$$

where

$$Q_t(z) = \frac{1}{t}\int_0^t Q_{t,x}(z)dx. \tag{4.1.35}$$

In many insurance applications, the 'value of the claim' reflects whether or not the claim satisfies a certain property, as will be apparent in what follows. In this case, the 'value of the claim' is either 0 which represents no contribution to the sum, or 1 which represents a contribution of 1 to the sum. Since $Q_{t,x}(0)$ is the probability that the 'value of the claim' is 0, the pgf $Q_{t,x}(z)$ may be expressed as

$$Q_{t,x}(z) = Q_{t,x}(0) + \{1 - Q_{t,x}(0)\}z, \tag{4.1.36}$$

and (4.1.35) becomes

$$Q_t(z) = Q_t(0) + \{1 - Q_t(0)\}z \tag{4.1.37}$$

where

$$Q_t(0) = \frac{1}{t}\int_0^t Q_{t,x}(0)dx. \tag{4.1.38}$$

Thus, if (4.1.37) holds, it follows from (4.1.34) and (4.1.23) that

$$E(z^{X_t}) = \int_0^\infty e^{\theta t\{Q_t(0)+\{1-Q_t(0)\}z-1\}} dK(\theta),$$

i.e.

$$E(z^{X_t}) = \int_0^\infty e^{\theta t\{1-Q_t(0)\}(z-1)} dK(\theta). \tag{4.1.39}$$

Comparison of (4.1.39) with the mixed Poisson pgf (3.2.1) reveals that (4.1.39) is a mixed Poisson pgf, with λ replaced by $t\{1-Q_t(0)\}$. Thus, X_t is simply a mixed Poisson pgf whose distribution may be analyzed using the results of chapter 3. The phenomenon above is referred to as mixed Poisson thinning, and is essentially discussed in Ross (1996, section 2.3) in connection with the ordinary Poisson process.

We now consider a quantity of interest in connection with insurance claim reserving.

Example 4.1.6 Incurred but not reported (IBNR) claims

We begin with the assumption that upon the incurral of a claim, there is a random amount of time L (called the reporting lag) until the claim is reported to the insurer. This time has df $W(x) = Pr(L \leq x),\ x \geq 0$, and let $\overline{W}(x) = 1 - W(x)$. Estimation of the IBNR reserve requires as input the number of IBNR claims X_t. First note that any claim incurred in the time period $(0,\ t)$ is either reported by time t or unreported (IBNR) at time t. If a claim occurs at time x, the claim is reported by time t with probability $Q_{t,x}(0) = W(t-x)$. Thus the claim is an IBNR claim with probability $\overline{W}(t-x)$. It follows from (4.1.38) that

$$Q_t(0) = \frac{1}{t}\int_0^t W(t-x)dx = \frac{1}{t}\int_0^t W(x)dx,$$

and thus

$$1 - Q_t(0) = \frac{1}{t}\int_0^t \overline{W}(x)dx.$$

Substitution into (4.1.39) yields the pgf of the number of IBNR claims, namely

$$E(z^{X_t}) = \int_0^\infty e^{\theta\left\{\int_0^t \overline{W}(x)dx\right\}(z-1)} dK(\theta). \tag{4.1.40}$$

We remark that X_t may also be interpreted as the number in the system at time t in an infinite server queue with mixed Poisson arrivals and service time df $W(x)$. If $\{N_t,\ t \geq 0\}$ is an ordinary Poisson process, then the queue is referred to as the M/G/∞ queue (e.g. Taylor and Karlin, 1998, pp. 564-5, or Ross 1996, p. 70). Clearly, one has

$$\lim_{t\to\infty} E(z^{X_t}) = \int_0^\infty e^{\theta E(L)(z-1)} dK(\theta), \tag{4.1.41}$$

and from (4.1.22) one may conclude upon comparison of the means of X_t and the number of incurred claims N_t, as $t \to \infty$, that one has

$$E(\# \text{ IBNR claims}) = E(\# \text{ incurred claims/unit time})E(\text{reporting lag}),$$

an intuitively reasonable result.

We now consider a more general model, which is in turn a special case of a much more general model discussed by Brown and Ross (1969) in the Poisson case.

Suppose that the number of incurred claims process is a compound mixed Poisson process. That is, times of claims incurral are the same as before, but the number of incurred claims at each claim incurral instant is a random variable with pgf $Q(z) = \sum_{n=1}^{\infty} q_n z^n$. The time to reporting each claim has df $W(x)$. Following the above reasoning one has

$$Q_{t,x}(z) = Q\{W(t-x) + \overline{W}(t-x)z\}, \tag{4.1.42}$$

the pgf of the number of claims not reported of those incurred at time x. The number of IBNR claims then has pgf (4.1.34), where from (4.1.42)

$$Q_t(z) = \frac{1}{t}\int_0^t Q\{W(x) + \overline{W}(x)z\}dx. \tag{4.1.43}$$

In a queueing context, the number of IBNR claims may be interpreted as the number in the system at time t in a batch arrival infinite server queue with mixed Poisson arrivals and service time df $W(x)$. In the case of ordinary Poisson arrivals, this is the $\mathrm{M^X/G/\infty}$ queue. In the special case with $W(x) = 1 - e^{-\beta x}$, this is the $\mathrm{M^X/M/\infty}$ queue, and (4.1.43) becomes

$$Q_t(z) = \sum_{n=0}^{\infty} q_n(t) z^n = \frac{1}{t}\int_0^t Q\{1 + e^{-\beta x}(z-1)\}dx. \tag{4.1.44}$$

Differentiation yields

$$\begin{aligned} \frac{d}{dz}Q_t(z) &= \frac{1}{t}\int_0^t e^{-\beta x}Q'\{1 + e^{-\beta x}(z-1)\}dx \\ &= \frac{1}{\beta t(1-z)}\int_0^t dQ\{1 + e^{-\beta x}(z-1)\}. \end{aligned}$$

That is, (4.1.44) satisfies

$$\frac{d}{dz}Q_t(z) = \frac{Q\{1 - e^{-\beta t} + e^{-\beta t}z\} - Q(z)}{\beta t(1-z)}. \tag{4.1.45}$$

The probabilities with pgf (4.1.44) may be obtained from (4.1.45). Let $Q\{1 - e^{-\beta t} + e^{-\beta t}z\} = \sum_{n=0}^{\infty} q_{n,t} z^n$, and we remark that the probabilities

$\{q_{n,t}; n = 0, 1, 2, \cdots\}$ are easily obtained for many choices of pgf's $Q(z)$. See Willmot and Drekic (2000) for details. Then from (4.1.45) one has

$$\frac{d}{dz}Q_t(z) = \frac{1-Q(z)}{\beta t(1-z)} - \frac{1-Q\{1-e^{-\beta t}+e^{-\beta t}z\}}{\beta t(1-z)}.$$

Using the tail generating function (2.6.5), equating coefficients of z^n yields

$$(n+1)q_{n+1}(t) = \frac{\sum_{k=n+1}^{\infty} q_k - \sum_{k=n+1}^{\infty} q_{k,t}}{\beta t},$$

i.e., with $q_0 = 0$,

$$q_n(t) = \frac{\sum_{k=0}^{n-1}(q_{k,t} - q_k)}{\beta t n}; \ n = 1, 2, 3, \cdots. \tag{4.1.46}$$

In order to obtain the distribution of the number of IBNR claims with pgf (4.1.34) where $Q_t(z)$ is given by (4.1.44), note that from (4.1.23), one has

$$\begin{aligned} E(z^{X_t}) &= \int_0^\infty e^{\theta t\{\sum_{n=0}^{\infty} q_n(t) z^n - 1\}} dK(\theta) \\ &= \int_0^\infty e^{\theta t\{1-q_0(t)\}\{\sum_{n=1}^{\infty} \frac{q_n(t)}{1-q_0(t)} z^n - 1\}} dK(\theta). \end{aligned}$$

That is,

$$E(z^{X_t}) = \int_0^\infty e^{\theta \beta(t)\{Q_t^*(z)-1\}} dK(\theta), \tag{4.1.47}$$

where

$$\beta(t) = t\sum_{n=1}^{\infty} q_n(t) \tag{4.1.48}$$

and

$$Q_t^*(z) = \sum_{n=1}^{\infty} q_n^*(t) z^n \tag{4.1.49}$$

with

$$q_n^*(t) = \frac{q_n(t)}{\sum_{k=1}^{\infty} q_k(t)}; \ n = 1, 2, 3, \cdots. \tag{4.1.50}$$

Substitution of (4.1.46) into (4.1.48) and (4.1.50) allows for explicit (numerical) evaluation of the components of the compound mixed Poisson pgf (4.1.47). Asymptotic evaluation of the probabilities with pgf (4.1.47) is discussed by Grandell (1997, section 8.3), and recursive numerical evaluation for a wide variety of choices of $K(\theta)$, including the ordinary Poisson case, by Grandell (1997, section 8.4).

The analysis is simpler in the truncated geometric case, with $Q(z) = (1-\phi)z/(1-\phi z)$. It is convenient to write

$$Q(z) = \frac{z}{1 - \frac{\phi}{1-\phi}(z-1)},$$

and substitution into (4.1.44) yields

$$\begin{aligned} Q_t(z) &= \frac{1}{t}\int_0^t \frac{1+e^{-\beta x}(z-1)}{1-\frac{\phi}{1-\phi}e^{-\beta x}(z-1)}dx \\ &= \frac{1}{t}\int_0^t \left\{1+\frac{1}{\phi\beta}\cdot\frac{\frac{\phi\beta}{1-\phi}e^{-\beta x}(z-1)}{1-\frac{\phi}{1-\phi}e^{-\beta x}(z-1)}\right\}dx \\ &= 1+\frac{1}{\phi\beta t}\ln\left\{\frac{1-\frac{\phi}{1-\phi}e^{-\beta t}(z-1)}{1-\frac{\phi}{1-\phi}(z-1)}\right\}. \end{aligned}$$

In the compound Poisson case (i.e. the $M^X/M/\infty$ queue) with $P_t(z) = e^{\lambda t(z-1)}$, it follows from the above and (4.1.34) that the number of IBNR claims has pgf

$$E(z^{X_t}) = \left\{\frac{1-\phi-\phi e^{-\beta t}(z-1)}{1-\phi z}\right\}^{\lambda/\phi\beta}. \tag{4.1.51}$$

It is easy to verify that (4.1.51) may be expressed as

$$E(z^{X_t}) = \left\{\frac{1-\phi_t}{1-\phi_t R_t(z)}\right\}^{\lambda/\phi\beta}, \tag{4.1.52}$$

where $\phi_t = 1 - e^{-\beta t}$ and

$$R_t(z) = \frac{1-\phi_t^*}{1-\phi_t^* z}, \tag{4.1.53}$$

with $\phi_t^* = \phi/\{\phi + e^{\beta t}(1-\phi)\}$. The representation (4.1.52) shows that X_t has a compound negative binomial-geometric distribution. Also, from (4.1.51),

$$\lim_{t\to\infty} E(z^{X_t}) = \left(\frac{1-\phi}{1-\phi z}\right)^{\lambda/\phi\beta} \tag{4.1.54}$$

which is a negative binomial pgf, and agrees with Chaudhry and Templeton (1983, p. 277). □

4.2 The general upper bound

In this section we derive general upper bounds on the tail $\overline{G}(x)$ defined by (4.1.2). These analytic results augment exact and asymptotic results.

We assume the existence of $\phi_1 \in (0,1)$ satisfying

$$a_{n+1} \leq \phi_1 a_n; \quad n = 0, 1, 2, \cdots, \tag{4.2.1}$$

which implies that $a_n \leq a_0 \phi_1^n$, i.e. the number of claims distribution is stochastically dominated by a modified geometric distribution with parameter ϕ_1. This dominance assumption is discussed in detail in section 2.6 and chapter 3. Further, we assume that $B(y)$ is a df satisfying

$$\int_0^\infty \left\{\overline{B}(y)\right\}^{-1} dF(y) = \frac{1}{\phi_1}, \tag{4.2.2}$$

where $\overline{B}(y) = 1 - B(y)$ is the associated survival function. Equation (4.2.2) is a generalization of the Lundberg adjustment equation (4.1.10) and is termed as a generalized adjustment equation. The generalization involves replacement of the exponential df $1 - e^{-\kappa y}$ by the df $B(y)$. In many situations the exponential case is convenient and appropriate, although there are situations where this is not true. In particular, if the distribution $F(y)$ has no moment generating function, the Lundberg adjustment equation (4.1.10) may not be used, and an alternative is necessary. Other choices of $B(y)$ which may be appropriate for a particular $F(y)$ include a Pareto df, and more generally, a member of the NWU class. These alternative choices as well as others will be discussed in the remainder of this and the following chapters.

Once a choice of $B(y)$ has been made which is suitable for the distribution $F(y)$ under consideration, it will be necessary in what follows to determine a dominating df $V(x)$ which satisfies

$$\overline{V}(x)\overline{B}(y) \leq \overline{V}(x+y), \ x \geq 0, \ y \geq 0, \tag{4.2.3}$$

where $\overline{V}(x) = 1 - V(x)$. The choice of $V(x)$ will normally be made as a result of reliability properties of $B(y)$, and will be discussed in more detail later. It is actually not necessary that $V(x)$ be a df but it is sufficient for our purposes. We remark that if (4.2.2) holds, then (4.2.3) implies that $\overline{V}(x) > 0$ for $x > 0$. To see this, note that since $F(0) = 0$ it follows from (4.2.2) that $\overline{B}(c) > 0$ for some c. Then (4.2.3) yields $\overline{V}(c) \geq \overline{V}(0)\overline{B}(c)$, and inductively, $\overline{V}(kc) \geq \overline{V}(0)\{\overline{B}(c)\}^k$ for any positive integer k. Since $\overline{V}(x)$ is monotone, one must have $\overline{V}(x) > 0$ for $x > 0$.

Of central importance in deriving an upper bound is the function

$$c(x,z) = \frac{\int_z^\infty \{\overline{B}(y)\}^{-1} dF(y)}{\overline{V}(x-z)\overline{F}(z)}, \quad 0 \leq z \leq x, \ \overline{F}(z) > 0. \tag{4.2.4}$$

Next, define $c_1(x)$ by

$$\frac{1}{c_1(x)} = \inf_{0 \leq z \leq x, \overline{F}(z) > 0} c(x,z), \quad x \geq 0. \tag{4.2.5}$$

Intuitively, $c_1(x)$ is a nondecreasing function which satisfies

$$\overline{F}(z) \leq \frac{c_1(x)}{\overline{V}(x-z)} \int_z^\infty \{\overline{B}(y)\}^{-1} dF(y), \quad 0 \leq z \leq x. \tag{4.2.6}$$

Moreover, if $c(x,z) \geq c_*(x)$, then $1/c_1(x) \geq c_*(x)$.

We are now in a position to state and prove the general upper bound, the importance of which will be made apparent through subsequent corollaries where $B(y)$ and $V(x)$ will be identified.

Theorem 4.2.1 Suppose that $\phi_1 \in (0,1)$ satisfies the dominance condition (4.2.1), and $B(y)$ is a df satisfying the generalized adjustment equation (4.2.2). If $V(x)$ is a df satisfying (4.2.3), then

$$\overline{G}(x) \leq \frac{1-p_0}{\phi_1 \overline{V}(0)} c_1(x), \quad x \geq 0, \tag{4.2.7}$$

where $c_1(x)$ satisfies (4.2.5), with $c(x,z)$ given by (4.2.4).

Proof: Define for $k = 0, 1, 2, \cdots$,

$$\overline{G}_k(z) = \sum_{m=0}^{k} a_m \left\{ \overline{F}^{*(m+1)}(z) - \overline{F}^{*m}(z) \right\}, \quad z \geq 0 \tag{4.2.8}$$

where $\overline{F}^{*0}(z) = 0$. We will show by induction on k that for $0 \leq z \leq x$,

$$\overline{G}_k(z) \leq \frac{1-p_0}{\phi_1} \frac{c_1(x)}{\overline{V}(x-z)}. \tag{4.2.9}$$

It follows from (4.2.6) that

$$\begin{aligned}
\overline{G}_0(z) &= (1-p_0)\,\overline{F}(z) \\
&\leq (1-p_0)\, c_1(x) \left\{\overline{V}(x-z)\right\}^{-1} \int_z^\infty \left\{\overline{B}(y)\right\}^{-1} dF(y) \\
&\leq (1-p_0)\, c_1(x) \left\{\overline{V}(x-z)\right\}^{-1} \int_0^\infty \left\{\overline{B}(y)\right\}^{-1} dF(y) \\
&= \frac{1-p_0}{\phi_1} \frac{c_1(x)}{\overline{V}(x-z)},
\end{aligned}$$

and (4.2.9) holds for $k = 0$. Now, for $m = 0, 1, 2, \cdots$, from the law of total probability

$$\overline{F}^{*(m+1)}(z) = \overline{F}(z) + \int_0^z \overline{F}^{*m}(z-y)\, dF(y),$$

and so

$$\begin{aligned}
&\int_0^z \overline{G}_k(z-y)dF(y) \\
= & \sum_{m=0}^{k} a_m \int_0^z \left\{\overline{F}^{*(m+1)}(z-y) - \overline{F}^{*m}(z-y)\right\} dF(y) \\
= & \sum_{m=0}^{k} a_m \left\{\overline{F}^{*(m+2)}(z) - \overline{F}(z) - \overline{F}^{*(m+1)}(z) + \overline{F}(z)\right\},
\end{aligned}$$

that is

$$\int_0^z \overline{G}_k(z-y)dF(y) = \sum_{m=0}^{k} a_m \left\{\overline{F}^{*(m+2)}(z) - \overline{F}^{*(m+1)}(z)\right\}. \tag{4.2.10}$$

Then, using (4.2.1),

$$\begin{aligned}
\overline{G}_{k+1}(z) &= a_0\overline{F}(z) + \sum_{m=1}^{k+1} a_m \left\{\overline{F}^{*(m+1)}(z) - \overline{F}^{*m}(z)\right\} \\
&\leq (1-p_0)\overline{F}(z) + \phi_1 \sum_{m=1}^{k+1} a_{m-1} \left\{\overline{F}^{*(m+1)}(z) - \overline{F}^{*m}(z)\right\} \\
&= (1-p_0)\,\overline{F}(z) + \phi_1 \sum_{m=0}^{k} a_m \left\{\overline{F}^{*(m+2)}(z) - \overline{F}^{*(m+1)}(z)\right\},
\end{aligned}$$

and from (4.2.10),

$$\overline{G}_{k+1}(z) \leq (1-p_0)\,\overline{F}(z) + \phi_1 \int_0^z \overline{G}_k(z-y)dF(y). \tag{4.2.11}$$

Assume, by the inductive hypothesis, that (4.2.9) holds for k, and from (4.2.11) and (4.2.3),

$$\begin{aligned}
\overline{G}_{k+1}(z) &\leq (1-p_0)\,\overline{F}(z) + \phi_1 \int_0^z \overline{G}_k(z-y)dF(y) \\
&\leq \frac{(1-p_0)\,c_1(x)}{\overline{V}(x-z)} \int_z^\infty \left\{\overline{B}(y)\right\}^{-1} dF(y) \\
&\quad + (1-p_0)\,c_1(x) \int_0^z \left\{\overline{V}(x+y-z)\right\}^{-1} dF(y) \\
&\leq \frac{(1-p_0)\,c_1(x)}{\overline{V}(x-z)} \int_0^\infty \left\{\overline{B}(y)\right\}^{-1} dF(y) \\
&= \frac{1-p_0}{\phi_1} \frac{c_1(x)}{\overline{V}(x-z)}.
\end{aligned}$$

Thus, by induction, (4.2.9) holds for $k = 0, 1, 2, \cdots$. Now from (4.2.8) and $a_m = a_{m+1} + p_{m+1}$, we obtain (since $\overline{F}^{*0}(z) = 0$)

$$\begin{aligned}\overline{G}_k(z) &= \sum_{m=0}^{k} a_m \overline{F}^{*(m+1)}(z) - \sum_{m=0}^{k} a_m \overline{F}^{*m}(z) \\ &= \sum_{m=0}^{k} a_{m+1} \overline{F}^{*(m+1)}(z) + \sum_{m=0}^{k} p_{m+1} \overline{F}^{*(m+1)}(z) - \sum_{m=1}^{k} a_m \overline{F}^{*m}(z) \\ &= \sum_{m=1}^{k+1} a_m \overline{F}^{*m}(z) + \sum_{m=1}^{k+1} p_m \overline{F}^{*m}(z) - \sum_{m=1}^{k} a_m \overline{F}^{*m}(z).\end{aligned}$$

To summarize, for $k = 0, 1, 2, \cdots$,

$$\overline{G}_k(z) = a_{k+1} \overline{F}^{*(k+1)}(z) + \sum_{m=1}^{k+1} p_m \overline{F}^{*m}(z), \quad z \geq 0. \tag{4.2.12}$$

We now show that $\lim_{k\to\infty} a_{k+1} \overline{F}^{*(k+1)}(z) = 0$. This is true since $a_{k+1} \overline{F}^{*(k+1)}(z) \leq a_{k+1}$ and $\lim_{k\to\infty} a_{k+1} = 0$ since $\{p_k; k = 0, 1, 2, \cdots\}$ is a probability distribution. Therefore, from (4.2.12) and (4.1.2),

$$\lim_{k\to\infty} \overline{G}_k(z) = \overline{G}(z), \quad z \geq 0. \tag{4.2.13}$$

Then by (4.2.13)

$$\overline{G}(z) = \lim_{k\to\infty} \overline{G}_k(z) \leq \frac{1-p_0}{\phi_1} \frac{c_1(x)}{\overline{V}(x-z)}; \quad 0 \leq z \leq x, \tag{4.2.14}$$

and (4.2.7) follows from (4.2.14) with $z = x$. □

The inductive proof of the above theorem is essentially due to Cai and Wu (1997). Also, the dominance condition (4.2.1) may be replaced in theorem 4.2.1 by the slightly weaker condition $a_n \leq \phi_1^n a_0$, as may be seen from theorem 7.1.2 and the discussion following it. The important special case when $B(y)$ is an exponential df is now given. As we will see later, it is a refinement of the well-known Lundberg bound in many cases.

Corollary 4.2.1 If $\phi_1 \in (0, 1)$ satisfies (4.2.1), and $\kappa > 0$ satisfies

$$\int_0^\infty e^{\kappa y} dF(y) = \frac{1}{\phi_1}, \tag{4.2.15}$$

then

$$\overline{G}(x) \leq \frac{1-p_0}{\phi_1} \alpha_1(x) e^{-\kappa x}, \quad x \geq 0 \tag{4.2.16}$$

where

$$\frac{1}{\alpha_1(x)} = \inf_{0 \le z \le x, \overline{F}(z) > 0} \int_z^\infty e^{\kappa y} dF(y) \Big/ \left\{ e^{\kappa z} \overline{F}(z) \right\}, \quad x \ge 0. \tag{4.2.17}$$

Proof: Let $B(y) = 1 - e^{-\kappa y}$. Then (4.2.3) holds with $V(y) = B(y)$, and theorem 4.2.1 applies. Then from (4.2.4)

$$c(x, z) = e^{\kappa x} \frac{\int_z^\infty e^{\kappa y} dF(y)}{e^{\kappa z} \overline{F}(z)},$$

and from (4.2.5) and (4.2.17)

$$\frac{1}{c_1(x)} = \frac{e^{\kappa x}}{\alpha_1(x)},$$

i.e. $c_1(x) = \alpha_1(x) e^{-\kappa x}$. □

It is worth noting that

$$\int_z^\infty e^{\kappa y} dF(y) \ge \int_z^\infty e^{\kappa z} dF(y) = e^{\kappa z} \overline{F}(z),$$

and therefore $1/\alpha_1(x) \ge 1$, i.e. $\alpha_1(x) \le 1$. Thus, (4.2.16) yields

$$\overline{G}(x) \le \frac{1 - p_0}{\phi_1} e^{-\kappa x}, \quad x \ge 0, \tag{4.2.18}$$

an extremely simple but useful bound. Improved choices for $\alpha_1(x)$ will be discussed in the next chapter.

As discussed previously, it is useful to consider generalizations of the exponential assumption in some situations, and the NWU class is a straightforward but useful generalization since it includes the Pareto distribution (to be discussed in section 6.2). The following special case is due to Cai and Garrido (1999).

Corollary 4.2.2 If $\phi_1 \in (0, 1)$ satisfies (4.2.1), and $B(y)$ is a NWU df satisfying (4.2.2), then

$$\overline{G}(x) \le \frac{1 - p_0}{\phi_1} c_1(x), \quad x \ge 0, \tag{4.2.19}$$

where $c_1(x)$ satisfies (4.2.5) with

$$c(x, z) = \frac{\int_z^\infty \{\overline{B}(y)\}^{-1} dF(y)}{\overline{B}(x - z) \overline{F}(z)}, \quad x \ge 0,\ 0 \le z \le x. \tag{4.2.20}$$

Proof: Since $B(y)$ is NWU, (4.2.3) holds with $V(y) = B(y)$, and theorem 4.2.1 applies. This proof assumes that $\overline{B}(0) = 1$, but the result holds without this restriction (Cai and Garrido, 1999). See also the discussion following (4.4.13). □

The next result is important since it gives an upper bound that can be computed easily, and generalizes (4.2.18).

Corollary 4.2.3 Suppose that $\phi_1 \in (0,1)$ satisfies (4.2.1) and $B(y)$ is a df satisfying (4.2.2). If $V(x)$ is a df satisfying (4.2.3), then

$$\overline{G}(x) \leq \frac{1-p_0}{\phi_1 \overline{V}(0)} \overline{V}(x), \quad x \geq 0. \tag{4.2.21}$$

In particular, if $\phi_1 \in (0,1)$ satisfies (4.2.1) and $B(y)$ is a NWU df satisfying (4.2.2), then

$$\overline{G}(x) \leq \frac{1-p_0}{\phi_1} \overline{B}(x), \quad x \geq 0. \tag{4.2.22}$$

Proof: Since $\overline{V}(y)$ is nonincreasing , it follows from (4.2.3) for $y \geq z$ that

$$\overline{V}(x-z)\overline{B}(y) \leq \overline{V}(x+y-z) \leq \overline{V}(x),$$

and so from (4.2.4),

$$c(x,z) \geq \frac{\int_z^\infty \{\overline{V}(x)\}^{-1} dF(y)}{\overline{F}(z)} = \{\overline{V}(x)\}^{-1}.$$

Thus, from (4.2.5), $1/c_1(x) \geq 1/\overline{V}(x)$, i.e. $c_1(x) \leq \overline{V}(x)$. Then (4.2.21) follows from theorem 4.2.1. Clearly, (4.2.22) is the special case where $V(x) = B(x)$ and follows from corollary 4.2.2. □

The above corollaries appear to reflect the most important applications of theorem 4.2.1, but the following results are also of interest.

Corollary 4.2.4 If $\phi_1 \in (0,1)$ satisfies (4.2.1), and $B(y)$ is a NWUC df satisfying (4.2.2), then

$$\overline{G}(x) \leq \frac{1-p_0}{\phi_1} \overline{B}_1(x), \quad x \geq 0, \tag{4.2.23}$$

with $B_1(y) = 1 - \overline{B}_1(y) = \displaystyle\int_0^y \overline{B}(t)dt \Big/ \int_0^\infty \overline{B}(t)dt$ the equilibrium df of $B(y)$.

Proof: Since $B(y)$ is NWUC, (4.2.3) holds with $V(y) = B_1(y)$. The result follows from corollary 4.2.3 and the fact that $\overline{B}_1(0) = 1$. □

For the larger NWUE class, we have the following.

Corollary 4.2.5 If $\phi_1 \in (0,1)$ satisfies (4.2.1), and $B(y)$ is a NWUE df satisfying (4.2.2), then

$$\overline{G}(x) \leq \frac{1-p_0}{\phi_1} \frac{\int_0^\infty y dB(y)}{x + \int_0^\infty y dB(y)}, \quad x \geq 0. \tag{4.2.24}$$

Proof: It follows from (2.4.3) that $\overline{B}(y) \leq \overline{V}(y)$ where

$$\overline{V}(x) = \frac{\int_0^\infty y dB(y)}{x + \int_0^\infty y dB(y)}.$$

It is not hard to see that $V(x)$ is a Pareto df which is DFR and hence NWU. Thus, $\overline{B}(y) \leq \overline{V}(y) \leq \overline{V}(x+y)/\overline{V}(x)$ and (4.2.3) is satisfied with $V(0) = 0$. The result follows from corollary 4.2.3. □

Improvements to (4.2.23) and (4.2.24) are easily obtained using theorem 4.2.1, rather than corollary 4.2.3.

As discussed in section 2.6, if the dominance condition (4.2.1) holds, then $\phi_1^{-1} \leq z_0$, and if $a_{n+1} \leq a_n/z_0$, then (4.2.2) becomes

$$\int_0^\infty \left\{\overline{B}(y)\right\}^{-1} dF(y) = z_0. \tag{4.2.25}$$

Conditions under which $a_{n+1} \leq a_n/z_0$ are given in section 2.6 and chapter 3. Intuitively, the right hand side of the generalized adjustment equation (4.2.2) can be made no larger than z_0, yielding (4.2.25), and so $\left\{\overline{B}(y)\right\}^{-1}$ is "maximized" with $\phi_1^{-1} = z_0$, i.e. $\overline{B}(y)$ is "minimized", and this leads to a best bound in theorem 4.2.1. Hence, the choice $\phi_1^{-1} = z_0$ is fundamentally important for the present application.

Markov's inequality may be applied in the present context as well, yielding

$$\begin{aligned}\overline{G}(x) &= \int_x^\infty dG(y) \leq \overline{B}(x) \int_x^\infty \left\{\overline{B}(y)\right\}^{-1} dG(y) \\ &\leq \overline{B}(x) \int_0^\infty \left\{\overline{B}(y)\right\}^{-1} dG(y),\end{aligned}$$

i.e. $\overline{G}(x) \leq E\left\{1/\overline{B}(X)\right\} \overline{B}(x), x \geq 0$. But from (4.2.2),

$$\begin{aligned}(1-p_0)/\phi_1 &= (1-p_0) E\left\{1/\overline{B}(Y_1)\right\} \\ &= \sum_{n=1}^\infty p_n E\left\{1/\overline{B}(Y_1)\right\}\end{aligned}$$

$$
\begin{aligned}
&\leq p_0 E\left\{1/\overline{B}(0)\right\} + \sum_{n=1}^{\infty} p_n E\left\{1/\overline{B}\left(Y_1 + Y_2 + \cdots + Y_n\right)\right\} \\
&= \sum_{n=0}^{\infty} p_n E\left\{1/\overline{B}(X) | N = n\right\} \\
&= E\left\{1/\overline{B}(X)\right\}.
\end{aligned}
$$

Thus, corollary 4.2.3 yields a tighter inequality then that obtained by Markov's inequality. Moreover, the latter is often inapplicable since $E\left\{1/\overline{B}(X)\right\} = \infty$ in many cases. To see this, suppose that $a_{n+1} \leq a_n/z_0$, and (4.2.25) holds with $\overline{B}(y) = e^{-\kappa y}$. Then from (4.1.3)

$$E\left\{1/\overline{B}(X)\right\} = E\left\{e^{\kappa X}\right\} = P\left\{E\left(e^{\kappa Y}\right)\right\} = P\left(z_0\right)$$

and $E\left\{1/\overline{B}(X)\right\} = \infty$ if $P\left(z_0\right) = \infty$. This is often the case, for example in the negative binomial case with $P(z) = \{(1-\phi)/(1-\phi z)\}^{\alpha}$ where $0 < \alpha \leq 1$ (example 2.5.1).

Willmot and Lin (1994) developed the general exponential bound of corollary 4.2.1 with coefficient $\alpha_1(x) = 1$. Willmot (1994) then introduced the idea of replacing $e^{\kappa y}$ in (4.2.15) by $1/\overline{B}(y)$ where $1 - \overline{B}(y)$ is a NWU df, and proved the slightly weaker version of corollary 4.2.2 with $\overline{B}(x-z)$ in the denominator of (4.2.20) replaced by $\overline{B}(x)/\overline{B}(z)$. The idea of replacing the exponential tail by a NWU tail is of central importance in applications. This allows us to deal with a larger class of claim size distributions and obtain sharper bounds. For example, if a claim size distribution is subexponential or has a finite number of moments, an exponential tail is not appropriate. In this case, we may use a Pareto tail, as discussed in section 6.2. Lin (1996) also derived a slightly weaker version of corollary 4.2.2 with $\overline{B}(y)\overline{B}(x-z)$ replaced by $\overline{B}(x+y-z)$, using a generalization of Wald's identity. This approach led to the derivation by Lin (1996) of lower bounds also, to be discussed in the next section. Cai and Wu (1997) proved Lin's result inductively as in the present proof, also using an inductive argument in the lower bound case. Cai and Garrido (1999) used the Wald-type approach to derive the present corollary 4.2.2, as well as the corresponding lower bound in the next section. Gerber (1994) and Rolski et al. (1999, section 4.5) use slightly different methods of proof. The present theorem 4.2.1 generalizes the result of Cai and Garrido (1999) by replacement of the NWU condition by (4.2.3). A derivation of the main upper bound in theorem 4.2.1 using the Wald-type approach is given in section 4.4.

4.3 The general lower bound

The results in this section are in some cases dual to the upper bound results, although in general the methods of proof are somewhat different.

Similar to the previous section, we begin with the following dominance assumption. We assume the existence of $\phi_2 \in (0,1)$ satisfying

$$a_{n+1} \geq \phi_2 a_n; \quad n = 0, 1, 2, \cdots. \tag{4.3.1}$$

As described in section 2.6 and chapter 3, the optimal choice of ϕ_2 is $1/z_0$ where z_0 is the radius of convergence of the pgf $P(z)$. We then assume that $B(y)$ is a df satisfying the generalized adjustment equation

$$\int_0^\infty \left\{\overline{B}(y)\right\}^{-1} dF(y) = \frac{1}{\phi_2}, \tag{4.3.2}$$

where $\overline{B}(y) = 1 - B(y)$. Following the logic of the previous section, we require a dominating df $V(x)$ to satisfy

$$\overline{V}(x)\overline{B}(y) \geq \overline{V}(x+y), \; x \geq 0, \; y \geq 0, \tag{4.3.3}$$

where $\overline{V}(x) = 1 - V(x)$. Define $c_2(x)$ by

$$\frac{1}{c_2(x)} = \sup_{0 \leq z \leq x, \overline{F}(z) > 0} c(x,z) \tag{4.3.4}$$

where $c(x,z)$ is defined by (4.2.4). The following theorem is a dual to theorem 4.2.1, and the proof is essentially due to Cai and Wu (1997).

Theorem 4.3.1 Suppose that $\phi_2 \in (0,1)$ satisfies the dominance condition (4.3.1), and $B(y)$ is a df satisfying the generalized adjustment equation (4.3.2). If $V(x)$ is a df satisfying (4.3.3), then

$$\overline{G}(x) \geq \frac{1-p_0}{\phi_2 \overline{V}(0)} c_2(x), \quad x \geq 0, \tag{4.3.5}$$

where $c_2(x)$ satisfies (4.3.4) with $c(x,z)$ given by (4.2.4).

Proof: If $c_2(x) = 0$ then (4.3.5) is trivially true. Thus assume that $c_2(x) > 0$. This implies that $c(x,0) = 1/\{\phi_2 \overline{V}(x)\} < \infty$, i.e. $\overline{V}(x) > 0$. Therefore, $\overline{V}(x-z) > 0$ for $0 \leq z \leq x$. Next, define

$$A_1(z) = 1 - \overline{A}_1(z) = \phi_2 \int_0^z \left\{\overline{B}(y)\right\}^{-1} dF(y), \tag{4.3.6}$$

and $A_1(z)$ is a df since (4.3.2) holds. For $k = 1, 2, 3, \cdots$, the sum of k independent random variables, each with df $A_1(z)$, has df $A_k(z) = 1 - \overline{A}_k(z)$ where, by the law of total probability,

$$\overline{A}_{k+1}(z) = \overline{A}_1(z) + \int_0^z \overline{A}_k(z-y) dA_1(y); \; k = 1, 2, 3, \cdots. \tag{4.3.7}$$

We shall show by induction that for $k = 0, 1, 2, \cdots$,

$$\overline{G}_k(z) \geq \frac{(1-p_0)\, c_2(x)\overline{A}_{k+1}(z)}{\phi_2 \overline{V}(x-z)}, \quad 0 \leq z \leq x, \tag{4.3.8}$$

where $\overline{G}_k(z)$ is given by (4.2.8). By (4.3.4),

$$\overline{F}(z) \geq \frac{c_2(x)}{\overline{V}(x-z)} \int_z^\infty \left\{\overline{B}(y)\right\}^{-1} dF(y), \quad 0 \leq z \leq x. \tag{4.3.9}$$

Then

$$\begin{aligned}
\overline{G}_0(z) &= (1-p_0)\,\overline{F}(z) \\
&\geq (1-p_0)\, c_2(x) \left\{\overline{V}(x-z)\right\}^{-1} \int_z^\infty \left\{\overline{B}(y)\right\}^{-1} dF(y) \\
&= \frac{1-p_0}{\phi_2} \frac{c_2(x)}{\overline{V}(x-z)} \overline{A}_1(z)
\end{aligned}$$

using (4.3.6), and (4.3.8) holds with $k = 0$. Using (4.3.1) and (4.2.8)

$$\begin{aligned}
\overline{G}_{k+1}(z) &= a_0 \overline{F}(z) + \sum_{m=1}^{k+1} a_m \left\{\overline{F}^{*(m+1)}(z) - \overline{F}^{*m}(z)\right\} \\
&\geq (1-p_0)\,\overline{F}(z) + \phi_2 \sum_{m=1}^{k+1} a_{m-1} \left\{\overline{F}^{*(m+1)}(z) - \overline{F}^{*m}(z)\right\} \\
&= (1-p_0)\,\overline{F}(z) + \phi_2 \sum_{m=0}^{k} a_m \left\{\overline{F}^{*(m+2)}(z) - \overline{F}^{*(m+1)}(z)\right\},
\end{aligned}$$

and using (4.2.10),

$$\overline{G}_{k+1}(z) \geq (1-p_0)\,\overline{F}(z) + \phi_2 \int_0^z \overline{G}_k(z-y)\,dF(y). \tag{4.3.10}$$

Now assume that (4.3.8) holds for k, and from (4.3.10), (4.3.7), and (4.3.3),

$$\begin{aligned}
&\overline{G}_{k+1}(z) \geq \overline{G}_0(z) + \phi_2 \int_0^z \overline{G}_k(z-y)\,dF(y) \\
\geq\ & \frac{1-p_0}{\phi_2} \frac{c_2(x)}{\overline{V}(x-z)} \overline{A}_1(z) \\
+\ & (1-p_0)\, c_2(x) \int_0^z \left\{\overline{V}(x+y-z)\right\}^{-1} \overline{A}_{k+1}(z-y)\,dF(y) \\
\geq\ & \frac{1-p_0}{\phi_2} \frac{c_2(x)}{\overline{V}(x-z)} \left\{\overline{A}_1(z) + \phi_2 \int_0^z \left\{\overline{B}(y)\right\}^{-1} \overline{A}_{k+1}(z-y)\,dF(y)\right\}
\end{aligned}$$

$$= \frac{1-p_0}{\phi_2} \frac{c_2(x)}{\overline{V}(x-z)} \left\{ \overline{A}_1(z) + \int_0^z \overline{A}_{k+1}(z-y)dA_1(y) \right\}$$

$$= \frac{1-p_0}{\phi_2} \frac{c_2(x)}{\overline{V}(x-z)} \overline{A}_{k+2}(z),$$

and (4.3.8) holds for $k = 0, 1, 2, \cdots$.

It follows from Ross (1996, pp. 99-100) that

$$m(x) = \sum_{k=1}^{\infty} A_k(x) < \infty,$$

implying that $\lim_{k \to \infty} A_k(x) = 0$. Thus,

$$\lim_{k \to \infty} \overline{A}_k(x) = 1. \tag{4.3.11}$$

Clearly, with $x = z$ one obtains from (4.3.8) that

$$\overline{G}_k(x) \geq \frac{1-p_0}{\phi_2 \overline{V}(0)} c_2(x) \overline{A}_{k+1}(x)$$

and, using (4.2.13) and (4.3.11),

$$\overline{G}(x) = \lim_{k \to \infty} \overline{G}_k(x) \geq \frac{1-p_0}{\phi_2 \overline{V}(0)} c_2(x) \lim_{k \to \infty} \overline{A}_{k+1}(x) = \frac{1-p_0}{\phi_2 \overline{V}(0)} c_2(x).$$

□

As in the upper bound case, the dominance condition (4.3.1) may be replaced in theorem 4.3.1 by the slightly weaker condition $a_n \geq \phi_2^n a_0$, as may be seen from theorem 7.1.2 and the discussion following it. The following corollary is a dual to corollary 4.2.1 and provides an exponential lower bound.

Corollary 4.3.1 If $\phi_2 \in (0, 1)$ satisfies (4.3.1), and $\kappa > 0$ satisfies

$$\int_0^{\infty} e^{\kappa y} dF(y) = \frac{1}{\phi_2}, \tag{4.3.12}$$

then

$$\overline{G}(x) \geq \frac{1-p_0}{\phi_2} \alpha_2(x) e^{-\kappa x}, \quad x \geq 0 \tag{4.3.13}$$

where

$$\frac{1}{\alpha_2(x)} = \sup_{0 \leq z \leq x, \overline{F}(z) > 0} \int_z^{\infty} e^{\kappa y} dF(y) \Big/ \left\{ e^{\kappa z} \overline{F}(z) \right\}, \quad x \geq 0. \tag{4.3.14}$$

Proof: Let $B(y) = 1 - e^{-\kappa y}$, and (4.3.3) holds with $V(x) = B(x)$, and theorem 4.3.1 applies. The rest of the proof is the same as that of corollary 4.2.1. □

Nonexponential lower bounds follow in a similar manner to the upper bounds. The following result, due to Cai and Garrido (1999), is a dual to corollary 4.2.2.

Corollary 4.3.2 If $\phi_2 \in (0,1)$ satisfies (4.3.1), and $B(y)$ is a NBU df satisfying (4.3.2), then

$$\overline{G}(x) \geq \frac{1-p_0}{\phi_2} c_2(x), \quad x \geq 0, \tag{4.3.15}$$

where $c_2(x)$ satisfies (4.3.4) with $c(x,z)$ given by (4.2.20).

Proof: Since $B(y)$ is NBU, $\overline{B}(0)\overline{B}(0) \geq \overline{B}(0)$, implying that $\overline{B}(0) = 1$, and (4.3.3) holds with $V(x) = B(x)$. Thus theorem 4.3.1 applies. □

There does not appear to be a dual to corollary 4.2.3, but the following is a simple and general lower bound.

Corollary 4.3.3 Suppose that $\phi_2 \in (0,1)$ satisfies (4.3.1), $B(y)$ is a df satisfying (4.3.2), and $V(x)$ is a df satisfying (4.3.3). If $m = \inf\{x;\ F(x) = 1\}$, then

$$\overline{G}(x) \geq \frac{1-p_0}{\phi_2 \overline{V}(0)} \gamma(x)\overline{V}(x), \quad x \geq 0, \tag{4.3.16}$$

where $\gamma(x) = \max\{\phi_2 \overline{F}(x), \overline{B}(m)\}$. In particular, if $\phi_2 \in (0,1)$ satisfies (4.3.1) and $B(y)$ is a NBU df satisfying (4.3.2), then

$$\overline{G}(x) \geq \frac{1-p_0}{\phi_2 \overline{V}(0)} \gamma(x)\overline{B}(x), \quad x \geq 0, \tag{4.3.17}$$

Proof: Since $\overline{V}(x)$ is nonincreasing in x, it follows that $\overline{V}(x-z) \geq \overline{V}(x)$ for $0 \leq z \leq x$. Thus, from (4.2.4) one has

$$c(x,z) \leq \frac{\int_z^\infty \{\overline{B}(y)\}^{-1} dF(y)}{\overline{V}(x)\overline{F}(z)}.$$

Therefore, if $\overline{B}(m) > 0$

$$c(x,z) \leq \frac{\int_z^m \{\overline{B}(m)\}^{-1} dF(y)}{\overline{V}(x)\overline{F}(z)} = \{\overline{V}(x)\overline{B}(m)\}^{-1},$$

and so from (4.3.4), $1/c_2(x) \le \{\overline{V}(x)\overline{B}(m)\}^{-1}$. Theorem 4.3.1 yields

$$\overline{G}(x) \ge \frac{1-p_0}{\phi_2 \overline{V}(0)} \overline{B}(m)\overline{V}(x), \quad x \ge 0, \tag{4.3.18}$$

which is trivially true if $\overline{B}(m) = 0$. Also,

$$c(x,z) \le \frac{\int_0^\infty \{\overline{B}(y)\}^{-1} dF(y)}{\overline{V}(x)\overline{F}(z)} = \{\phi_2 \overline{V}(x)\overline{F}(x)\}^{-1},$$

for $0 \le z \le x$, which implies using theorem 4.3.1 that

$$\overline{G}(x) \ge (1-p_0)\,\overline{F}(x)\overline{V}(x)/\overline{V}(0), \quad x \ge 0. \tag{4.3.19}$$

Equations (4.3.18) and (4.3.19) together imply (4.3.16). If $B(y)$ is NBU then $V(x) = B(x)$ and (4.3.16) becomes (4.3.17). □

In corollary 4.3.3, m is the right-end point of the region of support of the df $F(y)$. We remark that if $m = \infty$ then $\overline{B}(m) = 0$ and (4.3.18) is of no use, but $\overline{F}(x) > 0$, and (4.3.19) applies. If $m < \infty$ then $\overline{F}(x) = 0$ for $x > m$, but it is usually the case that $\overline{B}(m) > 0$. Thus (4.3.16) normally provides a non-zero lower bound.

As in the upper bound case, a lower bound is easily obtainable when $B(y)$ is NBUC.

Corollary 4.3.4 If $\phi_2 \in (0,1)$ satisfies (4.3.1), and $B(y)$ is a NBUC df satisfying (4.3.2), then

$$\overline{G}(x) \ge \frac{1-p_0}{\phi_2} \gamma(x)\overline{B}_1(x), \quad x \ge 0, \tag{4.3.20}$$

where $\gamma(x)$ is defined in corollary 4.3.3 and $\overline{B}_1(y) = \int_0^y \overline{B}(t)dt \,/ \int_0^\infty \overline{B}(t)dt$ is the equilibrium df of $B(y)$.

Proof: Since $B(y)$ is NBUC, (4.3.3) holds with $V(y) = B_1(y)$, and the result follows from corollary 4.3.3. □

4.4 A Wald-type martingale approach

In this section we briefly outline how the main bounds of section 4.2 and 4.3 may be obtained by identifying a martingale and a related stopping time, and then using optional stopping. The resulting equality is a generalization of Wald's identity (e.g. Grimmett and Stirzaker, 1992, p.467). The approach in this section is essentially due to Lin (1996) and Cai and Garrido (1999).

The inductive approach adopted in section 4.2 and 4.3 is very convenient due to its simplicity, whereas the present martingale approach provide a different type of insight which may often be employed in other more complex models. Consequently, both methods are of interest.

To begin the analysis, note that (4.2.8) and (4.2.13) imply that (with $\overline{F}^{*0}(x) = 0$)

$$\overline{G}(x) = \sum_{n=0}^{\infty} a_n \left\{ \overline{F}^{*(n+1)}(x) - \overline{F}^{*n}(x) \right\}, \quad x \geq 0. \tag{4.4.1}$$

Consider now the process $\{S_n,\ n = 0, 1, 2, \cdots\}$, where $S_0 = 0$ and $S_n = Y_1 + Y_2 + \cdots + Y_n$. Define $N(x) = \sup\{n;\ S_n \leq x\}$, i.e. $N(x)$ is the number of renewals in a renewal process with Y_i the time between the $(i-1)$st and ith renewals. Then $N(x)$ is a stopping time with

$$Pr\{N(x) = n\} = \overline{F}^{*(n+1)}(x) - \overline{F}^{*n}(x);\ n = 0, 1, 2, \cdots, \tag{4.4.2}$$

and

$$0 \leq S_{N(x)} \leq x < S_{N(x)+1}. \tag{4.4.3}$$

Furthermore, $N(x)$ is finite. This is because for fixed x, $\overline{F}^{*n}(x)$ is increasing in n from $\overline{F}^{*0}(x) = 0$ to $\lim_{n\to\infty} \overline{F}^{*n}(x) = 1$, as is clear from (4.3.11). Thus,

$$\lim_{n\to\infty} Pr\{N(x) \leq n\} = \sum_{n=0}^{\infty} \left\{ \overline{F}^{*(n+1)}(x) - \overline{F}^{*n}(x) \right\} = 1.$$

Obviously, the random sum X may be expressed as S_N, and from (4.4.1) and (4.4.2) it follows that

$$\overline{G}(x) = E\left\{a_{N(x)}\right\}, \quad x \geq 0. \tag{4.4.4}$$

We shall now derive the main upper bound of theorem 4.2.1. Since (4.2.1) holds one has $a_n \leq \phi_1^n a_0$, and hence (4.4.4) implies that

$$\overline{G}(x) \leq \frac{a_0}{\phi_1} E\left\{\phi_1^{N(x)+1}\right\}, \quad x \geq 0. \tag{4.4.5}$$

Next, since (4.2.2) holds, it follows that $\{Z_n,\ n = 0, 1, 2, \cdots\}$ is a martingale, where $Z_0 = 1$ and

$$Z_n = \phi_1^n \Big/ \prod_{i=1}^{n} \overline{B}(Y_i)\ ;\ n = 1, 2, \cdots. \tag{4.4.6}$$

It can be shown (e.g. Lin, 1996) that the optional sampling theorem applies to the finite stopping time $N(x) + 1$, i.e. $E\{Z_{N(x)+1}\} = E(Z_1) = 1$. Equivalently,

$$E\left\{ \phi_1^{N(x)+1} \Big/ \prod_{i=1}^{N(x)+1} \overline{B}(Y_i) \right\} = 1, \tag{4.4.7}$$

which is a generalized Wald's identity. Since (4.2.3) holds and $Y_i > 0$, one has

$$\overline{B}(Y_i) \leq \overline{V}(S_i) / \overline{V}(S_{i-1}); \quad i = 1, 2, 3, \cdots. \tag{4.4.8}$$

Therefore,

$$\prod_{i=1}^{N(x)} \overline{B}(Y_i) \leq \overline{V}(S_{N(x)}) / \overline{V}(0) \tag{4.4.9}$$

since $S_0 = 0$. It then follows from (4.4.7) that

$$\overline{V}(0) E \left\{ \phi_1^{N(x)+1} \Big/ \{\overline{B}(Y_{N(x)+1}) \overline{V}(S_{N(x)})\} \right\} \leq 1.$$

By conditioning on $N(x)$ and $S_{N(x)}$, it follows that

$$\overline{V}(0) E \left\{ \frac{\phi_1^{N(x)+1}}{\overline{V}(S_{N(x)})} \eta(x) \right\} \leq 1, \tag{4.4.10}$$

where

$$\eta(x) = E \left\{ \frac{1}{\overline{B}(Y_{N(x)+1})} \middle| \, N(x), S_{N(x)} \right\}. \tag{4.4.11}$$

Now, given $N(x)$ and $S_{N(x)}$, it is clear from (4.4.3) that $Y_{N(x)+1} > x - S_{N(x)}$. Therefore,

$$\begin{aligned} \eta(x) &= E \left\{ \frac{1}{\overline{B}(Y)} \middle| \, Y > x - S_{N(x)} \right\}. \\ &= \frac{\int_{x - S_{N(x)}}^{\infty} \{\overline{B}(y)\}^{-1} dF(y)}{\overline{F}(x - S_{N(x)})}. \end{aligned}$$

But from (4.2.4),

$$\eta(x) = \overline{V}(S_{N(x)}) c(x, x - S_{N(x)}), \tag{4.4.12}$$

and from (4.4.10),

$$\overline{V}(0) E \left\{ \phi_1^{N(x)+1} c(x, x - S_{N(x)}) \right\} \leq 1.$$

Obviously, (4.2.5) implies that $c(x, x - S_{N(x)}) \geq 1/c_1(x)$, and from (4.4.5),

$$\frac{\overline{V}(0)}{c_1(x)} \frac{\phi_1 \overline{G}(x)}{1 - p_0} \leq \frac{\overline{V}(0)}{c_1(x)} E \left\{ \phi_1^{N(x)+1} \right\} \leq 1,$$

i.e.

$$\overline{G}(x) \leq \frac{1 - p_0}{\phi_1 \overline{V}(0)} c_1(x), \quad x \geq 0. \tag{4.4.13}$$

It is not difficult to see that the term $\overline{V}(0)$ in the denominator may be omitted in the special case where $B(y)$ is NWU. In this case, (4.4.9) may clearly be replaced by $\prod_{i=1}^{N(x)} \overline{B}(Y_i) \leq \overline{B}(S_{N(x)})$.

The proof of the main lower bound in theorem 4.3.1 is similar and is therefore omitted.

5

Bounds based on reliability classifications

The general bounds given in theorems 4.2.1 and 4.3.1 require minimizing or maximizing the function $c(x, z)$ in (4.2.4), which may be awkward to evaluate. Simpler bounds are thus derived in the corollaries of these theorems and they are applicable for an arbitrary individual claim amount df $F(y)$. In this chapter we employ the results of chapter 4 with additional assumptions about the df $F(y)$. By taking advantage of reliability properties of $F(y)$, refinements of these bounds on the tail of the aggregate claims distribution are obtainable. Naturally, these refined bounds are sharper than those given in the corollaries and but still simple enough to use. The next section deals with first order assumptions about the df $F(y)$, by which we mean assumptions about the failure rate and the distribution function of the residual lifetime associated with $F(y)$. The following section deals with assumptions involving the equilibrium df $F_1(y)$.

5.1 First order properties

We begin with the case where the individual claim amount df $F(y)$ is NWU (NBU). In this case, a bound is obtained in terms of the tail of $F(y)$.

Theorem 5.1.1 If $a_{n+1} \leq (\geq)\ \phi a_n$ for $n = 0, 1, 2, \cdots$, and $F(y)$ is an absolutely continuous NWU (NBU) df, then

$$\overline{G}(x) \leq (\geq)\ (1 - p_0) \left\{\overline{F}(x)\right\}^{1-\phi}, \quad x \geq 0. \tag{5.1.1}$$

Proof: Since $F(y)$ is NWU (NBU), so is $B(y)$ where $B(y) = 1 - \{\overline{F}(y)\}^{1-\phi}$ or $\overline{B}(y) = \{\overline{F}(y)\}^{1-\phi}$. Now, by absolute continuity,

$$\phi \int_z^\infty \{\overline{F}(y)\}^{\phi-1} dF(y) = -\{\overline{F}(y)\}^\phi \Big|_z^\infty = \{\overline{F}(z)\}^\phi.$$

That is, with the above choice of $\overline{B}(y)$,

$$\frac{1}{\phi} = \frac{\overline{B}(z)}{\overline{F}(z)} \int_z^\infty \{\overline{B}(y)\}^{-1} dF(y).$$

Since $\overline{B}(x-z)\overline{B}(z) \leq (\geq)\ \overline{B}(x)$, one has from (4.2.4)

$$c(x,z) \geq (\leq)\ \frac{\overline{B}(z)}{\overline{B}(x)\overline{F}(z)} \int_z^\infty \{\overline{B}(y)\}^{-1} dF(y) = \frac{1}{\phi \overline{B}(x)}.$$

Thus, from (4.2.5) ((4.3.4)), $1/\{\phi\overline{B}(x)\} \leq (\geq)\ 1/c_1(x)(1/c_2(x))$ and the result follows from corollary 4.2.2 (4.3.2). □

We leave as an exercise the verification that if $p_n = (1-p_0)(1-\phi)\phi^{n-1}$ for $n = 1, 2, 3, \cdots$ and $\overline{F}(x) = e^{-\beta x}$, $x \geq 0$, then (5.1.1) is an equality. Thus, the bound in theorem 5.1.1 is sharp.

In what follows it will be convenient to weaken the results of theorems 4.2.1 and 4.3.1 slightly, in light of (4.2.3) and (4.3.3), by replacing $c(x,z)$ in (4.2.4) by $\tau(x,z)$ where

$$\tau(x,z) = \frac{1}{\overline{F}(z)} \int_z^\infty \{\overline{V}(x+y-z)\}^{-1} dF(y), \tag{5.1.2}$$

for $0 \leq z \leq x$, $\overline{F}(z) > 0$. In the notation of section 2.3, let T_z be the residual lifetime random variable given that $Y > z$. Thus, $Pr(T_z \leq y) = 1 - \overline{F}(z+y)/\overline{F}(z)$, and a change of variables from y to $y+z$ results in

$$\tau(x,z) = E\{1/\overline{V}(x+T_z)\}. \tag{5.1.3}$$

The results of this section are mainly due to Willmot and Lin (1997a), and those of the next mainly due to Willmot (1997b), both utilizing integration by parts. The stochastic ordering approach presented here is due to Cai and Garrido (1999), and is simpler and involves less extra conditions.

We now present a simplified (and slightly weakened) version of theorem 4.2.1.

Theorem 5.1.2 Suppose that $\phi_1 \in (0,1)$ satisfies (4.2.1), and $B(y)$ is a df satisfying (4.2.2). If $V(x)$ is a df satisfying (4.2.3) and $H_x(y)$ is a df satisfying

$$\overline{H}_x(y) \leq \inf_{0 \leq z \leq x, \overline{F}(z) > 0} \overline{F}(z+y)/\overline{F}(z); \quad y \geq 0, \tag{5.1.4}$$

then

$$\overline{G}(x) \leq \frac{1-p_0}{\phi_1 \overline{V}(0)} \left\{ \int_0^\infty \frac{dH_x(y)}{\overline{V}(x+y)} \right\}^{-1}, \quad x \geq 0. \tag{5.1.5}$$

Proof: For ease of notation, let $H_x(y) = Pr(W_x \leq y), y \geq 0$. Then, from (5.1.4), $Pr(W_x > y) \leq Pr(T_z > y)$ for $0 \leq z \leq x, \overline{F}(z) > 0$. Since $1/\overline{V}(x+y)$ is a nondecreasing function of y, it follows from Ross (1996, p. 405) that

$$\begin{aligned} c(x,z) &\geq \tau(x,z) = E\left\{1/\overline{V}(x+T_z)\right\} \\ &\geq E\left\{1/\overline{V}(x+W_x)\right\} = \int_0^\infty \left\{\overline{V}(x+y)\right\}^{-1} dH_x(y). \end{aligned}$$

Therefore, from (4.2.5),

$$1/c_1(x) \geq \int_0^\infty \left\{\overline{V}(x+y)\right\}^{-1} dH_x(y),$$

or

$$c_1(x) \leq 1/\int_0^\infty \left\{\overline{V}(x+y)\right\}^{-1} dH_x(y).$$

and the result follows from theorem 4.2.1. □

The following dual to theorem 5.1.2 follows.

Theorem 5.1.3 Suppose that $\phi_2 \in (0,1)$ satisfies (4.3.1), and $B(y)$ is a df satisfying (4.3.2). If $V(x)$ is a df satisfying (4.3.3) and $H_x(y)$ is a df satisfying

$$\overline{H}_x(y) \geq \sup_{0 \leq z \leq x, \overline{F}(z) > 0} \overline{F}(z+y)/\overline{F}(z), \quad y \geq 0, \tag{5.1.6}$$

then

$$\overline{G}(x) \geq \frac{1-p_0}{\phi_2 \overline{V}(0)} \left\{ \int_0^\infty \frac{dH_x(y)}{\overline{V}(x+y)} \right\}^{-1}, \quad x \geq 0. \tag{5.1.7}$$

Proof: Clearly, as in the proof of theorem 5.1.2,

$$\begin{aligned} c(x,z) &\leq \tau(x,z) = E\left\{1/\overline{V}(x+T_z)\right\} \\ &\leq E\left\{1/\overline{V}(x+W_x)\right\} = \int_0^\infty \left\{\overline{V}(x+y)\right\}^{-1} dH_x(y). \end{aligned}$$

Therefore, from (4.3.4), $1/c_2(x) \leq \int_0^\infty \left\{\overline{V}(x+y)\right\}^{-1} dH_x(y)$, and the result follows from theorem 4.3.1. □

We now identify $H_x(y)$ for various classes of distributions for $F(y)$.

Corollary 5.1.1 Suppose that $\phi \in (0,1)$ satisfies $a_{n+1} \leq (\geq)\ \phi a_n$ for

$n = 0, 1, 2, \cdots$, and $B(y)$ is a df satisfying $\phi^{-1} = \int_0^\infty \{\overline{B}(y)\}^{-1} dF(y)$. If $V(x)$ is a df satisfying $\overline{V}(x)\overline{B}(y) \leq (\geq)\ \overline{V}(x+y)$ for $x \geq 0$ and $y \geq 0$, and $F(y)$ is NWU (NBU), then

$$\overline{G}(x) \leq (\geq)\ \frac{1-p_0}{\phi\overline{V}(0)} \left\{ \int_0^\infty \frac{dF(y)}{\overline{V}(x+y)} \right\}^{-1}, \quad x \geq 0. \tag{5.1.8}$$

Proof: Since $F(y)$ is NWU (NBU), $\overline{F}(z+y)/\overline{F}(z) \geq (\leq)\ \overline{F}(y)$ and theorem 5.1.2 (5.1.3) is satisfied by $H_x(y) = F(y)$. □

In the case where $F(y)$ is IFR (DFR), we also have an upper (lower) bound as follows.

Corollary 5.1.2 Suppose that $\phi \in (0,1)$ satisfies $a_{n+1} \leq (\geq)\ \phi a_n$ for $n = 0, 1, 2, \cdots$, and $B(y)$ is a df satisfying $\phi^{-1} = \int_0^\infty \{\overline{B}(y)\}^{-1} dF(y)$. If $V(x)$ is a df satisfying $\overline{V}(x)\overline{B}(y) \leq (\geq)\ \overline{V}(x+y)$ for $x \geq 0$ and $y \geq 0$, and $F(y)$ is IFR with $\overline{F}(x) > 0$ (DFR), then

$$\overline{G}(x) \leq (\geq)\ \frac{1-p_0}{\phi\overline{V}(0)} \overline{F}(x) \Big/ \int_x^\infty \{\overline{V}(y)\}^{-1} dF(y), \quad x \geq 0. \tag{5.1.9}$$

Proof: Since $F(y)$ is IFR (DFR), it follows that $\overline{F}(z+y)/\overline{F}(z)$ is nonincreasing (nondecreasing) in z, and so for $0 \leq z \leq x, \overline{F}(z+y)/\overline{F}(z) \geq (\leq)\ \overline{F}(x+y)/\overline{F}(x)$. Therefore, theorem 5.1.2 (5.1.3) is satisfied by $H_x(y) = 1 - \overline{F}(x+y)/\overline{F}(x)$, the residual lifetime df. Since $dH_x(y) = dF(x+y)/\overline{F}(x)$ and

$$\frac{1}{\overline{F}(x)} \int_0^\infty \{\overline{V}(x+y)\}^{-1} dF(x+y) = \frac{1}{\overline{F}(x)} \int_x^\infty \{\overline{V}(y)\}^{-1} dF(y),$$

the result follows. □

The assumption that $\overline{F}(x) > 0$ is needed in the IFR case in corollary 5.1.2 so that the result is well defined, but is not needed in the DFR case since one must have $\overline{F}(x) > 0$ if $x > 0$.

The UBA (UWA) case is considered next.

Corollary 5.1.3 Suppose that $\phi \in (0,1)$ satisfies $a_{n+1} \leq (\geq)\ \phi a_n$ for $n = 0, 1, 2, \cdots$, and $B(y)$ is a df satisfying $\phi^{-1} = \int_0^\infty \{\overline{B}(y)\}^{-1} dF(y)$. If $V(x)$ is a df satisfying $\overline{V}(x)\overline{B}(y) \leq (\geq)\ \overline{V}(x+y)$ for $x \geq 0$ and $y \geq 0$, and $F(y)$ is UBA (UWA) with mean residual lifetime $r(y) = \int_y^\infty \overline{F}(t)dt/\overline{F}(y)$, then

$$\overline{G}(x) \leq (\geq)\ \frac{1-p_0}{\phi\overline{V}(0)} \left\{ \frac{1}{r(\infty)} \int_0^\infty \frac{e^{-y/r(\infty)}}{\overline{V}(x+y)} dy \right\}^{-1}; \quad x \geq 0. \tag{5.1.10}$$

Proof: Since $F(y)$ is UBA (UWA), $\overline{F}(z+y)/\overline{F}(z) \geq (\leq)\ e^{-y/r(\infty)}$ and theorem 5.1.2 (5.1.3) is satisfied by $H_x(y) = e^{-y/r(\infty)}$. □

We remark that if $F(y)$ is IFR (DFR), then each of corollaries 5.1.2 and 5.1.3 applies. But the choice of $H_x(y)$ employed in corollary 5.1.2 results in (5.1.4) ((5.1.6)) being an equality. Therefore, corollary 5.1.2 gives a better bound in this case.

We now consider the bounded failure rate case.

Corollary 5.1.4 Suppose that $\phi \in (0,1)$ satisfies $a_{n+1} \leq (\geq)\ \phi a_n$ for $n = 0, 1, 2, \cdots$, and $B(y)$ is a df satisfying $\phi^{-1} = \int_0^\infty \{\overline{B}(y)\}^{-1} dF(y)$. If $V(x)$ is a df satisfying $\overline{V}(x)\overline{B}(y) \leq (\geq)\ \overline{V}(x+y)$ for $x \geq 0$ and $y \geq 0$, and $F(y)$ is absolutely continuous with failure rate $\mu(y) = -\frac{d}{dy}\ln \overline{F}(y)$ which satisfies $\mu(y) \leq (\geq)\ \mu$ where $\mu \in (0,\infty)$, then

$$\overline{G}(x) \leq (\geq)\ \frac{1-p_0}{\phi \overline{V}(0)} \left\{ \int_0^\infty \frac{\mu e^{-\mu y} dy}{\overline{V}(x+y)} \right\}^{-1}, \quad y \geq 0. \tag{5.1.11}$$

Proof: One has

$$\frac{\overline{F}(z+y)}{\overline{F}(z)} = e^{-\int_0^y \mu(z+t)dt} \geq (\leq)\ e^{-\int_0^y \mu dt} = e^{-\mu y}$$

and theorem 5.1.2 (5.1.3) is satisfied by $H_x(y) = 1 - e^{-\mu y}$. □

The corollaries in this section involve identification of $H_x(y)$ for various reliability classifications of the distribution of $F(y)$. The following example illustrates how a suitable $H_x(y)$ may be constructed if the failure rate is nonmonotone.

Example 5.1.1 Burr distribution

Suppose that $\overline{F}(y) = \left(\frac{\beta^2}{\beta^2+y^2}\right)^\alpha, \quad y \geq 0$, where $\beta > 0$ and $\alpha > 0$. Then $F(y)$ is the df of a Burr distribution with arbitrary parameters α and β and $\gamma = 2$. The failure rate is

$$\mu(y) = -\frac{d}{dy}\ln \overline{F}(y) = \frac{2\alpha y}{\beta^2 + y^2}, \quad y \geq 0.$$

It is not hard to see that $\mu(0) = \mu(\infty) = 0$, and since

$$\mu'(y) = \frac{2\alpha(\beta^2 - y^2)}{(\beta^2 + y^2)^2},$$

it follows that $\mu(y)$ increases to its maximum value of $\mu(\beta) = \alpha/\beta$ at $y = \beta$, and then decreases. Thus, one could use corollary 5.1.4 with $\mu = \alpha/\beta$ and choose $H_x(y) = 1 - e^{-\frac{\alpha}{\beta}y}$ to find an upper bound for $\overline{G}(x)$. A better inequality may be obtained as follows, however. Fix $y \geq 0$ and consider

$$Q_y(z) = \frac{\overline{F}(y+z)}{\overline{F}(z)} = \left\{ \frac{\beta^2 + z^2}{\beta^2 + (y+z)^2} \right\}^\alpha .$$

One finds that

$$\frac{\partial}{\partial z} Q_y(z) = \frac{2\alpha y(z^2 + yz - \beta^2) Q_y(z)}{(\beta^2 + z^2)\{\beta^2 + (y+z)^2\}},$$

from which it is not hard to see that $Q_y(z)$ attains a minimum at

$$z(y) = \frac{\sqrt{y^2 + 4\beta^2} - y}{2} = \frac{2\beta^2}{\sqrt{y^2 + 4\beta^2} + y}.$$

To simplify $Q_y\{z(y)\}$, one finds after a little algebra that

$$\beta^2 + z(y)^2 = \frac{2\beta^2 \sqrt{y^2 + 4\beta^2}}{\sqrt{y^2 + 4\beta^2} + y}.$$

Similarly,

$$z(y) + y = \frac{\sqrt{y^2 + 4\beta^2} + y}{2} = \frac{2\beta^2}{\sqrt{y^2 + 4\beta^2} - y}$$

and

$$\beta^2 + \{z(y) + y\}^2 = \frac{2\beta^2 \sqrt{y^2 + 4\beta^2}}{\sqrt{y^2 + 4\beta^2} - y}.$$

Thus,

$$\begin{aligned} Q_y\{z(y)\} &= \left\{ \frac{\beta^2 + z(y)^2}{\beta^2 + \{y + z(y)\}^2} \right\}^\alpha \\ &= \left\{ \frac{\sqrt{y^2 + 4\beta^2} - y}{\sqrt{y^2 + 4\beta^2} + y} \right\}^\alpha = \left\{ \frac{2\beta}{y + \sqrt{y^2 + 4\beta^2}} \right\}^{2\alpha}. \end{aligned}$$

To summarize, $Q_y(z) \geq Q_y\{z(y)\}$ which may be restated as $\overline{F}(y+z)/\overline{F}(z) \geq \overline{H}(y)$ where

$$\overline{H}(y) = \left\{ \frac{2\beta}{y + \sqrt{y^2 + 4\beta^2}} \right\}^{2\alpha}.$$

Now, $\overline{H}(0) = 1$, $\overline{H}(\infty) = 0$, and $\overline{H}(y)$ is decreasing in y. Thus, $1 - \overline{H}(y)$ is a df with failure rate which satisfies

$$-\frac{d}{dy} \ln \overline{H}(y) = \frac{2\alpha}{\sqrt{y^2 + 4\beta^2}} \leq \frac{\alpha}{\beta}.$$

This implies that $\overline{H}(y) \geq e^{-\frac{\alpha}{\beta}y}$ and thus $\overline{H}(y)$ is a better choice than $e^{-\frac{\alpha}{\beta}y}$ in theorem 5.1.2 since it leads to a tighter bound. Suitable choices of $\overline{B}(y)$ and $\overline{V}(y)$ are given by Willmot and Lin (1997a) in this situation. □

5.2 Bounds based on equilibrium properties

In the previous section bounds were constructed based on properties of the df $F(y)$, yielding results for DFR (IFR) and NWU (NBU) classes, among others. In this section, we consider classes based on properties of $F_1(y) = \int_0^y \overline{F}(t)dt/E(Y)$ or equivalently the mean residual lifetime $r(y)$ from section 2.3. Applications to classes such as IMRL(DMRL), NWUC (NBUC), and NWUE (NBUE) follow.

We will normally require that $1/\overline{V}(y)$ be a convex function of y. Thus, if $V(y)$ is twice differentiable with failure rate

$$\mu_V(y) = -\frac{d}{dy}\ln \overline{V}(y), \quad y \geq 0, \tag{5.2.1}$$

it follows that $1/\overline{V}(y) = \exp\left\{\int_0^y \mu_V(t)dt\right\}$. Since a sufficient condition for convexity is a nonnegative second derivative, this means that $1/\overline{V}(y)$ is convex if $\mu_V'(y) + \{\mu_V(y)\}^2 \geq 0$, or equivalently $\frac{d}{dy}\left\{\frac{1}{\mu_V(y)}\right\} \leq 1$. Integrating both sides from 0 to y yields $\frac{1}{\mu_V(y)} - \frac{1}{\mu_V(0)} \leq y$. In other words, $\mu_V(y) \geq \frac{\mu_V(0)}{1+\mu_V(0)y}$, $y \geq 0$. This in turn implies that $\overline{V}(y) \leq \frac{1}{1+\mu_V(0)y}$, $y \geq 0$. A sufficient condition for convexity is $\mu_V'(y) \geq -\left\{\frac{\mu_V(0)}{1+\mu_V(0)y}\right\}^2$, since integration yields $\mu_V(y) \geq \frac{\mu_V(0)}{1+\mu_V(0)y}$ from which $\mu_V'(y) + \{\mu_V(y)\}^2 \geq 0$ follows. Clearly, $1/\overline{V}(y)$ is convex if $1 - \overline{V}(y)$ is IFR. Other examples include the exponential tail $\overline{V}(y) = e^{-\kappa y}$ and the Pareto tail $\overline{V}(y) = (1+\kappa y)^{-m}$ where $m \geq 1$. In particular, with $m = 1$ it follows that $\kappa = \mu_V(0)$ and we obtain the Pareto tail $\overline{V}(y) = \{1 + \mu_V(0)y\}^{-1}$, which from the above reasoning, may be regarded as the 'weakest' distribution such that $1/\overline{V}(y)$ is convex. If $1/\overline{V}(y)$ is convex we immediately obtain the following result (Lin, 1996).

Theorem 5.2.1 Suppose that $\phi_1 \in (0,1)$ satisfies (4.2.1), and $B(y)$ is a df satisfying (4.2.2). If $V(x)$ is a df satisfying (4.2.3) with $1/\overline{V}(x)$ convex, then

$$\overline{G}(x) \leq \frac{1-p_0}{\phi_1 \overline{V}(0)} \overline{V}\left(x + r_*(x)\right), \quad x \geq 0 \tag{5.2.2}$$

where

$$r_*(x) = \inf_{0 \leq z \leq x, \overline{F}(z) > 0} r(z), \; x \geq 0, \tag{5.2.3}$$

and $r(y)$ defined by (2.3.2) is the mean residual lifetime of the df $F(y)$.

Proof: Using (2.3.2), the mean associated with the residual lifetime df $1 - \overline{F}(z+y)/\overline{F}(z)$ is

$$z + r(z) = \frac{1}{\overline{F}(z)} \int_z^\infty y dF(y).$$

As a function of $y, 1/\overline{V}(x+y-z)$ is convex, and from (5.1.3) Jensen's inequality yields for $0 \le z \le x$

$$\begin{aligned} \tau(x,z) &= E\{1/\overline{V}(x+T_z)\} \\ &\geq \frac{1}{\overline{V}\{x+r(z)\}} \geq \frac{1}{\overline{V}\{x+r_*(x)\}} \end{aligned}$$

since $\overline{V}(y)$ is nonincreasing. Therefore, from (4.2.5),

$$1/c_1(x) \geq 1/\overline{V}\{x+r_*(x)\} \text{ or } c_1(x) \leq \overline{V}\{x+r_*(x)\}.$$

The result follows from theorem 4.2.1. □

Clearly, theorem 5.2.1 is a refinement of corollary 4.2.3 since

$$\overline{V}(x+r_*(x)) \leq \overline{V}(x).$$

Theorem 5.2.2 Suppose that $\phi_1 \in (0,1)$ satisfies (4.2.1), and $B(y)$ is a df satisfying (4.2.2). If $V(x)$ is a df satisfying (4.2.3) with $1/\overline{V}(x)$ convex, and $H_x(y) = 1 - \overline{H}_x(y)$ is a df satisfying

$$\int_y^\infty \overline{H}_x(t)dt \leq \inf_{0 \le z \le x, \overline{F}(z)>0} \int_y^\infty \frac{\overline{F}(t+z)}{\overline{F}(z)}dt, \quad y \geq 0, \tag{5.2.4}$$

then

$$\overline{G}(x) \leq \frac{1-p_0}{\phi_1 \overline{V}(0)} \left\{ \int_0^\infty \frac{dH_x(y)}{\overline{V}(x+y)} \right\}^{-1}, \quad x \geq 0. \tag{5.2.5}$$

Proof: As in the proof of theorem 5.1.2, let $H_x(y) = Pr(W_x \le y)$ and $1 - \overline{F}(y+z)/\overline{F}(z) = Pr(T_z \le y)$. Then (5.2.4) implies that $\int_y^\infty Pr(W_x > t)dt \le \int_y^\infty Pr(T_z > t)dt$. But from Shaked and Shanthikumar (1994, pp. 83-5), this implies that

$$E\{1/\overline{V}(x+W_x)\} \leq E\{1/\overline{V}(x+T_z)\}$$

by convexity. Therefore, from (4.2.4) and (5.1.3),

$$c(x,z) \geq \tau(x,z) \geq E\{1/\overline{V}(x+W_x)\}$$

and from (4.2.5), $1/c_1(x) \geq E\{1/\overline{V}(x+W_x)\}$, and the result follows from theorem 4.2.1. □

In terms of the equilibrium df's $F_1(y) = 1 - \overline{F}_1(y) = \int_0^y \overline{F}(t)dt/E(Y)$ and $H_{1,x}(y) = 1 - \overline{H}_{1,x}(y) = \int_0^y \overline{H}_x(t)dt/m_x$ with $m_x = \int_0^\infty y dH_x(y)$, (5.2.4) is equivalent to

$$\overline{H}_{1,x}(y) \leq \frac{E(Y)}{m_x} \inf_{0 \le z \le x, \overline{F}(z)>0} \overline{F}_1(y+z)/\overline{F}(z); \quad y \geq 0, \tag{5.2.6}$$

Turning to the lower bound case, we have the following dual to theorem 5.2.2.

Theorem 5.2.3 Suppose that $\phi_2 \in (0,1)$ satisfies (4.3.1), and $B(y)$ is a df satisfying (4.3.2). If $V(x)$ is a df satisfying (4.3.3) with $1/\overline{V}(x)$ convex, and $H_x(y) = 1 - \overline{H}_x(y)$ is a df satisfying

$$\int_y^\infty \overline{H}_x(t)dt \geq \sup_{0 \leq z \leq x, \overline{F}(z)>0} \int_y^\infty \frac{\overline{F}(t+z)}{\overline{F}(z)} dt, \quad y \geq 0, \tag{5.2.7}$$

then

$$\overline{G}(x) \geq \frac{1-p_0}{\phi_2 \overline{V}(0)} \left\{ \int_0^\infty \frac{dH_x(y)}{\overline{V}(x+y)} \right\}^{-1}, \quad x \geq 0. \tag{5.2.8}$$

Proof: Simply reverse the inequalities in theorem 5.2.2. □

As with (5.2.6), (5.2.7) is equivalent to

$$\overline{H}_{1,x}(y) \geq \frac{E(Y)}{m_x} \sup_{0 \leq z \leq x, \overline{F}(z)>0} \overline{F}_1(y+z)/\overline{F}(z); \quad y \geq 0. \tag{5.2.9}$$

It is clear that theorems 5.2.2 and 5.2.3 give the same bounds as the first order theorems 5.1.2 and 5.1.3 respectively, but for larger classes of distributions. With the extra condition that $1/\overline{V}(y)$ is convex, bounds are obtainable based on stochastic ordering of the equilibrium df's of $1-\overline{F}(z+y)/\overline{F}(z)$ and $H_x(y)$ rather than the df's themselves (as is clear from (5.2.6) and (5.2.9)). As is evident in what follows, this allows for bounds which are applicable for larger classes of claim size distributions than was the case using the first order results. The key to the use of theorems 5.2.2 and 5.2.3 is the identification of a suitable $H_x(y)$.

Corollary 5.2.1 Suppose that $\phi \in (0,1)$ satisfies $a_{n+1} \leq (\geq)\ \phi a_n$; $n = 0, 1, 2, \cdots$, and $B(y)$ is a df satisfying $\phi^{-1} = \int_0^\infty \left\{\overline{B}(y)\right\}^{-1} dF(y)$. If $V(x)$ is a df satisfying $\overline{V}(x)\overline{B}(y) \leq (\geq)\ \overline{V}(x+y)$ for $x \geq 0$ and $y \geq 0$ with $1/\overline{V}(x)$ convex, and $F(y)$ is NWUC (NBUC), then

$$\overline{G}(x) \leq (\geq) \frac{1-p_0}{\phi \overline{V}(0)} \left\{ \int_0^\infty \frac{dF(y)}{\overline{V}(x+y)} \right\}^{-1}, \quad x \geq 0. \tag{5.2.10}$$

Proof: Since $F(y)$ is NWUC (NBUC) $\overline{F}_1(y+z) \geq (\leq)\ \overline{F}(z)\overline{F}_1(y)$ and theorem 5.2.2. (5.2.3) is satisfied with $H_x(y) = F(y)$. □

The above corollary generalizes corollary 5.1.1 with the additional condition on $\overline{V}(y)$, and the following generalizes corollary 5.1.2.

Corollary 5.2.2. Suppose that $\phi \in (0,1)$ satisfies $a_{n+1} \leq (\geq)\ \phi a_n;\ n = 0, 1, 2, \cdots$, and $B(y)$ is a df satisfying $\phi^{-1} = \int_0^\infty \{\overline{B}(y)\}^{-1} dF(y)$. If $V(x)$ is a df satisfying $\overline{V}(x)\overline{B}(y) \leq (\geq)\ \overline{V}(x+y)$ for $x \geq 0$ and $y \geq 0$ with $1/\overline{V}(x)$ convex, and $F(y)$ is DMRL with $\overline{F}(x) > 0$ (IMRL), then

$$\overline{G}(x) \leq (\geq)\ \frac{1-p_0}{\phi\overline{V}(0)}\overline{F}(x) \Big/ \int_x^\infty \{\overline{V}(y)\}^{-1} dF(y), \quad x \geq 0. \tag{5.2.11}$$

Proof: Using (2.3.4), one has $E(Y)\overline{F}_1(y+z)/\overline{F}(z) = r(z)\overline{F}_1(y+z)/\overline{F}_1(z)$. Since $F(y)$ is DMRL (IMRL) is equivalent to $F_1(y)$ is IFR (DFR) (for example, from (2.3.5)), it follows that $\overline{F}_1(y+z)/\overline{F}_1(z)$ is nonincreasing (nondecreasing) in z. Therefore, for $0 \leq z \leq x$,

$$\frac{E(Y)\overline{F}_1(y+z)}{\overline{F}(z)} = r(z)\frac{\overline{F}_1(y+z)}{\overline{F}_1(z)} \geq (\leq)\ r(x)\frac{\overline{F}_1(y+x)}{\overline{F}_1(x)}.$$

Then theorem 5.2.2 (theorem 5.2.3) is satisfied with $H_x(y) = 1 - \overline{F}(x+y)/\overline{F}(x)$ since $m_x = r(x)$ and from example 2.3.3, $\overline{H}_{1,x}(y) = \overline{F}_1(x+y)/\overline{F}_1(x)$. This choice of $\overline{H}_{1,x}(y)$ obviously satisfies (5.2.6) ((5.2.9)) by the above inequality. □

The next bound involves the mean residual lifetime.

Corollary 5.2.3 Suppose that $\phi \in (0,1)$ satisfies $a_{n+1} \leq (\geq)\ \phi a_n;\ n = 0, 1, 2, \cdots$, and $B(y)$ is a df satisfying $\phi^{-1} = \int_0^\infty \{\overline{B}(y)\}^{-1} dF(y)$. If $V(x)$ is a df satisfying $\overline{V}(x)\overline{B}(y) \leq (\geq)\ \overline{V}(x+y)$ for $x \geq 0$ and $y \geq 0$ with $1/\overline{V}(x)$ convex, and the mean residual lifetime $r(y) = \int_y^\infty \overline{F}(x)dx/\overline{F}(y)$ of $F(y)$ satisfies $r(y) \geq (\leq)\ r$ where $r \in (0, \infty)$, then

$$\overline{G}(x) \leq (\geq)\ \frac{1-p_0}{\phi\overline{V}(0)}\left\{\frac{1}{r}\int_0^\infty \frac{e^{-y/r}}{\overline{V}(x+y)}dy\right\}^{-1}; \quad x \geq 0. \tag{5.2.12}$$

Proof: It follows from (2.3.4) and (2.3.6) that for $0 \leq z \leq x, \overline{F}(z) > 0$,

$$\begin{aligned} E(Y)\overline{F}_1(y+z)/\overline{F}(z) &= r(z)\overline{F}_1(y+z)/\overline{F}_1(z) \\ &= r(z)e^{-\int_z^{y+z}\{r(t)\}^{-1}dt} \\ &\geq (\leq)\ re^{-\int_z^{y+z} r^{-1}dt} \\ &= re^{-y/r}. \end{aligned}$$

Therefore, theorem 5.2.2 (theorem 5.2.3) is satisfied by $H_x(y) = 1 - e^{-y/r}$. □

If $r(y) \geq r$, then from (5.2.3), $r_*(x) \geq r$, and since, $\overline{V}(y)$ is nonincreasing, it follows from theorem 5.2.1 that

$$\overline{G}(x) \leq \frac{1-p_0}{\phi_1 \overline{V}(0)} \overline{V}(x+r), \quad x \geq 0. \tag{5.2.13}$$

However, since $H_x(y) = 1 - e^{-y/r}$ is the df of an exponential random variable with mean r, Jensen's inequality yields for convex $1/\overline{V}(y)$

$$\frac{1}{r}\int_0^\infty \frac{e^{-y/r}}{\overline{V}(x+y)} dy \geq \frac{1}{\overline{V}(x+r)}.$$

Therefore, (5.2.12) is sharper than (5.2.13).

Two important special cases are now given.

Corollary 5.2.4 Suppose that $\phi \in (0,1)$ satisfies $a_{n+1} \leq (\geq)\ \phi a_n;\ \ n = 0, 1, 2, \cdots$, and $B(y)$ is a df satisfying $\phi^{-1} = \int_0^\infty \left\{\overline{B}(y)\right\}^{-1} dF(y)$. If $V(x)$ is a df satisfying $\overline{V}(x)\overline{B}(y) \leq (\geq)\ \overline{V}(x+y)$ for $x \geq 0$ and $y \geq 0$ with $1/\overline{V}(x)$ convex, and $F(y)$ is NWUE (NBUE), then

$$\overline{G}(x) \leq (\geq)\ \frac{1-p_0}{\phi \overline{V}(0)} \left\{\frac{1}{E(Y)} \int_0^\infty \frac{e^{-y/E(Y)}}{\overline{V}(x+y)} dy\right\}^{-1}; \quad x \geq 0. \tag{5.2.14}$$

Proof: Corollary 5.2.3 holds with $r = E(Y)$. □

Corollary 5.2.5 Suppose that $\phi \in (0,1)$ satisfies $a_{n+1} \leq (\geq)\ \phi a_n;\ n = 0, 1, 2, \cdots$, and $B(y)$ is a df satisfying $\phi^{-1} = \int_0^\infty \left\{\overline{B}(y)\right\}^{-1} dF(y)$. If $V(x)$ is a df satisfying $\overline{V}(x)\overline{B}(y) \leq (\geq)\ \overline{V}(x+y)$ for $x \geq 0$ and $y \geq 0$ with $1/\overline{V}(x)$ convex, and $F(y)$ is UBAE (UWAE), then

$$\overline{G}(x) \leq (\geq)\ \frac{1-p_0}{\phi \overline{V}(0)} \left\{\frac{1}{r(\infty)} \int_0^\infty \frac{e^{-y/r(\infty)}}{\overline{V}(x+y)} dy\right\}^{-1}; \quad x \geq 0. \tag{5.2.15}$$

Proof: Corollary 5.2.3 holds with $r = r(\infty)$. □

6
Parametric Bounds

In this chapter we consider in more detail particular choices of $\overline{B}(y)$ which are convenient and useful as bounds. We begin with the most important case: exponential bounds. An exponential bound on the tail probability is often considered for a distribution whose moment generating function exists. In this case, a simple exponential upper bound can be obtained by applying Markov's inequality but it is often unsatisfactory, as was demonstrated in section 4.2. To derive a better bound, one needs to take advantage of distributional properties of the distribution under consideration. For example, Brown (1990) considers upper and lower bounds on the tail of compound geometric distributions where these bounds are expressed in terms of the first two moments of the individual claim amount distribution. Other types of exponential bounds on the tail of compound geometric distributions are obtainable in the context of the waiting time distribution of a G/G/1 queue or the probability of ruin in insurance risk theory. These bounds will be discussed in detail in the next chapter. In the next section, we consider various simple and tight exponential bounds on the tail of compound distributions based on the reliability classification of the individual claim amount df $F(x)$. Section 6.2 deals with Pareto-type bounds. Exponential bounds are not applicable to distributions whose moment generating function does not exist. In this case, a Pareto-type bound is often appropriate. The advantage of a Pareto-type bound is that it only requires a finite number of moments of $F(x)$. The fact that the product of two NWU tails is again an NWU tail enables us to construct other bounds. We discuss this approach in section 6.3.

6.1 Exponential bounds

The choice $\overline{B}(y) = e^{-\kappa y}$ is considered in this section, where κ is a parameter to be determined. Since both (4.2.3) and (4.3.3) are satisfied by $\overline{V}(x) = \overline{B}(x) = e^{-\kappa x}$, this is useful both for upper and lower bounds. The general upper bound in this case is given in corollary 4.2.1, and the corresponding lower bound in corollary 4.3.1.

It follows from the generalized adjustment equation (4.2.2) that the parameter $\kappa > 0$ is the solution to

$$\int_0^\infty e^{\kappa y} dF(y) = \frac{1}{\phi}, \tag{6.1.1}$$

which is referred to as the Lundberg adjustment equation, and is clearly a statement involving the moment generating function of $F(y)$. A general upper bound, obtained by Willmot and Lin (1994), is now stated.

Corollary 6.1.1 Suppose that $\phi \in (0,1)$ satisfies $a_{n+1} \leq \phi a_n$ for $n = 0, 1, 2, \cdots$, and $\kappa > 0$ satisfies (6.1.1). Then

$$\overline{G}(x) \leq \frac{1-p_0}{\phi} e^{-\kappa x}, \quad x \geq 0. \tag{6.1.2}$$

Proof: Corollary 4.2.3 applies with $\overline{B}(x) = e^{-\kappa x}$. □

Similarly, we have the following general lower bound.

Corollary 6.1.2 Suppose that $\phi \in (0,1)$ satisfies $a_{n+1} \geq \phi a_n$ for $n = 0, 1, 2, \cdots$, and $\kappa > 0$ satisfies (6.1.1). Then

$$\overline{G}(x) \geq \frac{1-p_0}{\phi} \gamma(x) e^{-\kappa x}, \quad x \geq 0, \tag{6.1.3}$$

where $m = \inf\{x;\ F(x) = 1\}$ and $\gamma(x) = \max\{\phi \overline{F}(x), e^{-\kappa m}\}$.

Proof: Corollary 4.3.3 applies with $\overline{B}(x) = e^{-\kappa x}$. □

Turning now to the case where further restrictions are placed on the individual claim amount distribution, we have the following bounds.

Corollary 6.1.3 Suppose that $\phi \in (0,1)$ satisfies $a_{n+1} \leq (\geq)\ \phi a_n$ for $n = 0, 1, 2, \cdots$, and $\kappa > 0$ satisfies (6.1.1). If $H_x(y)$ is a df satisfying (5.1.4) ((5.1.6)), then

$$\overline{G}(x) \leq (\geq) \frac{1-p_0}{\phi \int_0^\infty e^{\kappa y} dH_x(y)} e^{-\kappa x}, \quad x \geq 0. \tag{6.1.4}$$

Proof: Theorem 5.1.2 (5.1.3) applies with $\overline{V}(x) = \overline{B}(x) = e^{-\kappa x}$. □

Since $1/\overline{V}(y) = e^{\kappa y}$ is convex, (6.1.4) is also obtained from theorem 5.2.2 (theorem 5.2.3) with the condition on $\overline{V}(y)$ replaced by $\kappa > 0$ satisfying (6.1.1). Moreover, we obtain the following simple results.

Corollary 6.1.4 Suppose that $\phi \in (0,1)$ satisfies $a_{n+1} \leq (\geq)\ \phi a_n$ for $n = 0, 1, 2, \cdots$, and $\kappa > 0$ satisfies (6.1.1). If $F(y)$ is NWUC (NBUC) then

$$\overline{G}(x) \leq (\geq)\ (1 - p_0)\, e^{-\kappa x}, \quad x \geq 0. \tag{6.1.5}$$

Proof: Corollary 5.2.1 applies with $\overline{V}(x) = \overline{B}(x) = e^{-\kappa x}$. □

We remark that the bounds obtained in corollary 6.1.4 match the true value of $\overline{G}(x)$ at $x = 0$. This implies that (6.1.5) is the best possible bound of the form $\overline{G}(x) \leq (\geq)\ Ce^{-\kappa x}$, since with $x = 0$, one obtains $1 - p_0 = \overline{G}(0) \leq (\geq)\ C$, implying that $(1 - p_0)e^{-\kappa x} \leq (\geq)\ Ce^{-\kappa x}$. Furthermore, as mentioned following (4.2.25), the best choice of κ is obtained if $1/\phi = z_0$ in (6.1.1). Also, in the geometric-exponential case, the upper and lower bounds coincide and we have

$$\overline{G}(x) = (1 - p_0)\, e^{-\kappa x}. \tag{6.1.6}$$

Corollary 6.1.5 Suppose that $\phi \in (0,1)$ satisfies $a_{n+1} \leq (\geq)\ \phi a_n$ for $n = 0, 1, 2, \cdots$, and $\kappa > 0$ satisfies (6.1.1). If $F(y)$ is DMRL with $\overline{F}(x) > 0$ (IMRL), then

$$\overline{G}(x) \leq (\geq)\ \frac{1 - p_0}{\phi} \overline{F}(x) \Big/ \int_x^{\infty} e^{\kappa y} dF(y), \quad x \geq 0. \tag{6.1.7}$$

Proof: Corollary 5.2.2 applies with $\overline{V}(x) = \overline{B}(x) = e^{-\kappa x}$. □

Corollary 6.1.6 Suppose that $\phi \in (0,1)$ satisfies $a_{n+1} \leq (\geq)\ \phi a_n$ for $n = 0, 1, 2, \cdots$, and $\kappa > 0$ satisfies (6.1.1). If the mean residual lifetime $r(y) = \int_y^{\infty} \overline{F}(x)dx \Big/ \overline{F}(y)$ of $F(y)$ satisfies $r(y) \geq (\leq)\ r$ where $r \in (0, \infty)$, then

$$\overline{G}(x) \leq (\geq)\ \frac{(1 - p_0)\,(1 - \kappa r)}{\phi} e^{-\kappa x}, \quad x \geq 0. \tag{6.1.8}$$

Proof: Corollary 5.2.3 applies with $\overline{V}(x) = \overline{B}(x) = e^{-\kappa x}$. □

Corollary 6.1.7 Suppose that $\phi \in (0,1)$ satisfies $a_{n+1} \leq (\geq)\ \phi a_n$ for $n = 0, 1, 2, \cdots$, and $\kappa > 0$ satisfies (6.1.1). If $F(y)$ is NWUE (NBUE), then

$$\overline{G}(x) \leq (\geq)\ \frac{(1 - p_0)\,\{1 - \kappa E(Y)\}}{\phi} e^{-\kappa x}, \quad x \geq 0. \tag{6.1.9}$$

Proof: Corollary 6.1.6 applies with $r = E(Y)$. □

Corollary 6.1.8 Suppose that $\phi \in (0,1)$ satisfies $a_{n+1} \leq (\geq)\ \phi a_n$ for $n = 0, 1, 2, \cdots$, and $\kappa > 0$ satisfies (6.1.1). If $F(y)$ is UBAE (UWAE), then

$$\overline{G}(x) \leq (\geq)\ \frac{(1-p_0)\{1-\kappa r(\infty)\}}{\phi} e^{-\kappa x}, \quad x \geq 0. \tag{6.1.10}$$

Proof: Corollary 6.1.6 applies with $r = r(\infty)$. □

There is one other technical result which is implicit in corollary 6.1.4, but for later reference we now make it explicit.

Proposition 6.1.1 Suppose $t > 0$, $x \geq 0$, and $\int_0^\infty e^{ty} dF(y) < \infty$. Then if $F(y)$ is NWUC

$$\inf_{0 \leq z \leq x, \overline{F}(z) > 0} \frac{\int_z^\infty e^{ty} dF(y)}{e^{tz}\overline{F}(z)} = \int_0^\infty e^{ty} dF(y), \tag{6.1.11}$$

whereas if $F(y)$ is NBUC

$$\sup_{0 \leq z \leq x, \overline{F}(z) > 0} \frac{\int_z^\infty e^{ty} dF(y)}{e^{tz}\overline{F}(z)} = \int_0^\infty e^{ty} dF(y). \tag{6.1.12}$$

Proof: Let $Pr(T_z \leq y) = 1 - \overline{F}(y+z)/\overline{F}(z)$, and thus

$$\frac{\int_z^\infty e^{ty} dF(y)}{e^{tz}\overline{F}(z)} = \int_0^\infty e^{ty} dPr(T_z \leq y) = E(e^{tT_z}).$$

If $F(y)$ is NWUC (NBUC), $\overline{F}_1(y) \leq (\geq)\ \overline{F}_1(y+z)/\overline{F}(z)$, i.e.

$$\int_y^\infty \overline{F}(x)dx \leq (\geq) \int_y^\infty \frac{\overline{F}(x+z)}{\overline{F}(z)} dx.$$

In other words, $\int_y^\infty Pr(Y > x)dx \leq (\geq) \int_y^\infty Pr(T_z > x)dx$, with $F(y) = Pr(Y \leq y)$. Since e^{ty} is convex, it follows from Shaked and Shanthikumar (1994, pp. 83-5) that $E(e^{tT_z}) \geq (\leq)\ E(e^{tY})$. Thus,

$$\inf_{0 \leq z \leq x, \overline{F}(z) > 0} E(e^{tT_z}) \geq E(e^{tY}) \left(\sup_{0 \leq z \leq x, \overline{F}(z) > 0} E(e^{tT_z}) \leq E(e^{tY}) \right).$$

But when $z = 0$, T_z and Y have the same distribution function $F(y)$ and thus

$$\inf_{0 \leq z \leq x, \overline{F}(z) > 0} E(e^{tT_z}) = E(e^{tY}) \left(\sup_{0 \leq z \leq x, \overline{F}(z) > 0} E(e^{tT_z}) = E(e^{tY}) \right).$$

Hence (6.1.11) ((6.1.12)) holds. □

Evidently, if (6.1.1) holds, then proposition 6.1.1 implies that if $F(y)$ is NWUC (NBUC) then $\alpha_1(x) = \phi$ ($\alpha_2(x) = \phi$) where $\alpha_1(x)$ ($\alpha_2(x)$) is given by (4.2.17) ((4.3.14)). Further use will be made of proposition 6.1.1 later.

6.2 Pareto bounds

The bounds of the previous section are very useful, but require the existence of the moment generating function of the df $F(y)$. In cases where no such $\kappa > 0$ exists satisfying (6.1.1), often because no moment generating function exists, an alternative is to consider a moment based bound.

Therefore, we assume that moments exist up to order m, i.e.

$$E\left(Y^j\right) = \int_0^\infty y^j dF(y) < \infty, \quad j \leq m. \tag{6.2.1}$$

Since the Pareto distribution is DFR and hence NWU, it is convenient to choose the Pareto tail $\overline{B}(y) = \overline{V}(y) = (1+\kappa y)^{-m}$, where $\kappa > 0$ satisfies

$$\int_0^\infty (1+\kappa y)^m \, dF(y) = \frac{1}{\phi_1}. \tag{6.2.2}$$

It is convenient if m is a positive integer, in which case (6.2.2) becomes, after a binomial expansion,

$$\sum_{j=1}^{m} \binom{m}{j} E(Y^j)\kappa^j = \frac{1}{\phi_1} - 1, \tag{6.2.3}$$

a polynomial in κ of degree m. If $m = 1$ then one obtains $\kappa = (1 - \phi_1)/\left\{\phi_1 E(Y)\right\}$ whereas if $m = 2$ one obtains

$$\kappa = \frac{\sqrt{\frac{E(Y^2)}{\phi_1} - \mathrm{Var}(Y)} - E(Y)}{E(Y^2)}.$$

Now, for $m \geq 1, 1/\overline{V}(y) = (1+\kappa y)^m$ is convex, and Jensen's inequality yields, from (6.2.2),

$$\frac{1}{\phi_1} = \int_0^\infty (1+\kappa y)^m \, dF(y) \geq \left\{1 + \kappa E(Y)\right\}^m,$$

which may be restated as $\kappa \leq \left\{\phi_1^{-1/m} - 1\right\} \big/ E(Y)$.

We obtain the following general bound.

Corollary 6.2.1 If $\phi_1 \in (0,1)$ satisfies (4.2.1), and $\kappa > 0$ and $m > 0$ satisfy (6.2.2), then

$$\overline{G}(x) \leq \frac{1-p_0}{\phi_1}(1+\kappa x)^{-m}, \quad x \geq 0. \tag{6.2.4}$$

Proof: Corollary 4.2.3 applies with $\overline{B}(y) = (1+\kappa y)^{-m}$. □

It should be noted that if $\phi_1 \in (0,\ 1)$ satisfies (4.2.1) and $E(Y^m) < \infty$, one has $E(X^m) = \int_0^\infty x^m dG(x) < \infty$, and Markov's inequality yields

$$\overline{G}(x) = \int_x^\infty dG(y) \leq \int_x^\infty y^m dG(y) \Big/ x^m \leq \int_0^\infty y^m dG(y) \Big/ x^m,$$

i.e.

$$\overline{G}(x) \leq E(X^m)/x^m, \quad x \geq 0. \tag{6.2.5}$$

The inequality (6.2.4) is tighter than (6.2.5) for all x satisfying

$$\frac{1-p_0}{\phi_1}(1+\kappa x)^{-m} \leq E(X^m)x^{-m},$$

or equivalently

$$\frac{1}{x} \geq \left\{\frac{1-p_0}{\phi_1 E(X^m)}\right\}^{1/m} - \kappa. \tag{6.2.6}$$

Therefore, (6.2.4) is tighter than (6.2.5) for x sufficiently small, or for all $x \geq 0$ if the right hand side of (6.2.6) is not strictly positive.

Improved inequalities may also be obtained with further restrictions. As noted, $1/\overline{V}(y) = (1+\kappa y)^m$ is convex if $m \geq 1$, and refinements are possible.

Corollary 6.2.2 Suppose $\phi_1 \in (0,1)$ satisfies (4.2.1), and (6.2.2) holds with $\kappa > 0$ and $m \geq 1$. Then

$$\overline{G}(x) \leq \frac{1-p_0}{\phi_1}\{1+\kappa x+\kappa r_*(x)\}^{-m}, \quad x \geq 0, \tag{6.2.7}$$

where $r_*(x) = \inf_{0 \leq z \leq x, \overline{F}(z) > 0} r(z)$, and $r(y)$ defined by (2.3.2) is the mean residual lifetime of the df $F(y)$.

Proof: Theorem 5.2.1 holds with $\overline{B}(y) = \overline{V}(y) = (1+\kappa y)^{-m}$. □

Corollary 6.2.2, a refinement of corollary 6.2.1, may also be itself refined in many cases. Continuing to utilize convexity of $1/\overline{V}(y)$, we obtain the following.

Corollary 6.2.3 Suppose $\phi_1 \in (0,1)$ satisfies (4.2.1), and (6.2.2) holds with $\kappa > 0$ and $m \in \{1,2,3,\cdots\}$. If $F(y)$ is NWUC, then

$$\overline{G}(x) \leq \frac{1-p_0}{\phi_1}\left\{(1+\kappa x)^m + \sum_{j=1}^{m}\binom{m}{j}\kappa^j E(Y^j)(1+\kappa x)^{m-j}\right\}^{-1}, \quad x \geq 0. \tag{6.2.8}$$

Proof: Corollary 5.2.1 applies with $\overline{B}(y) = \overline{V}(y) = (1+\kappa y)^{-m}$ to give

$$\overline{G}(x) \leq \frac{1-p_0}{\phi_1}\left\{\int_0^\infty (1+\kappa x+\kappa y)^m dF(y)\right\}^{-1}, \quad x \geq 0. \tag{6.2.9}$$

Substitution of $(1+\kappa x+\kappa y)^m = \sum_{j=0}^{m}\binom{m}{j}(\kappa y)^j(1+\kappa x)^{m-j}$ into (6.2.9) yields (6.2.8) with the help of (6.2.1). □

It is worth noting that (6.2.9) holds for $m \geq 1$ (not necessarily an integer) if $\phi_1 \in (0,1)$ satisfies (4.2.1), (6.2.2) holds with $\kappa > 0$ and $m \geq 1$, and $F(y)$ is NWUC.

We also have the following.

Corollary 6.2.4 Suppose $\phi_1 \in (0,1)$ satisfies (4.2.1), and (6.2.2) holds with $\kappa > 0$ and $m \in \{1,2,3,\cdots\}$. If the mean residual lifetime $r(y) = \int_y^\infty \overline{F}(x)dx \big/ \overline{F}(y)$ of $F(y)$ satisfies $r(y) \geq r$ where $r \in (0,\infty)$, then

$$\overline{G}(x) \leq \frac{1-p_0}{\phi_1(\kappa r)^m m!}\left\{\sum_{j=0}^{m}\frac{1}{j!}\left(\frac{1+\kappa x}{\kappa r}\right)^j\right\}^{-1}, \quad x \geq 0. \tag{6.2.10}$$

Proof: From corollary 5.2.3 with $\overline{B}(y) = \overline{V}(y) = (1+\kappa y)^{-m}$, one obtains

$$\overline{G}(x) \leq \frac{1-p_0}{\phi_1}\left\{\frac{1}{r}\int_0^\infty (1+\kappa x+\kappa y)^m e^{-y/r}dy\right\}^{-1}, \quad x \geq 0.$$

A change of variable of integration from y to $t = (1+\kappa x+\kappa y)/(\kappa r)$ results in

$$\frac{1}{r}\int_0^\infty (1+\kappa x+\kappa y)^m e^{-y/r}dy = (\kappa r)^m e^{\frac{1+\kappa x}{\kappa r}}\int_{\frac{1+\kappa x}{\kappa r}}^\infty t^m e^{-t}dt,$$

i.e.

$$\overline{G}(x) \leq \frac{(1-p_0)\,e^{-\frac{1+\kappa x}{\kappa r}}}{\phi_1(\kappa r)^m}\left\{\int_{\frac{1+\kappa x}{\kappa r}}^\infty t^m e^{-t}dt\right\}^{-1}, \quad x \geq 0. \tag{6.2.11}$$

From (2.1.12) and (2.1.13), one obtains

$$\int_{\frac{1+\kappa x}{\kappa r}}^\infty t^m e^{-t}dt = (m!)e^{-\frac{1+\kappa x}{\kappa r}}\sum_{j=0}^{m}\frac{1}{j!}\left(\frac{1+\kappa x}{\kappa r}\right)^j,$$

and substitution into (6.2.11) yields (6.2.10). □

We remark that if $\phi_1 \in (0,1)$ satisfies (4.2.1), and (6.2.2) holds with $\kappa > 0$ and $m \geq 1$, (not necessarily an integer) and $r(y) \geq r$ where $r \in (0, \infty)$, then from (6.2.11) one obtains

$$\overline{G}(x) \leq \frac{(1-p_0)\, e^{-\frac{1+\kappa x}{\kappa r}}}{\phi_1(\kappa r)^m \Gamma(m+1)} \left\{1 - \Gamma\left(m+1, \frac{1+\kappa x}{\kappa r}\right)\right\}^{-1}, \quad x \geq 0, \quad (6.2.12)$$

where

$$\Gamma(\alpha, x) = \int_0^x \frac{t^{\alpha-1} e^{-t}}{\Gamma(\alpha)} dt \qquad (6.2.13)$$

is the incomplete gamma function . Also, from (2.3.18), a sufficient condition for $r(y) \geq r$ is that $-\frac{d}{dy} \ln \overline{F}(y) \leq 1/r$. In addition, as mentioned following corollary 5.2.3, (6.2.10) is a refinement of (6.2.7) with $r_*(x)$ replaced by r. The following special cases are also of interest.

Corollary 6.2.5 Suppose $\phi_1 \in (0,1)$ satisfies (4.2.1), and (6.2.2) holds with $\kappa > 0$ and $m \in \{1, 2, 3, \cdots\}$. If $F(y)$ is NWUE, then

$$\overline{G}(x) \leq \frac{1-p_0}{\phi_1 \{\kappa E(Y)\}^m m!} \left\{\sum_{j=0}^{m} \frac{1}{j!} \left(\frac{1+\kappa x}{\kappa E(Y)}\right)^j\right\}^{-1}, \quad x \geq 0. \qquad (6.2.14)$$

Proof: Corollary 6.2.4 holds with $r = E(Y)$. □

Corollary 6.2.6 Suppose $\phi_1 \in (0,1)$ satisfies (4.2.1), and (6.2.2) holds with $\kappa > 0$ and $m \in \{1, 2, 3, \cdots\}$. If $F(y)$ is UBAE, then

$$\overline{G}(x) \leq \frac{1-p_0}{\phi_1 \{\kappa r(\infty)\}^m m!} \left\{\sum_{j=0}^{m} \frac{1}{j!} \left(\frac{1+\kappa x}{\kappa r(\infty)}\right)^j\right\}^{-1}, \quad x \geq 0. \qquad (6.2.15)$$

Proof: Corollary 6.2.4 holds with $r = r(\infty)$. □

6.3 Product based bounds

In the previous two sections exponential and Pareto tails were considered as two important choices for $\overline{B}(x)$. There are situations where neither are appropriate, however, and other approaches to construction of bounds should be used. A useful technique involves multiplication of NWU tails, as illustrated in the following two examples.

Example 6.3.1 The generalized inverse Gaussian distribution

The generalized inverse Gaussian distribution has density

$$f(y) = Cy^{\lambda-1}e^{-\theta y-\beta/y}, \quad y > 0, \tag{6.3.1}$$

where $\theta > 0, \beta > 0$, and $-\infty < \lambda < \infty$. The normalizing constant C is given by $C^{-1} = 2(\beta/\theta)^{\lambda/2}K_\lambda\left(2\sqrt{\theta\beta}\right)$ where $K_\lambda(x)$ is a modified Bessel function (Jorgensen, 1982). The inverse Gaussian distribution is the special case $\lambda = -1/2$. If Y has pdf (6.3.1) then $E\left(e^{\theta Y}\right) = C\int_0^\infty y^{\lambda-1}e^{-\beta/y}dy$ which is infinite if $\lambda > 0$. Therefore, if $\lambda > 0$ one can always find $\kappa > 0$ satisfying

$$\frac{1}{\phi_1} = \int_0^\infty e^{\kappa y}f(y)dy, \tag{6.3.2}$$

where $0 < \phi_1 < 1$. This means that an exponential upper bound may always be obtained if $\lambda > 0$. If $\lambda < 0$ however, this may not be the case, and this situation is examined further. Consider the inverse gamma pdf

$$h(y) = \frac{\beta^{-\lambda}y^{\lambda-1}e^{-\beta/y}}{\Gamma(-\lambda)}, \quad y > 0. \tag{6.3.3}$$

Then (6.3.1) may be expressed as

$$f(y) = C\beta^\lambda\Gamma(-\lambda)e^{-\theta y}h(y), \quad y > 0, \tag{6.3.4}$$

and it is clear that $E\left(e^{tY}\right) < \infty$ for $t \leq \theta, E\left(e^{tY}\right) = \infty$ for $t > \theta$, and using (6.3.4) one has $E\left(e^{\theta Y}\right) = C\beta^\lambda\Gamma(-\lambda)$. This implies that one can find $\kappa > 0$ satisfying (6.3.2) if $C\beta^\lambda\Gamma(-\lambda) \geq 1/\phi_1$ and an exponential bound is available. On the other hand, if $\phi_1 C\beta^\lambda\Gamma(-\lambda) < 1$, no $\kappa > 0$ may be found satisfying (6.3.2). In this situation let $\phi_1^* = \phi_1 C\beta^\lambda\Gamma(-\lambda)$ and so $\phi_1^* < 1$. Now, consider $\overline{B}(y) = e^{-\theta y}(1+\kappa y)^{-\alpha}$ where $\kappa > 0$ and $\alpha > 0$ are unspecified for the moment. Then equation (4.2.2) becomes, with these choices of $\overline{B}(y)$ and $F(y)$,

$$\frac{1}{\phi_1} = \int_0^\infty e^{\theta y}(1+\kappa y)^\alpha f(y)dy,$$

and using (6.3.4) one obtains

$$\frac{1}{\phi_1} = C\beta^\lambda\Gamma(-\lambda)\int_0^\infty (1+\kappa y)^\alpha h(y)dy,$$

i.e.

$$\frac{1}{\phi_1^*} = \int_0^\infty (1+\kappa y)^\alpha h(y)dy. \tag{6.3.5}$$

Therefore, we will require $\kappa > 0$ and $\alpha > 0$ to satisfy (6.3.5). Then, since $h(y)$ has moments up to order $-\lambda$, we must choose $\alpha \in (0, -\lambda)$ and calculate κ from (6.3.5). As in the previous section, it is convenient if α is a positive integer.

Since $1 - \overline{B}(y)$ is NWU, it follows from corollary 4.2.3 that

$$\overline{G}(x) \leq \frac{1-p_0}{\phi_1}(1+\kappa x)^{-\alpha} e^{-\theta x}, \quad x \geq 0, \tag{6.3.6}$$

valid if $\phi_1^* = \phi_1 C\beta^\lambda \Gamma(-\lambda) < 1$, and $\kappa \in (0, \infty)$ and $\alpha \in (0, -\lambda)$ satisfy (6.3.5).

It is possible to refine (6.3.6). The generalized inverse Gaussian failure rate $\mu(y) = -\frac{d}{dy} \ln \int_y^\infty f(t)dt$ is bounded from above (Jorgensen, 1982, p. 102), i.e. $\mu(y) \leq \mu < \infty$. Therefore, from corollary 5.1.4 with $V(y) = B(y)$, one obtains

$$\overline{G}(x) \leq \frac{1-p_0}{\phi_1} \left\{ \int_0^\infty e^{\theta(x+y)} \{1 + \kappa(x+y)\}^\alpha \mu e^{-\mu y} dy \right\}^{-1}.$$

To simplify the integral, first note that

$$\int_0^\infty e^{\theta y} \overline{F}(y) dy = \frac{\int_0^\infty e^{\theta y} f(y) dy - 1}{\theta} = \frac{C\beta^\lambda \Gamma(-\lambda) - 1}{\theta} < \infty.$$

Therefore,

$$\int_0^\infty e^{\theta y - \mu y} dy \leq \int_0^\infty e^{\theta y} \overline{F}(y) dy < \infty,$$

which implies that $\mu > \theta$. Then

$$\begin{aligned}
&\int_0^\infty \mu e^{-\mu y} e^{\theta(x+y)} \{1 + \kappa(x+y)\}^\alpha dy \\
&= \mu e^{\mu x} \int_0^\infty e^{-(\mu-\theta)(x+y)} \{1 + \kappa(x+y)\}^\alpha dy \\
&= \mu e^{\mu x} \int_x^\infty e^{-(\mu-\theta)y} (1+\kappa y)^\alpha dy \\
&= \mu e^{\mu x} \int_{(\mu-\theta)(x+\kappa^{-1})}^\infty e^{\frac{\mu-\theta}{\kappa} - t} \left(\frac{\kappa t}{\mu - \theta} \right)^\alpha \frac{dt}{\mu - \theta} \\
&= \frac{\mu \kappa^\alpha e^{\frac{\mu-\theta}{\kappa}}}{(\mu-\theta)^{\alpha+1}} e^{\mu x} \int_{(\mu-\theta)(x+\kappa^{-1})}^\infty t^\alpha e^{-t} dt
\end{aligned}$$

with the change of variables from y to $t = (\mu - \theta)(y + \kappa^{-1})$. With the help of the incomplete gamma function (6.2.13), one obtains for $x \geq 0$

$$\overline{G}(x) \leq \frac{(1-p_0)(\mu-\theta)^{\alpha+1}}{\phi_1 \mu \kappa^\alpha e^{(\mu-\theta)/\kappa} \Gamma(\alpha+1)} e^{-\mu x} \{1 - \Gamma(\alpha+1, (\mu-\theta)(x+1/\kappa))\}^{-1}. \tag{6.3.7}$$

Thus, (6.3.7) is a refinement of (6.3.6), and a simpler but weaker refinement is obtainable using Jensen's inequality, as long as $1/\overline{V}(y) = e^{\theta y}(1+\kappa y)^\alpha$ is convex.

□

Example 6.3.2

Consider the distribution with tail $\overline{F}(y) = e^{-\theta y}\overline{H}(y)$ where $\theta > 0$ and $H(y) = 1 - \overline{H}(y)$ is an absolutely continuous DFR df with pdf $h(y) = -\overline{H}'(y)$, finite mean m, and failure rate $\mu_H(y) = -\frac{d}{dy}\ln\overline{H}(y)$ which satisfies $\lim_{y\to\infty}\mu_H(y) = 0$. Then $f(y) = -\overline{F}'(y) = e^{-\theta y}\left\{\theta\overline{H}(y) + h(y)\right\}$. Now, for $t > 0$, $\lim_{y\to\infty} e^{ty}\overline{H}(y) = \infty$ since $\int_0^y \{t - \mu_H(s)\}\,ds \to \infty$ as $y \to \infty$ because $\lim_{y\to\infty}\mu_H(y) = 0$. But this implies that $E\left(e^{tY}\right) = \int_0^\infty e^{ty}f(y)dy = \infty$ for $t > \theta$. Also $E\left(e^{\theta Y}\right) = \theta\int_0^\infty \overline{H}(y)dy + \int_0^\infty h(y)dy = \theta m + 1$. Therefore, no $\kappa > 0$ may be found satisfying $\phi_1^{-1} = \int_0^\infty e^{\kappa y}f(y)dy$ if $\phi_1^{-1} > 1 + \theta m$, i.e. if $\phi_1^* = \phi_1(1+\theta m) < 1$. Let $\overline{H}_1(y) = \int_y^\infty \overline{H}(s)ds/m$, and $H_1(y) = 1 - \overline{H}_1(y)$ is the equilibrium df of $H(y)$. Since $H(y)$ is DFR, $H_1(y)$ is DFR, and so is the df $H_*(y) = 1 - \overline{H}_*(y) = \{H(y) + \theta m H_1(y)\}/\,(1+\theta m)$ since it is a mixture of two DFR df's. So, therefore, is $B(y) = 1 - \overline{B}(y)$ where $\overline{B}(y) = e^{-\theta y}\left\{\overline{H}_*(y)\right\}^{1-\phi_1^*}$. We now show that (4.2.2) is satisfied with this choice of $\overline{B}(y)$. Note that the pdf $h_*(y) = H_*'(y)$ satisfies

$$h_*(y) = \frac{h(y) + \theta\overline{H}(y)}{1+\theta m} = \frac{e^{\theta y}f(y)}{1+\theta m}. \tag{6.3.8}$$

Thus, for $z \geq 0$,

$$\begin{aligned}
\int_z^\infty \left\{\overline{B}(y)\right\}^{-1} dF(y) &= \int_z^\infty e^{\theta y}\left\{\overline{H}_*(y)\right\}^{\phi_1^*-1} f(y)dy \\
&= (1+\theta m)\int_z^\infty \left\{\overline{H}_*(y)\right\}^{\phi_1^*-1} h_*(y)dy \\
&= -\frac{1+\theta m}{\phi_1^*}\left\{\overline{H}_*(y)\right\}^{\phi_1^*}\Big|_{y=z}^{\infty}.
\end{aligned}$$

In other words, for $z \geq 0$, we obtain

$$\int_z^\infty \left\{\overline{B}(y)\right\}^{-1} dF(y) = \frac{\left\{\overline{H}_*(z)\right\}^{\phi_1^*}}{\phi_1}. \tag{6.3.9}$$

Clearly, (6.3.9) with $z = 0$ is (4.2.2). The intention here is to use corollary 4.2.2. Consequently, from (6.3.9), we obtain

$$\phi_1\overline{B}(z)\int_z^\infty \left\{\overline{B}(y)\right\}^{-1} dF(y) = e^{-\theta z}\overline{H}_*(z).$$

Therefore, using (6.3.8),

$$\phi_1\overline{B}(z)\int_z^\infty \left\{\overline{B}(y)\right\}^{-1} dF(y) = e^{-\theta z}\int_z^\infty h_*(y)dy = \frac{e^{-\theta z}}{1+\theta m}\int_z^\infty e^{\theta y}f(y)dy$$

which may be expressed as

$$\phi_1^* \overline{B}(z) \int_z^\infty \left\{\overline{B}(y)\right\}^{-1} dF(y) = e^{-\theta z} \int_z^\infty e^{\theta y} dF(y). \tag{6.3.10}$$

Now, $0 \le e^{\theta y}\overline{F}(y) = e^{\theta y}\int_y^\infty f(s)ds \le \int_y^\infty e^{\theta s} f(s)ds$ which must go to 0 as $y \to \infty$ since $\int_0^\infty e^{\theta s} f(s)ds = 1 + \theta m < \infty$. Thus, $\lim_{y\to\infty} e^{\theta y}\overline{F}(y)dy = 0$ and integration by parts yields

$$\begin{aligned} \int_z^\infty e^{\theta y} dF(y) &= -e^{\theta y}\overline{F}(y)\Big|_z^\infty + \theta \int_z^\infty e^{\theta y}\overline{F}(y)dy \\ &= e^{\theta z}\overline{F}(z) + \theta e^{\theta z} \int_z^\infty e^{\theta(y-z)}\overline{F}(y)dy. \end{aligned}$$

In other words

$$e^{-\theta z} \int_z^\infty e^{\theta y} dF(y) = \overline{F}(z) + \theta \int_0^\infty e^{\theta y}\overline{F}(y+z)dy. \tag{6.3.11}$$

Now, $F(y)$ is DFR and hence NWU, implying that $\overline{F}(y+z) \ge \overline{F}(y)\overline{F}(z)$. Therefore

$$e^{-\theta z} \int_z^\infty e^{\theta y} dF(y) \ge \overline{F}(z)\left\{1 + \theta \int_0^\infty e^{\theta y}\overline{F}(y)dy\right\} = \overline{F}(z)\int_0^\infty e^{\theta y} dF(y)$$

using (6.3.11) with $z = 0$. Thus for $z \ge 0$,

$$e^{-\theta z} \int_z^\infty e^{\theta y} dF(y) \ge E\left(e^{\theta Y}\right)\overline{F}(z) = (1+\theta m)\overline{F}(z), \tag{6.3.12}$$

with equality when $z = 0$. Combining (6.3.10) and (6.3.12) results in

$$\phi_1^* \overline{B}(z) \int_z^\infty \left\{\overline{B}(y)\right\}^{-1} dF(y) \ge (1+\theta m)\overline{F}(z).$$

Thus, since $B(y)$ is NWU, (4.2.20) yields

$$c(x,z) \ge \frac{\overline{B}(z)}{\overline{B}(x)\overline{F}(z)} \int_z^\infty \left\{\overline{B}(y)\right\}^{-1} dF(y) \ge \frac{1+\theta m}{\phi_1^* \overline{B}(x)} = \frac{1}{\phi_1 \overline{B}(x)}.$$

It then follows from (4.2.5) that $c_1(x) \le \phi_1 \overline{B}(x)$, and from corollary 4.2.2 that $\overline{G}(x) \le (1-p_0)\,\overline{B}(x)$. To summarize,

$$\overline{G}(x) \le (1-p_0)\, e^{-\theta x} \left\{ \frac{\overline{H}(x) + \theta \int_x^\infty \overline{H}(t)dt}{1+\theta m} \right\}^{1-\phi_1(1+\theta m)}, \quad x \ge 0, \tag{6.3.13}$$

a result which is valid if $\phi_1(1+\theta m) < 1$. □

Each of the two examples in this section has the property that $E(e^{tY}) < \infty$ for $t \leq \theta$, and $E(e^{tY}) = \infty$ for $t > \theta$. This leads to a bound of the form $\overline{G}(x) \leq Ce^{-\theta x}\overline{B}_1(x),\ x \geq 0$, where $\overline{B}_1(x)$ is a nonexponential tail chosen to reflect properties of the normalized density $e^{\theta x}f(x)/E(e^{\theta Y})$. Densities with this property may be viewed as intermediate between the heavy tailed subexponential distributions and those with a tail decaying exponentially fast. Asymptotic properties are discussed in Embrechts and Goldie (1982). See also Embrechts et al (1997, p. 57), and references therein.

7

Compound geometric and related distributions

In this chapter we consider the two special cases where the number of claims distribution $\{p_n; n = 0, 1, 2, \cdots\}$ is a geometric distribution (possibly modified or truncated at 0), and a negative binomial distribution. Compound geometric distributions play an important role in reliability, queueing, regenerative processes, and insurance applications. For example, the equilibrium waiting time in the G/G/1 queue has a compound geometric distribution and so does the maximal aggregate loss of the surplus process under the classical and renewal risk models. For discussions of these applications see Gertsbakh (1984), Kalashnikov (1994b, 1997a), and Panjer and Willmot (1992). Brown (1990) discussed moment-based bounds on the tails of compound geometric distributions. Bounds in the context of G/G/1 queues are given in Kingman (1964, 1970), Ross (1974), and Stoyan (1983, p. 83). Similar results in the insurance risk setting are given in Gerber (1973, 1979) and Taylor (1976). Their bounds are largely based on the service time distribution or the claim amount distribution of the underlying aggregate claims process, as opposed to the 'individual claim amount distribution' in the present compound geometric context. In the next section, various bounds on the tails of compound geometric distributions are derived using the results in chapter 4. These bounds are based on the individual claim amount df $F(y)$. Section 7.2 gives a detailed analysis when the number of claims distribution is a discrete compound geometric distribution. Here the secondary distribution (termed the individual claim amount distribution in the present context but called secondary distribution when the number of claims itself has a compound distribution) is also a counting distribution. We apply the results for the compound geometric distributions to insurance risk theory, which is the subject of section 7.3. Compound negative binomial distributions are discussed in section 7.4, and are widely used in modeling aggregate claims arising from automobile insurance, where the number of claims follows a mixed Poisson with a gamma mixing distribution (see Lemaire, 1995). In accident insurance, the negative binomial distribution is of use since it is also a compound Poisson with logarithmic series secondary distribution. See Panjer and Willmot (1992, section 6.8), for example.

7.1 Compound modified geometric distributions

We consider the special case of example 2.5.1 where $0 \leq p_0 < 1$ and

$$p_n = (1-p_0)(1-\phi)\phi^{n-1}; \quad n = 1, 2, 3, \cdots, \tag{7.1.1}$$

with $0 < \phi < 1$. That is, $\{p_n; n = 0, 1, 2, \cdots\}$ is a modified geometric distribution. Then

$$a_n = \sum_{k=n+1}^{\infty} p_k = (1-p_0)\,\phi^n; \quad n = 0, 1, 2, \cdots, \tag{7.1.2}$$

and it follows that

$$a_{n+1} = \phi a_n; \quad n = 0, 1, 2, \cdots \tag{7.1.3}$$

in this special case. Since each of $a_{n+1} \leq \phi a_n$ and $a_{n+1} \geq \phi a_n$ hold, bounds for the tail of the aggregate claims, $\overline{G}(x)$, are determined in terms of the df $B(y)$ with $\overline{B}(y) = 1 - B(y)$ which satisfies

$$\int_0^{\infty} \left\{\overline{B}(y)\right\}^{-1} dF(y) = \frac{1}{\phi}. \tag{7.1.4}$$

A unique and useful feature of the distribution (7.1.1) is the fact that simple upper and lower bounds for $\overline{G}(x)$ are obtainable in the exponential case. That is, if $\overline{B}(x) = e^{-\kappa x}$, then $\overline{V}(x) = \overline{B}(x) = e^{-\kappa x}$ may be chosen, as discussed in section 6.1. Thus, the adjustment coefficient $\kappa > 0$ is chosen to satisfy

$$\int_0^{\infty} e^{\kappa y} dF(y) = \frac{1}{\phi}. \tag{7.1.5}$$

Upper and lower bounds for $\overline{G}(x)$ are defined in terms of $\alpha_1(x)$ and $\alpha_2(x)$ where

$$\frac{1}{\alpha_1(x)} = \inf_{0 \leq z \leq x, \overline{F}(z) > 0} \int_z^{\infty} e^{\kappa y} dF(y) \Big/ \left\{e^{\kappa z}\overline{F}(z)\right\}, \quad x \geq 0, \tag{7.1.6}$$

and

$$\frac{1}{\alpha_2(x)} = \sup_{0 \le z \le x, \overline{F}(z) > 0} \int_z^\infty e^{\kappa y} dF(y) \Big/ \left\{ e^{\kappa z} \overline{F}(z) \right\}, \quad x \ge 0. \tag{7.1.7}$$

The following general bound has been obtained by many different techniques. Other references, in addition to those in chapter 4, include Taylor (1976) and Kalashnikov (1997b).

Corollary 7.1.1 If p_n satisfies (7.1.1) for $n = 1, 2, 3, \cdots$ with $0 < \phi < 1$, and $\kappa > 0$ satisfies (7.1.5), then

$$\frac{1 - p_0}{\phi} \alpha_2(x) e^{-\kappa x} \le \overline{G}(x) \le \frac{1 - p_0}{\phi} \alpha_1(x) e^{-\kappa x}, \quad x \ge 0, \tag{7.1.8}$$

where $\alpha_1(x)$ and $\alpha_2(x)$ are given by (7.1.6) and (7.1.7) respectively.

Proof: Since (7.1.3) holds, each of corollaries 4.2.1 and 4.3.1 holds. □

Evidently, (7.1.8) gives fairly precise information about the quantity $e^{\kappa x} \overline{G}(x)$. This is supplemented by the fact that if $F(y)$ is non-arithmetic,

$$\lim_{x \to \infty} e^{\kappa x} \overline{G}(x) = \frac{(1 - p_0)(1 - \phi)}{\kappa \phi^2 \int_0^\infty y e^{\kappa y} dF(y)}, \tag{7.1.9}$$

a result which follows from (4.1.8) and (4.1.9) with $C = (1 - p_0)(1 - \phi)/\phi, \alpha = 0$, and $\tau = \phi$.

The following simple bounds hold, and require no restriction on the df $F(y)$.

Corollary 7.1.2 If p_n satisfies (7.1.1) for $n = 1, 2, 3, \cdots$ with $0 < \phi < 1$, and $\kappa > 0$ satisfies (7.1.5), then

$$\frac{1 - p_0}{\phi} \gamma(x) e^{-\kappa x} \le \overline{G}(x) \le \frac{1 - p_0}{\phi} e^{-\kappa x}, \quad x \ge 0, \tag{7.1.10}$$

where $\gamma(x)$ is defined in corollary 6.1.2.

Proof: Each of corollaries 6.1.1 and 6.1.2 holds since (7.1.3) holds. □

In situations where the mean residual lifetime is monotone, we have the following results.

Corollary 7.1.3 Suppose that p_n satisfies (7.1.1) for $n = 1, 2, 3, \cdots$ with $0 < \phi < 1$, and $\kappa > 0$ satisfies (7.1.5). Then if $F(y)$ is IMRL

$$\frac{(1 - p_0) \overline{F}(x)}{\phi \int_x^\infty e^{\kappa y} dF(y)} \le \overline{G}(x) \le (1 - p_0) e^{-\kappa x}, \quad x \ge 0. \tag{7.1.11}$$

Proof: Since IMRL implies NWUC, both corollaries 6.1.4 and 6.1.5 apply. □

Corollary 7.1.4 Suppose that p_n satisfies (7.1.1) for $n = 1, 2, 3, \cdots$ with $0 < \phi < 1$, and $\kappa > 0$ satisfies (7.1.5). Then if $F(y)$ is DMRL with $\overline{F}(x) > 0$

$$(1 - p_0)\, e^{-\kappa x} \leq \overline{G}(x) \leq \frac{(1 - p_0)\, \overline{F}(x)}{\phi \int_x^\infty e^{\kappa y} dF(y)}, \quad x \geq 0. \tag{7.1.12}$$

Proof: Since DMRL implies NBUC, both corollaries 6.1.4 and 6.1.5 apply. □

We remark that if $F(y)$ is an exponential df the upper and lower bounds in both corollaries 7.1.3 and 7.1.4 coincide. Thus, these bounds give an exact expression for $\overline{G}(x)$.

If $F(y)$ has a bounded mean residual lifetime, a simple bound results.

Corollary 7.1.5 Suppose that p_n satisfies (7.1.1) for $n = 1, 2, 3, \cdots$ with $0 < \phi < 1$, and $\kappa > 0$ satisfies (7.1.5). If the mean residual lifetime $r(y) = \int_y^\infty \overline{F}(x)dx \Big/ \overline{F}(y)$ satisfies $r_1 \leq r(y) \leq r_2$ for $y \geq 0$ where $0 \leq r_1 < \infty$ and $0 < r_2 < \infty$, then

$$\frac{(1 - p_0)\,(1 - \kappa r_2)}{\phi} e^{-\kappa x} \leq \overline{G}(x) \leq \frac{(1 - p_0)\,(1 - \kappa r_1)}{\phi} e^{-\kappa x}, \quad x \geq 0. \tag{7.1.13}$$

Proof: Corollary 6.1.6 applies immediately. □

If $F(y)$ is NWUE (NBUE) then (7.1.13) applies with $r_1 = E(Y) (r_2 = E(Y))$, and if $F(y)$ is UBAE (UWAE) with $r_1 = r(\infty)$ $(r_2 = r(\infty))$. Also, from (2.3.18), it follows that a sufficient condition for $r_1 \leq r(y) \leq r_2$ is that

$$1/r_2 \leq -\frac{d}{dy} \ln \overline{F}(y) \leq 1/r_1. \tag{7.1.14}$$

We remark that when $F(y) = 1 - e^{-y/E(Y)}$, it is easy to see from (7.1.5) that $\kappa = (1 - \phi)/E(Y)$, and from corollary 7.1.5 with $r_1 = r_2 = E(Y)$, one must have

$$\overline{G}(x) = (1 - p_0)\, e^{-\frac{1-\phi}{E(Y)} x}, \quad x \geq 0. \tag{7.1.15}$$

Similarly, in this case $\overline{F}(x) / \left\{\phi \int_x^\infty e^{\kappa y} dF(y)\right\} = e^{-\kappa x}$, and so both sides of the inequalities in corollaries 7.1.3 and 7.1.4 also reduce to (7.1.15). That is, the bounds in corollaries 7.1.3, 7.1.4, and 7.1.5 are sharp.

The general bound in corollary 7.1.1 may often be refined. First note that, from (7.1.6), $\alpha_1(x)$ is a nondecreasing function which satisfies (with $z = x$)

$$\overline{F}(x) \leq \alpha_1(x) e^{-\kappa x} \int_x^\infty e^{\kappa y} dF(y), \quad x \geq 0. \tag{7.1.16}$$

Similarly, from (7.1.7) it follows that $\alpha_2(x)$ is nonincreasing and satisfies

$$\overline{F}(x) \geq \alpha_2(x)e^{-\kappa x}\int_x^\infty e^{\kappa y}dF(y), \quad x \geq 0. \tag{7.1.17}$$

The following is essentially due to Wu (1996).

Theorem 7.1.1 If p_n satisfies (7.1.1) for $n = 1, 2, 3, \cdots$ with $0 < \phi < 1$, and $\kappa > 0$ satisfies (7.1.5), then for $x \geq 0$,

$$\overline{G}(x) \leq \frac{1-p_0}{\phi}\alpha_1(x)e^{-\kappa x} - (1-p_0)\left\{\alpha_1(x)e^{-\kappa x}\int_x^\infty e^{\kappa y}dF(y) - \overline{F}(x)\right\}, \tag{7.1.18}$$

and

$$\overline{G}(x) \geq \frac{1-p_0}{\phi}\alpha_2(x)e^{-\kappa x} + (1-p_0)\left\{\overline{F}(x) - \alpha_2(x)e^{-\kappa x}\int_x^\infty e^{\kappa y}dF(y)\right\}. \tag{7.1.19}$$

Proof: It follows from (4.2.13) and (4.2.8) with $k = \infty$ that

$$\overline{G}(x) = \sum_{m=0}^\infty a_m\left\{\overline{F}^{*(m+1)}(x) - \overline{F}^{*m}(x)\right\} \tag{7.1.20}$$

and (4.2.10) with $k = \infty$ becomes

$$\int_0^x \overline{G}(x-y)dF(y) = \sum_{m=0}^\infty a_m\left\{\overline{F}^{*(m+2)}(x) - \overline{F}^{*(m+1)}(x)\right\}.$$

Thus, since $a_{m+1} = \phi a_m$ from (7.1.3), it follows that

$$\begin{aligned}
\phi\int_0^x \overline{G}(x-y)dF(y) &= \sum_{m=0}^\infty a_{m+1}\left\{\overline{F}^{*(m+2)}(x) - \overline{F}^{*(m+1)}(x)\right\} \\
&= \sum_{m=1}^\infty a_m\left\{\overline{F}^{*(m+1)}(x) - \overline{F}^{*m}(x)\right\} \\
&= \overline{G}(x) - a_0\overline{F}(x).
\end{aligned}$$

That is

$$\overline{G}(x) = \phi\int_0^x \overline{G}(x-y)dF(y) + (1-p_0)\overline{F}(x). \tag{7.1.21}$$

Therefore from (7.1.8) and (7.1.5),

$$\begin{aligned}
\overline{G}(x) &\leq (1-p_0)\int_0^x \alpha_1(x-y)e^{-\kappa(x-y)}dF(y) + (1-p_0)\overline{F}(x) \\
&\leq (1-p_0)\alpha_1(x)e^{-\kappa x}\int_0^x e^{\kappa y}dF(y) + (1-p_0)\overline{F}(x) \\
&= (1-p_0)\alpha_1(x)e^{-\kappa x}\left\{\frac{1}{\phi} - \int_x^\infty e^{\kappa y}dF(y)\right\} + (1-p_0)\overline{F}(x).
\end{aligned}$$

Similarly, (7.1.19) holds. □

Refinements to corollaries 7.1.2 through 7.1.5 may be obtained similarly. Nonexponential bounds are also obtained in the same way. We have the following result involving the mean $E(X) = \int_0^\infty x dG(x)$.

Corollary 7.1.6 Suppose that p_n satisfies (7.1.1) for $n = 1, 2, 3, \cdots$ with $0 < \phi < 1$. Then if $E(Y) = \int_0^\infty y dF(y) < \infty$,

$$\overline{G}(x) \leq E(X) / \left\{ x + \frac{\phi}{1 - p_0} E(X) \right\}, \quad x \geq 0. \tag{7.1.22}$$

Proof: Since $E(Y) < \infty$, $\phi^{-1} = \int_0^\infty (1 + \kappa y) dF(y)$ is satisfied by $\kappa = (1 - \phi) / \{\phi E(Y)\}$, and from corollary 6.2.1 with $m = 1$ one obtains

$$\overline{G}(x) \leq \frac{(1 - p_0) E(Y)}{\phi E(Y) + (1 - \phi) x}, \quad x \geq 0. \tag{7.1.23}$$

Since (7.1.2) holds, it follows that

$$\sum_{n=0}^{\infty} n p_n = \sum_{n=0}^{\infty} a_n = (1 - p_0) \sum_{n=0}^{\infty} \phi^n = (1 - p_0) / (1 - \phi),$$

and so $E(X) = E(Y)(1 - p_0)/(1 - \phi)$. Thus, $E(Y) = E(X)(1 - \phi)/(1 - p_0)$, and substitution into (7.1.23) yields (7.1.22) upon rearrangement. □

It is clear that (7.1.22) is a refinement of $\overline{G}(x) \leq E(X)/x$, which is Markov's inequality (i.e. (6.2.5) with $m = 1$). Also, if $p_0 \geq 1 - \phi$, (7.1.22) is a refinement of

$$\overline{G}(x) \leq \frac{E(X)}{x + E(X)}, \quad x \geq 0. \tag{7.1.24}$$

See Cai and Garrido (1998) for further refinements. The next result, essentially due to Lin (1996) and Cai and Wu (1997), is of interest since it may be used to show that the general upper and lower bounds of sections 4.2 and 4.3 hold for a larger class of counting distributions (an example of which is given in theorem 7.2.5) than those which satisfy the dominance conditions (4.2.1) and (4.3.1).

Theorem 7.1.2 Suppose that $a_n \leq (\geq)\ K\phi^n$ for $n = 1, 2, 3, \cdots$, where $\phi \in (0,\ 1)$. Then,

$$\overline{G}(x) \leq (\geq)\ \frac{K}{\phi} \psi_\phi(x) - (K - a_0) \overline{F}(x) \tag{7.1.25}$$

where

$$\psi_\phi(x) = \sum_{n=1}^{\infty} (1 - \phi) \phi^n \overline{F}^{*n}(x) \tag{7.1.26}$$

is a compound geometric tail.

Proof: One has from (4.2.8) and (4.2.13)

$$\begin{aligned}
\overline{G}(x) &= \sum_{n=0}^{\infty} a_n \left\{\overline{F}^{*(n+1)}(x) - \overline{F}^{*n}(x)\right\} \\
&\leq (\geq) \; a_0\overline{F}(x) + \frac{K}{\phi}\sum_{n=1}^{\infty} \phi^{n+1} \left\{\overline{F}^{*(n+1)}(x) - \overline{F}^{*n}(x)\right\} \\
&= a_0\overline{F}(x) + \frac{K}{\phi}\left\{\sum_{n=1}^{\infty}(1-\phi)\phi^n\overline{F}^{*n}(x) - \phi\overline{F}(x)\right\},
\end{aligned}$$

from which (7.1.25) follows. □

If $K = a_0$, it follows that

$$\overline{G}(x) \leq (\geq) \; \frac{1-p_0}{\phi}\psi_\phi(x), \quad x \geq 0.$$

By applying the general upper and lower bounds of theorem 4.2.1 and theorem 4.3.1 to the compound geometric tail $\psi_\phi(x)$, one obtains $\psi_\phi(x) \leq c_1(x)$ and $\psi_\phi(x) \geq c_2(x)$ where $c_1(x)$ and $c_2(x)$ are as in theorem 4.2.1 and theorem 4.3.1 respectively. Therefore, combining the above one obtains

$$\overline{G}(x) \leq \frac{1-p_0}{\phi\overline{V}(0)}c_1(x) \text{ and } \overline{G}(x) \geq \frac{1-p_0}{\phi\overline{V}(0)}c_2(x).$$

To summarize, the dominance conditions (4.2.1) and (4.3.1) in theorem 4.2.1 and theorem 4.3.1 may be replaced by the slightly weaker conditions $a_n \leq \phi_1^n a_0$ and $a_n \geq \phi_2^n a_0$ respectively, for $n = 0, 1, 2, \cdots$. This is also evident from the martingale approach of section 4.4.

Finally a simple evaluation procedure holds when $F(y)$ is the df of an Erlang mixture.

Example 7.1.1 Mixture of Erlangs

If $F'(y) = \sum_{k=1}^{r} q_k \frac{\beta(\beta y)^{k-1}e^{-\beta y}}{(k-1)!}$, then from example 4.1.1,

$$\overline{G}(x) = e^{-\beta x}\sum_{j=0}^{\infty}\overline{C}_j\frac{(\beta x)^j}{j!}, \quad x \geq 0,$$

where $\overline{C}_j = \sum_{n=j+1}^{\infty} c_n$, and $C(z) = \sum_{n=0}^{\infty} c_n z^n = P\{Q(z)\}$. In this case $P(z) = p_0 + (1-p_0)\frac{(1-\phi)z}{1-\phi z}$ from (7.1.1). Also (e.g. Feller, 1968, p. 265)

$$\sum_{j=0}^{\infty}\overline{C}_j z^j = \sum_{j=0}^{\infty}\sum_{n=j+1}^{\infty} c_n z^j = \sum_{n=1}^{\infty}\sum_{j=0}^{n-1} c_n z^j = \sum_{n=1}^{\infty} c_n\left(\frac{1-z^n}{1-z}\right),$$

and since $1 - z^0 = 0$,

$$\sum_{j=0}^{\infty} \overline{C}_j z^j = \frac{1 - C(z)}{1 - z}.$$

Thus,

$$\begin{aligned}
\sum_{j=0}^{\infty} \overline{C}_j z^j &= \frac{1 - p_0 - (1 - p_0)\frac{(1-\phi)Q(z)}{1-\phi Q(z)}}{1 - z} \\
&= (1 - p_0)\frac{1 - \phi Q(z) - (1 - \phi)Q(z)}{(1 - z)\{1 - \phi Q(z)\}} \\
&= (1 - p_0)\frac{1 - Q(z)}{1 - z}\frac{1}{1 - \phi Q(z)}.
\end{aligned}$$

The coefficients $\{\overline{C}_0, \overline{C}_1, \cdots\}$ may sometimes be obtained by expansion of the above generating function in powers of z (as when $r \leq 2$, for example). Alternatively, one has

$$\sum_{j=0}^{\infty} \overline{C}_j z^j = \phi Q(z) \sum_{j=0}^{\infty} \overline{C}_j z^j + (1 - p_0)\frac{1 - Q(z)}{1 - z},$$

from which it follows that for $j = 1, 2, 3, \cdots$,

$$\overline{C}_j = \phi \sum_{i=1}^{j} q_i \overline{C}_{j-i} + (1 - p_0) \sum_{i=j+1}^{\infty} q_i$$

where it is assumed that $q_i = 0$ if $i \notin \{1, 2, \cdots, r\}$. The $\overline{C}_j$'s may be calculated recursively, beginning with $\overline{C}_0 = 1 - p_0$. □

7.2 Discrete compound geometric distributions

It is clear from the discussion in sections 2.6 and 4.2 that best upper (lower) bounds may be obtained when $\{p_n;\ n = 0, 1, 2, \cdots\}$ is D-IMRL (D-DMRL). This is implied if $\{p_n;\ n = 0, 1, 2, \cdots\}$ is log-convex (log-concave). As discussed by Van Harn (1978), another larger class of distributions which contains the log-convex class is the class of discrete compound geometric distributions.

The following theorem lends further support to the view that the class of discrete compound geometric distributions may be viewed as a reliability class.

Theorem 7.2.1 If $\{p_n;\ n = 0, 1, 2, \cdots\}$ is a discrete compound geometric distribution, then $\{p_n;\ n = 0, 1, 2, \cdots\}$ is DS-NWU.

Proof: Brown (1990) and Cai and Kalashnikov (2000) have shown that compound geometric distributions are NWU. Therefore, with $a_n = Pr(N > n) = p_{n+1} + p_{n+2} + \cdots$,

$$\begin{aligned} a_{m+n+1} &= Pr(N > m + n + 1) \\ &\geq Pr(N > m + \frac{1}{2})Pr(N > n + \frac{1}{2}) \\ &= Pr(N > m)Pr(N > n) \\ &= a_m a_n \end{aligned}$$

and the result is proved. □

We may write the pgf of the discrete compound geometric distribution as

$$P(z) = \sum_{n=0}^{\infty} p_n z^n = \frac{1-p}{1-pQ(z)}, \quad |z| < z_0, \tag{7.2.1}$$

where $0 < p < 1$ and $Q(z) = \sum_{n=0}^{\infty} q_n z^n$ is a pgf. For notational convenience, let $\overline{Q}_n = \sum_{j=n+1}^{\infty} q_j$ for $n = 0, 1, 2, \cdots$. It is not difficult to see that the radius of convergence z_0 is the lesser of the radius of convergence of $Q(z)$ and the value $\tau > 1$ (if it exists) satisfying $Q(\tau) = 1/p$. Furthermore, it is possible to assume without loss of generality that $Q(0) = 0$. In other words, (7.2.1) may be expressed as

$$P(z) = \frac{1-a_0}{1-a_0 K(z)} \tag{7.2.2}$$

where

$$0 < a_0 = \frac{p(1-q_0)}{1-pq_0} < 1, \tag{7.2.3}$$

and

$$K(z) = \sum_{n=1}^{\infty} k_n z^n = \frac{Q(z)-q_0}{1-q_0}. \tag{7.2.4}$$

Theorem 7.2.1 may be refined with additional assumptions about $\{q_n;\ n = 0, 1, 2, \cdots\}$.

Theorem 7.2.2 Suppose that $\{p_n;\ n = 0, 1, 2, \cdots\}$ is a discrete compound geometric distribution with pgf (7.2.1), and $Q(z)$ is the pgf of a D-DFR distribution. Then, $\{p_n;\ n = 0, 1, 2, \cdots\}$ is DS-DFR.

Proof: It is clear from (7.2.4) that $k_n = q_n/(1-q_0)$ and thus $\overline{K}_n = \overline{Q}_n/(1-q_0)$ where $\overline{K}_n = \sum_{j=n+1}^{\infty} k_j$. Thus $\{k_n;\ n = 1, 2, 3, \cdots\}$ is D-DFR since $\{q_n;\ n = 0, 1, 2, \cdots\}$ is D-DFR. Now, since $a_n = \sum_{j=n+1}^{\infty} p_j$, it follows from Feller (1968, p. 265) and (7.2.2) that

$$A(z) = \sum_{n=0}^{\infty} a_n z^n = \frac{1 - P(z)}{1 - z} = a_0 \frac{1 - W(z)}{1 - z},$$

where

$$W(z) = \sum_{n=1}^{\infty} w_n z^n = \frac{(1 - a_0)K(z)}{1 - a_0 K(z)},$$

implying that $a_n = a_0 \sum_{j=n+1}^{\infty} w_j$. But since $\{k_n;\ n = 1, 2, 3, \cdots\}$ is D-DFR, it follows from theorem 3.5 of Shanthikumar (1988) that $\{w_n;\ n = 1, 2, 3, \cdots\}$ is D-DFR. Since $a_n = a_0 \sum_{j=n+1}^{\infty} w_j$, $\{p_n;\ n = 0, 1, 2, \cdots\}$ is D-DFR. Furthermore, it follows from (7.2.2) that $p_1 = p_0 a_0 k_1 = p_0(1 - p_0)k_1 \leq p_0(1 - p_0)$. Thus, from the discussion following (2.5.5), $\{p_n;\ n = 0, 1, 2, \cdots\}$ is DS-DFR. $\square$

For further discussion of reliability properties of discrete compound geometric distributions, see Van Harn (1978), Shanthikumar (1988), Hansen and Frenk (1991), and Willmot and Cai (1999).

If, as is usually the case, simple expressions for p_n are difficult to obtain, the following approach is useful. Since (7.2.1) may be reexpressed as

$$P(z) = pP(z)Q(z) + (1 - p),$$

it follows by equating coefficients of z^n that

$$p_n = p \sum_{k=0}^{n} q_k p_{n-k}; \quad n = 1, 2, 3, \cdots. \tag{7.2.5}$$

Equation (7.2.5) is a discrete defective renewal equation (e.g. Karlin and Taylor, 1975, p. 81), a formula which may be adopted to allow for recursive numerical computation of the probabilities $\{p_n;\ n = 0, 1, 2, \cdots\}$, e.g. Klugman, Panjer, and Willmot (1998). This may be supplemented by an asymptotic formula (e.g. Karlin and Taylor, 1975, p. 81, or Willmot, 1989a). If $Q(z)$ has radius of convergence exceeding τ where $\tau > 1$ satisfies $Q(\tau) = 1/p$, then $z_0 = \tau$ as mentioned above and

$$p_n \sim \frac{1 - p}{pQ'(z_0)} \left(\frac{1}{z_0}\right)^{n+1}, \quad n \to \infty. \tag{7.2.6}$$

It follows from (7.2.6) and Willmot (1989a) that the tail probabilities satisfy

$$a_n \sim \frac{1 - p}{p(z_0 - 1)Q'(z_0)} \left(\frac{1}{z_0}\right)^{n+1}, \quad n \to \infty, \tag{7.2.7}$$

again valid if $Q(z_0) = 1/p$. We will now derive a general upper bound on a_n (Willmot and Cai, 1999).

Theorem 7.2.3 Suppose that $\{p_n;\ n = 0, 1, 2, \cdots\}$ is a discrete compound geometric distribution with pgf (7.2.1). Then

$$a_n \leq \gamma_n \left(\frac{1}{z_0}\right)^{n+1}; \quad n = 0, 1, 2, \cdots, \tag{7.2.8}$$

where

$$\frac{1}{\gamma_n} = \inf_{m \in \{0,1,2,\cdots,n\}, \overline{Q}_m > 0} \left\{ \frac{\sum_{j=m+1}^{\infty} q_j z_0^j}{z_0^{m+1} \overline{Q}_m} \right\}. \tag{7.2.9}$$

Proof: If $z_0 = 1$ then $\gamma_n = 1$ from (7.2.9) and (7.2.8) is trivially true. It is clear from the discussion following (7.2.1) and (7.2.4) that the radius of convergence z_0 satisfies $K(z_0) \leq 1/a_0$ or $a_0 K(z_0) \leq 1$. Also, it follows from (7.2.4) that $k_n = q_n/(1 - q_0)$ and $\overline{K}_n = \overline{Q}_n/(1 - q_0)$ where $\overline{K}_n = k_{n+1} + k_{n+2} + \cdots$. Therefore, (7.2.9) implies that γ_n is nondecreasing and satisfies

$$\overline{K}_n \leq \gamma_n \sum_{j=n+1}^{\infty} k_j z_0^{j-n-1}; \quad n = 0, 1, 2, \cdots, \tag{7.2.10}$$

with equality if $\overline{K}_n = 0$. We shall prove (7.2.8) by induction on n. For $n = 0$, (7.2.9) yields

$$\frac{1}{\gamma_0} = \frac{\sum_{j=1}^{\infty} q_j z_0^j}{z_0 \overline{Q}_0} = \frac{\sum_{j=1}^{\infty} k_j z_0^j}{z_0 \overline{K}_0} = \frac{K(z_0)}{z_0}.$$

Thus, $\gamma_0/z_0 = 1/K(z_0) \geq a_0$, i.e. (7.2.8) holds when $n = 0$. Also, from (7.2.2) and (2.5.1), one has

$$A(z) = \sum_{n=0}^{\infty} a_n z^n = \frac{1 - a_0 K(z) - (1 - a_0)}{(1 - z)\{1 - a_0 K(z)\}} = \frac{a_0\{1 - K(z)\}}{(1 - z)\{1 - a_0 K(z)\}},$$

which may be reexpressed as $A(z) = a_0 K(z) A(z) + a_0\{1 - K(z)\}/(1 - z)$. Thus, equating coefficients of z^n yields the identity

$$a_n = a_0 \sum_{j=1}^{n} k_j a_{n-j} + a_0 \overline{K}_n; \quad n = 0, 1, 2, \cdots. \tag{7.2.11}$$

If $a_j \leq \gamma_j z_0^{-j-1}$ for $j = 0, 1, 2, \cdots, n-1$, then (7.2.10) and (7.2.11) imply that

$$\begin{aligned} a_n &\leq a_0 \sum_{j=1}^{n} k_j \gamma_{n-j} z_0^{j-n-1} + a_0 \gamma_n \sum_{j=n+1}^{\infty} k_j z_0^{j-n-1} \\ &\leq a_0 \gamma_n z_0^{-n-1} \left\{ \sum_{j=1}^{n} k_j z_0^j + \sum_{j=n+1}^{\infty} k_j z_0^j \right\} \\ &= \gamma_n z_0^{-n-1} \{a_0 K(z_0)\} \end{aligned}$$

and (7.2.8) holds for n as well. Thus, (7.2.8) holds for all $n = 0, 1, 2, \cdots$. □

The next result follows immediately.

Corollary 7.2.1 Suppose that $\{p_n;\ n = 0, 1, 2, \cdots\}$ is a discrete compound geometric distribution with pgf (7.2.1). Then

$$a_n \leq \left(\frac{1}{z_0}\right)^{n+1}; \quad n = 0, 1, 2, \cdots. \tag{7.2.12}$$

Proof: Since

$$\frac{\sum_{k=m+1}^{\infty} q_k z_0^k}{z_0^{m+1} \overline{Q}_m} \geq \frac{\sum_{k=m+1}^{\infty} q_k z_0^{m+1}}{z_0^{m+1} \overline{Q}_m} = 1,$$

it follows from (7.2.9) that $1/\gamma_n \geq 1$, i.e. $\gamma_n \leq 1$. Thus, from (7.2.8), $a_n \leq \gamma_n z_0^{-n-1} \leq z_0^{-n-1}$. □

Evidently, if $z_0 > 1$, corollary 7.2.1 gives a stochastic ordering between the discrete compound geometric random variable and a geometrically distributed random variable with parameter $1/z_0$.

In some cases, it is possible to improve (7.2.12), as is now demonstrated. First note that summation by parts yields

$$\sum_{j=m+1}^{\infty} q_j z_0^j = \overline{Q}_m z_0^{m+1} + (z_0 - 1) \sum_{j=m+1}^{\infty} \overline{Q}_j z_0^j. \tag{7.2.13}$$

We may now prove the following corollary.

Corollary 7.2.2 If $\{p_n;\ n = 0, 1, 2, \cdots\}$ is a discrete compound geometric distribution with pgf (7.2.1) where $Q(z)$ is the pgf of a DS-NWU distribution, then

$$a_n \leq \frac{1}{Q(z_0)} \left(\frac{1}{z_0}\right)^{n+1}; \quad n = 0, 1, 2, \cdots. \tag{7.2.14}$$

Proof. If $z_0 = 1$ then (7.2.14) is trivially true. If $z_0 > 1$ then from (7.2.13), since $\overline{Q}_{j+m+1} \geq \overline{Q}_j \overline{Q}_m$

$$\begin{aligned}
\sum_{j=m+1}^{\infty} q_j z_0^j &= \overline{Q}_m z_0^{m+1} + z_0^{m+1}(z_0 - 1) \sum_{j=0}^{\infty} \overline{Q}_{j+m+1} z_0^j \\
&\geq \overline{Q}_m z_0^{m+1} \left\{ 1 + (z_0 - 1) \sum_{j=0}^{\infty} \overline{Q}_j z_0^j \right\} \\
&= \overline{Q}_m z_0^{m+1} \left\{ 1 + (z_0 - 1) \frac{Q(z_0) - 1}{z_0 - 1} \right\} \\
&= \overline{Q}_m z_0^{m+1} \{ Q(z_0) \}.
\end{aligned}$$

Thus, from (7.2.9), $1/\gamma_n \geq Q(z_0)$ or $\gamma_n \leq 1/Q(z_0)$ and the result follows from theorem 7.2.2. □

Clearly, if $\tau > 1$ satisfies $Q(\tau) = 1/p$, then $z_0 = \tau$ and (7.2.14) becomes $a_n \leq p z_0^{-n-1}$. In the special case where $Q(z) = q_0/\{1 - (1 - q_0)z\}$, it is easy to see that $z_0 = (1 - pq_0)/(1 - q_0)$ and $a_n = p z_0^{-n-1}$, and (7.2.14) is an equality. We consider the larger D-NWU class next.

Corollary 7.2.3 If $\{p_n;\ n = 0, 1, 2, \cdots\}$ is a discrete compound geometric distribution with pgf (7.2.1) where $Q(z)$ is the pgf of a D-NWU distribution, then

$$a_n \leq \frac{1}{Q(z_0) + q_0(z_0 - 1)} \left(\frac{1}{z_0} \right)^n; \quad n = 0, 1, 2, \cdots. \tag{7.2.15}$$

Proof. If $z_0 = 1$ then (7.2.15) is trivially true. For $z_0 > 1$, it follows from (7.2.13) and $\overline{Q}_{j+m} \geq \overline{Q}_j \overline{Q}_m$ that

$$\begin{aligned}
\sum_{j=m+1}^{\infty} q_j z_0^j &= \overline{Q}_m z_0^{m+1} + z_0^{m}(z_0 - 1) \sum_{j=1}^{\infty} \overline{Q}_{j+m} z_0^j \\
&\geq \overline{Q}_m z_0^{m+1} + \overline{Q}_m z_0^{m+1} \left(1 - \frac{1}{z_0} \right) \sum_{j=1}^{\infty} \overline{Q}_j z_0^j \\
&= \overline{Q}_m z_0^{m+1} \left\{ 1 + \left(1 - \frac{1}{z_0} \right) \left(\sum_{j=0}^{\infty} \overline{Q}_j z_0^j - 1 + q_0 \right) \right\} \\
&= \overline{Q}_m z_0^{m+1} \left\{ 1 + \frac{1}{z_0}(z_0 - 1) \left(\frac{Q(z_0) - 1}{z_0 - 1} - 1 + q_0 \right) \right\} \\
&= \overline{Q}_m z_0^{m+1} \left\{ \frac{z_0 + Q(z_0) - 1 + (z_0 - 1)(q_0 - 1)}{z_0} \right\} \\
&= \overline{Q}_m z_0^{m} \{ Q(z_0) + q_0(z_0 - 1) \}.
\end{aligned}$$

Thus, (7.2.9) implies that $1/\gamma_n \geq \{Q(z_0) + q_0(z_0 - 1)\}/z_0$, and (7.2.15) follows from (7.2.8). □

The following lower bound is a dual to theorem 7.2.3 (Willmot and Cai, 1999).

Theorem 7.2.4 Suppose that $\{p_n;\ n = 0, 1, 2, \cdots\}$ is a discrete compound geometric distribution with pgf (7.2.1), and $\tau > 1$ satisfies $Q(\tau) = 1/p$. Then $z_0 = \tau$ and

$$a_n \geq \beta_n \left(\frac{1}{z_0}\right)^{n+1}; \quad n = 0, 1, 2, \cdots, \tag{7.2.16}$$

where

$$\frac{1}{\beta_n} = \sup_{m \in \{0,1,2,\cdots,n\}, \overline{Q}_m > 0} \left\{ \frac{\sum_{k=m+1}^{\infty} q_k z_0^k}{z_0^{m+1} \overline{Q}_m} \right\}. \tag{7.2.17}$$

Proof: Clearly, (7.2.4) yields $K(z_0) = 1/a_0$ with $z_0 = \tau$ since $P(z) < \infty$ for $z < z_0$ and $P(z) = \infty$ for $z \geq z_0$. Also, β_n is nonincreasing and satisfies (as in (7.2.10))

$$\overline{K}_n \geq \beta_n \sum_{j=n+1}^{\infty} k_j z_0^{j-n-1}; \quad n = 0, 1, 2, \cdots . \tag{7.2.18}$$

The rest of the proof is the same as that of theorem 7.2.3 with the inequalities reversed and γ_n replaced by β_n. □

In what follows it is convenient to define

$$\mu_Q = \sum_{n=0}^{\infty} n q_n = \sum_{n=0}^{\infty} \overline{Q}_n \tag{7.2.19}$$

to be the mean associated with $\{q_n;\ n = 0, 1, 2, \cdots\}$. Furthermore, define

$$q_n^* = \frac{\overline{Q}_n}{\mu_Q}; \quad n = 0, 1, 2, \cdots, \tag{7.2.20}$$

to be the associated equilibrium distribution, and

$$\overline{Q}_n^* = \sum_{j=n+1}^{\infty} q_j^* = \frac{1}{\mu_Q} \sum_{j=n+1}^{\infty} \overline{Q}_j; \quad n = 0, 1, 2, \cdots \tag{7.2.21}$$

to be the equilibrium tail. The following optimal bound is essentially a discrete analogue of corollary 6.1.5 in the geometric case with $\{q_n;\ n =$

$0, 1, 2, \cdots\}$ a discrete version of the NWUC (NBUC) class.

Theorem 7.2.5 Suppose that $\{p_n;\ n = 0, 1, 2, \cdots\}$ is a discrete compound geometric distribution with pgf (7.2.1). If the pgf $Q(z) = \sum_{n=0}^{\infty} q_n z^n$ is such that the distribution $\{q_n;\ n = 0, 1, 2, \cdots\}$ satisfies

$$\overline{Q}^*_{m+j} \geq (\leq) \frac{\overline{Q}_m \overline{Q}^*_j}{1 - q_0}; \quad j, m = 0, 1, 2, \cdots, \tag{7.2.22}$$

and $\tau > 1$ satisfies $Q(\tau) = 1/p$, then $z_0 = \tau$ and

$$a_n \leq (\geq)\, a_0 \left(\frac{1}{z_0}\right)^n; \quad n = 0, 1, 2, \cdots \ . \tag{7.2.23}$$

Proof: Summation by parts yields

$$\sum_{j=m+1}^{\infty} \overline{Q}_j z_0^j = \mu_Q \left\{ \overline{Q}^*_m z_0^{m+1} + (z_0 - 1) \sum_{j=m+1}^{\infty} \overline{Q}^*_j z_0^j \right\}, \tag{7.2.24}$$

as with (7.2.13). Therefore, combining (7.2.13) and (7.2.24) results in

$$\begin{aligned}
\frac{\sum_{j=m+1}^{\infty} q_j z_0^j}{z_0^{m+1} \overline{Q}_m} &= 1 + \frac{z_0 - 1}{z_0^{m+1} \overline{Q}_m} \sum_{j=m+1}^{\infty} \overline{Q}_j z_0^j \\
&= 1 + \frac{\mu_Q (z_0 - 1)}{z_0^{m+1} \overline{Q}_m} \left\{ \overline{Q}^*_m z_0^{m+1} + (z_0 - 1) \sum_{j=m+1}^{\infty} \overline{Q}^*_j z_0^j \right\}. \\
&= 1 + \frac{\mu_Q (z_0 - 1) \overline{Q}^*_m}{\overline{Q}_m} + \frac{\mu_Q (z_0 - 1)^2}{z_0 \overline{Q}_m} \sum_{j=1}^{\infty} \overline{Q}^*_{m+j} z_0^j \\
&\geq (\leq) \ 1 + \frac{\mu_Q (z_0 - 1) \overline{Q}^*_m}{\overline{Q}_m} + \frac{\mu_Q (z_0 - 1)^2}{z_0 (1 - q_0)} \sum_{j=1}^{\infty} \overline{Q}^*_j z_0^j.
\end{aligned}$$

But from (7.2.24) with $m = 0$ and $\overline{Q}_0 = 1 - q_0$, one has

$$\begin{aligned}
&\frac{\sum_{j=m+1}^{\infty} q_j z_0^j}{z_0^{m+1} \overline{Q}_m} \\
\geq (\leq) &\ 1 + \frac{\mu_Q (z_0 - 1) \overline{Q}^*_m}{\overline{Q}_m} + \frac{z_0 - 1}{z_0 (1 - q_0)} \left\{ \sum_{j=1}^{\infty} \overline{Q}_j z_0^j - \mu_Q \overline{Q}^*_0 z_0 \right\} \\
= &\ 1 + \mu_Q (z_0 - 1) \left\{ \frac{\overline{Q}^*_m}{\overline{Q}_m} - \frac{\overline{Q}^*_0}{\overline{Q}_0} \right\} + \frac{z_0 - 1}{z_0 (1 - q_0)} \left\{ \sum_{j=0}^{\infty} \overline{Q}_j z_0^j - \overline{Q}_0 \right\}.
\end{aligned}$$

Since (7.2.22) holds with $j = 0$, it follows (again with $\overline{Q}_0 = 1 - q_0$) that

$$\begin{aligned}\frac{\sum_{j=m+1}^{\infty} q_j z_0^j}{z_0^{m+1}\overline{Q}_m} &\geq (\leq) \; 1 + \frac{z_0 - 1}{z_0(1-q_0)}\left\{\frac{Q(z_0)-1}{z_0-1} - (1-q_0)\right\} \\ &= 1 + \frac{Q(z_0)-1}{z_0(1-q_0)} - \frac{z_0-1}{z_0} \\ &= \frac{Q(z_0) - 1 + (1-q_0)}{z_0(1-q_0)} \\ &= \frac{Q(z_0)-q_0}{z_0(1-q_0)}.\end{aligned}$$

But since (7.2.4) holds, this means that

$$\frac{\sum_{j=m+1}^{\infty} q_j z_0^j}{z_0^{m+1}\overline{Q}_m} \geq (\leq) \; \frac{K(z_0)}{z_0} = \frac{1}{a_0 z_0}$$

since $Q(z_0) = 1/p$ using (7.2.3). Therefore from theorem 7.2.3 (7.2.4), $\gamma_n \leq a_0 z_0$ ($\beta_n \geq a_0 z_0$) and the result follows from (7.2.8) ((7.2.16)). □

The bound (7.2.23) is the same (optimal) bound which results from $a_{n+1} \leq (\geq) \; (1/z_0)a_n$, and results in the same bound for the associated compound distribution, as follows from theorem 7.1.2.

Corollary 7.2.4 Suppose that $\{p_n;\; n = 0, 1, 2, \cdots\}$ is a discrete compound geometric distribution with pgf (7.2.1). If $Q(z)$ is the pgf of a D-IMRL (D-DMRL) distribution, and $\tau > 1$ satisfies $Q(\tau) = 1/p$, then $z_0 = \tau$ and

$$a_n \leq (\geq) \; a_0 \left(\frac{1}{z_0}\right)^n; \quad n = 0, 1, 2, \cdots. \tag{7.2.25}$$

Proof: Clearly, the result follows from theorem 7.2.5 if it can be demonstrated that (7.2.22) holds. As discussed in section 2.5, if $\{q_n; n = 0, 1, 2, \cdots\}$ is D-IMRL (D-DMRL) then $\{q_n^*; n = 0, 1, 2, \cdots\}$ is DS-DFR (DS-IFR) and hence D-DFR (D-IFR). Therefore $\overline{Q}^*_{m+j}/\overline{Q}^*_j$ is nondecreasing (nonincreasing) in j for fixed m. This implies that $\overline{Q}^*_{m+j}/\overline{Q}^*_j \geq (\leq) \; \overline{Q}^*_m/\overline{Q}^*_0$ and (7.2.22) holds if $\overline{Q}^*_m/\overline{Q}^*_0 \geq (\leq) \; \overline{Q}_m/\overline{Q}_0$, or equivalently, $\overline{Q}^*_m/\overline{Q}_m \geq (\leq) \; \overline{Q}^*_0/\overline{Q}_0$. But from (2.5.9), the mean residual lifetime satisfies, in an obvious notation,

$$r_{Q,m} = \frac{\sum_{k=m}^{\infty} \overline{Q}_k}{\overline{Q}_m} = 1 + \mu_Q \frac{\overline{Q}^*_m}{\overline{Q}_m}.$$

Since $\{q_n;\ n = 0, 1, 2, \cdots\}$ is D-IMRL (D-DMRL), $r_{Q,m} \geq (\leq)\ r_{Q,0}$ which is equivalent to $\overline{Q}_m^*/\overline{Q}_m \geq (\leq)\ \overline{Q}_0^*/\overline{Q}_0$. $\square$

In the special case when $\{q_n;\ n = 0, 1, 2, \cdots\}$ is D-DFR, then $\{p_n;\ n = 0, 1, 2, \cdots\}$ is D-DFR (e.g. Shanthikumar, 1988, and Willmot and Cai, 1999), and (7.2.25) follows easily as in section 2.6.

Corollary 7.2.5 Suppose that $\{p_n;\ n = 0, 1, 2, \cdots\}$ is a discrete compound geometric distribution with pgf (7.2.1). If the pgf $Q(z) = \sum_{n=0}^{\infty} q_n z^n$ is such that the distribution $\{q_n;\ n = 0, 1, 2, \cdots\}$ is D-NBU (D-NWU with $q_0 = 0$) and $\tau > 1$ satisfies $Q(\tau) = 1/p$, then $z_0 = \tau$ and

$$a_n \geq (\leq)\ a_0 \left(\frac{1}{z_0}\right)^n; \quad n = 0, 1, 2, \cdots. \tag{7.2.26}$$

Proof: In the D-NBU case $\overline{Q}_{m+j} \leq \overline{Q}_m \overline{Q}_j$ holds, and with $m = j = 0$ it follows that $\overline{Q}_0 \geq 1$, i.e. $\overline{Q}_0 = 1$ and $q_0 = 0$. Then (7.2.22) may be expressed as $\sum_{n=j+m+1}^{\infty} q_n^* \leq (\geq)\ \overline{Q}_m \sum_{n=j+1}^{\infty} q_n^*$, i.e. $\sum_{n=j+1}^{\infty} \overline{Q}_{n+m} \leq (\geq)\ \overline{Q}_m \sum_{n=j+1}^{\infty} \overline{Q}_n$, clearly true. The result follows from theorem 7.2.5. The D-NWU case also follows from corollary 7.2.3. $\square$

The following corollary holds for the compound tail $\overline{G}(x) = \sum_{n=1}^{\infty} p_n \overline{F}^{*n}(x)$.

Corollary 7.2.6 Suppose that $\{p_n;\ n = 0, 1, 2, \cdots\}$ is a discrete compound geometric distribution with pgf (7.2.1) and $z_0 > 1$. Then

$$\overline{G}(x) \leq \gamma_\infty \sum_{n=1}^{\infty} \left(1 - \frac{1}{z_0}\right) \left(\frac{1}{z_0}\right)^n \overline{F}^{*n}(x), \quad x \geq 0, \tag{7.2.27}$$

where $\gamma_\infty = \lim_{n\to\infty} \gamma_n$ and γ_n satisfies (7.2.9). If, in addition z_0 satisfies $Q(z_0) = 1/a_0$, then

$$\overline{G}(x) \geq \beta_\infty \sum_{n=1}^{\infty} \left(1 - \frac{1}{z_0}\right) \left(\frac{1}{z_0}\right)^n \overline{F}^{*n}(x), \quad x \geq 0, \tag{7.2.28}$$

where $\beta_\infty = \lim_{n\to\infty} \beta_n$ and β_n satisfies (7.2.17).

Proof: Since γ_n is nondecreasing, it follows from (7.2.8) that $a_n \leq \gamma_n \left(\frac{1}{z_0}\right)^{n+1} \leq \gamma_\infty \left(\frac{1}{z_0}\right)^{n+1}$, and from theorem 7.1.2 with $\phi = 1/z_0$ and $K = \gamma_\infty/z_0$, and the comment following it, (7.2.27) holds. Similarly,

(7.2.28) holds. □

The results of this section are essentially discrete analogues of some of the compound geometric bounds of section 7.1. It is not hard to see that more general bounds of the type considered there may be obtained. Also, the compound geometric bounds of (7.2.27) and (7.2.28) may be used in conjunction with the results of the previous section.

Example 7.2.1 A discrete ruin model

The following ruin model follows that of Shiu (1989) and Willmot (1993). See also Gerber (1988), Michel (1989), and references therein. Time is measured in discrete units $0, 1, 2, \cdots$. In each time period, the probability of a claim is $q \in (0,1)$ and the probability of no claim is $1-q$. Claims in different time periods are independent and identically distributed, positive, and integer valued, with common probability function $\{f_1, f_2, \cdots\}$. Let the mean be $\mu = \sum_{n=1}^{\infty} n f_n \in (1, 1/q)$, and $\tilde{f}(z) = \sum_{n=1}^{\infty} f_n z^n$. Premiums are payable at the rate of 1 per unit time, and if x is the initial surplus allocated to the portfolio of business, the ruin probability is defined to be

$$\psi_x = Pr\{S_k > x + k; \text{ for some integer } k\}$$

where S_k is the aggregate claims arising from k time periods. Then (e.g. Willmot, 1993),

$$\sum_{x=0}^{\infty} \psi_x z^x = \frac{1}{1-z}\{1 - P(z)\} \tag{7.2.29}$$

where

$$P(z) = \frac{1 - q\mu}{1 - q\mu\tilde{b}(z)} \tag{7.2.30}$$

with

$$\tilde{b}(z) = \frac{\tilde{f}(z) - 1}{\mu(z-1)}. \tag{7.2.31}$$

Clearly, $\{\psi_0, \psi_1, \cdots\}$ is a compound geometric tail, and the results of this section apply. But $\tilde{b}(0) = 1/\mu \neq 0$. Therefore, it is necessary to put $P(z)$ in the form (7.2.1) with $a_0 = \psi_0$. Clearly,

$$\psi_0 = 1 - P(0) = 1 - \frac{1 - q\mu}{1 - q} = \frac{q(\mu - 1)}{1 - q}. \tag{7.2.32}$$

Now, let $\overline{F}_n = \sum_{k=n+1}^{\infty} f_k$, and it follows that

$$\mu = \sum_{n=0}^{\infty} \overline{F}_n = \overline{F}_0 + \sum_{n=1}^{\infty} \overline{F}_n = 1 + \sum_{n=1}^{\infty} \overline{F}_n.$$

Therefore, let

$$q_n = \frac{\overline{F}_n}{\mu - 1} = \frac{\sum_{k=n+1}^{\infty} f_k}{\mu - 1}; \quad n = 1, 2, 3, \cdots, \tag{7.2.33}$$

and $\{q_n;\ n = 1, 2, 3, \cdots\}$ is a probability distribution with pgf

$$Q(z) = \sum_{n=1}^{\infty} q_n z^n = \frac{\sum_{n=0}^{\infty} \overline{F}_n z^n - 1}{\mu - 1} = \frac{\mu \tilde{b}(z) - 1}{\mu - 1}. \tag{7.2.34}$$

Therefore, $\tilde{b}(z) = \frac{1}{\mu} + \left(1 - \frac{1}{\mu}\right) Q(z)$. Then (7.2.30) may be expressed as

$$\begin{aligned} P(z) &= \frac{1 - q\mu}{1 - q\mu\tilde{b}(z)} \\ &= \frac{(1-q)(1-\psi_0)}{1 - q\mu\left\{\frac{1}{\mu} + \left(\frac{\mu - 1}{\mu}\right) Q(z)\right\}} \\ &= \frac{(1-q)(1-\psi_0)}{(1-q) - q(\mu - 1)Q(z)}, \end{aligned}$$

i.e.

$$P(z) = \frac{1 - \psi_0}{1 - \psi_0 Q(z)}. \tag{7.2.35}$$

Now, if $z_0 > 1$ satisfies $Q(z_0) = 1/\psi_0$, or equivalently $\tilde{f}(z_0) = 1 + (z_0 - 1)/q$, then (e.g. Willmot, 1989a)

$$\psi_x \sim \frac{1 - \psi_0}{\psi_0 (z_0 - 1) Q'(z_0)} \left(\frac{1}{z_0}\right)^{x+1}, \quad x \to \infty. \tag{7.2.36}$$

To put (7.2.36) in a simpler form, note that

$$Q(z) = \frac{\frac{\tilde{f}(z) - 1}{z - 1} - 1}{\mu - 1} = \frac{\tilde{f}(z) - z}{(\mu - 1)(z - 1)} = \frac{q}{\psi_0(1-q)} \frac{\tilde{f}(z) - z}{z - 1}.$$

That is,

$$\psi_0 (z - 1) Q(z) = \frac{q}{1 - q}\{\tilde{f}(z) - z\},$$

and differentiation yields

$$\psi_0 Q(z) + \psi_0 (z - 1) Q'(z) = \frac{q}{1 - q}\{\tilde{f}'(z) - 1\}.$$

Since $Q(z_0) = 1/\psi_0$, substitution of $z = z_0$ yields

$$\psi_0(z_0 - 1)Q'(z_0) = \frac{q}{1-q}\{\tilde{f}'(z_0) - 1\} - 1 = \frac{q\tilde{f}'(z_0) - 1}{1-q}.$$

Therefore, it follows using (7.2.32) and (7.2.36) that

$$\psi_x \sim \frac{1 - q\mu}{q\tilde{f}'(z_0) - 1}\left(\frac{1}{z_0}\right)^{x+1}, \quad x \to \infty. \tag{7.2.37}$$

Also, corollary 7.2.1 yields

$$\psi_x \leq \left(\frac{1}{z_0}\right)^{x+1}, \quad x = 0, 1, 2, \cdots, \tag{7.2.38}$$

a result which may be refined to

$$\psi_x \leq (\geq)\ \frac{q(\mu - 1)}{1-q}\left(\frac{1}{z_0}\right)^{x}; \quad x = 0, 1, 2, \cdots, \tag{7.2.39}$$

if $\{q_n;\ n = 1, 2, 3, \cdots\}$ satisfies (7.2.22). □

Example 7.2.2 The M/G/1 equilibrium queue length distribution
In this queue discipline, arrivals to the queue occur according to a Poisson process with rate λ, and the service time df is $H(x) = 1 - \overline{H}(x)$ with mean $\mu = \int_0^\infty \overline{H}(x)dx < 1/\lambda$. Define the mixed Poisson pgf

$$Q(z) = \sum_{n=0}^{\infty} q_n z^n = \int_0^\infty e^{\lambda x(z-1)} dH_1(x) \tag{7.2.40}$$

where $H_1(x) = 1 - \overline{H}_1(x) = \int_0^x \overline{H}(t)dt/\mu$ is the equilibrium df of $H(x)$. The equilibrium queue length distribution has pgf

$$G(z) = \sum_{n=0}^{\infty} g_n z^n = (1 - \rho) + \rho z V(z) \tag{7.2.41}$$

where $\rho = \lambda\mu < 1$ and

$$V(z) = \sum_{n=0}^{\infty} v_n z^n = \frac{(1-\rho)Q(z)}{1 - \rho Q(z)}. \tag{7.2.42}$$

See Willmot and Lin (1994) and references therein. It is convenient to introduce the compound geometric pgf

$$P(z) = \sum_{n=0}^{\infty} p_n z^n = \frac{1-\rho}{1 - \rho Q(z)} \tag{7.2.43}$$

and tail probabilities $a_n = p_{n+1} + p_{n+2} + \cdots$. Now,

$$\frac{1 - V(z)}{1 - z} = \frac{1 - \rho Q(z) - (1-\rho)Q(z)}{(1-z)\{1 - \rho Q(z)\}} = \frac{1 - Q(z)}{(1-z)\{1 - \rho Q(z)\}},$$

and it follows from (2.6.5) that

$$A(z) = \sum_{n=0}^{\infty} a_n z^n = \frac{1 - \rho Q(z) - (1-\rho)}{(1-z)\{1 - \rho Q(z)\}} = \frac{\rho\{1 - Q(z)\}}{(1-z)\{1 - \rho Q(z)\}}.$$

Consequently,

$$A(z) = \rho \left\{ \frac{1 - V(z)}{1 - z} \right\}, \tag{7.2.44}$$

from which it follows that

$$a_n = \rho \sum_{j=n+1}^{\infty} v_j; \quad n = 0, 1, 2, \cdots. \tag{7.2.45}$$

Now, it follows from (7.2.41) that

$$g_n = \rho v_{n-1}; \quad n = 1, 2, 3, \cdots, \tag{7.2.46}$$

and so the tail probabilities satisfy

$$\overline{G}_n = \sum_{j=n+1}^{\infty} g_j = \rho \sum_{j=n+1}^{\infty} v_{j-1} = \rho \sum_{j=n}^{\infty} v_j = a_{n-1}; \quad n = 1, 2, \cdots. \tag{7.2.47}$$

Theorem 7.2.1 implies that $a_{m+n+1} \geq a_m a_n$ and thus $\overline{G}_{m+n+2} \geq \overline{G}_{m+1}\overline{G}_{n+1}$. Since $\overline{G}_0 = \rho < 1$ it follows that $\overline{G}_{m+n} \geq \overline{G}_m \overline{G}_n$ for all $m \geq 0$ and $n \geq 0$. That is, the distribution $\{g_n;\ n = 0, 1, 2, \cdots\}$ is D-NWU. See Stoyan (1983, p. 96) for related results.

If $H(x)$ is IMRL the stronger result that $\{g_n;\ n = 0, 1, 2, \cdots\}$ is DS-DFR holds. To demonstrate this result, note that if $H(x)$ is IMRL then $H_1(x)$ is DFR, and from Grandell (1997, pp. 134-5) and (7.2.40) that $\{q_n;\ n = 0, 1, 2, \cdots\}$ is DS-DFR, and hence D-DFR. Thus, theorem 7.2.2 and (7.2.43) imply that $\{p_n;\ n = 0, 1, 2, \cdots\}$ is DS-DFR and hence D-DFR, i.e. a_{n+1}/a_n is nondecreasing in n for $n = 0, 1, 2, \cdots$. But from (7.2.47), $\overline{G}_{n+1}/\overline{G}_n$ is nondecreasing for $n = 1, 2, 3, \cdots$. Thus, it remains to prove that $\overline{G}_2/\overline{G}_1 \geq \overline{G}_1/\overline{G}_0$ and that $\overline{G}_1 \geq \overline{G}_0^2$ in order to prove that $\{g_n;\ n = 0, 1, 2, \cdots\}$ is DS-DFR, as is clear from the discussion following (2.5.5).

To prove that $\overline{G}_2/\overline{G}_1 \geq \overline{G}_1/\overline{G}_0$, note that since $\{q_n;\ n = 0, 1, 2, ...\}$ is DS-DFR, it follows that $q_1 \leq q_0(1-q_0)$, i.e. that $k_1 \leq q_0$ using (7.2.4). Thus, $p_0(1-p_0)k_1 \leq p_0(1-p_0)q_0$. In the proof of theorem 7.2.2, one obtained $p_1 = p_0(1-p_0)k_1$, and thus $p_1 \leq p_0(1-p_0)q_0$ or $p_1/(1-p_0) \leq p_0 q_0 =$

$q_0(1-\rho)/(1-\rho q_0)$ from (7.2.43). In turn, this implies that $1-p_1/(1-p_0) \geq 1-q_0(1-\rho)/(1-\rho q_0)$, i.e.

$$\frac{a_1}{a_0} \geq \frac{(1-\rho q_0)-q_0(1-\rho)}{1-\rho q_0} = \frac{1-q_0}{1-\rho q_0}.$$

But

$$a_0 = 1-p_0 = 1-\frac{1-\rho}{1-\rho q_0} = \frac{\rho(1-q_0)}{1-\rho q_0}.$$

Thus, $a_1/a_0 \geq a_0/\rho$, i.e. $\overline{G}_2/\overline{G}_1 \geq \overline{G}_1/\overline{G}_0$ using (7.2.47) and $\overline{G}_0 = \rho$.

In order to show that $\overline{G}_1 \geq \overline{G}_0^2$, note that since $H(x)$ is IMRL and hence NWUE, it follows from (2.4.1) that $\overline{H}_1(x) \geq e^{-x/\mu}$. Since $1-e^{-\lambda x}$ is increasing in x, It follows from Ross (1996, p. 405) that

$$\int_0^\infty (1-e^{-\lambda x})dH_1(x) \geq \int_0^\infty (1-e^{-\lambda x})\mu^{-1}e^{-x/\mu}dx,$$

which may be restated using (7.2.40) as $1-q_0 \geq 1-\frac{\mu^{-1}}{\lambda+\mu^{-1}}$, i.e. $q_0 \leq 1/(1+\rho)$. Thus $1 \geq q_0(1+\rho)$ and $1-\rho \geq q_0(1-\rho^2)$. That is, $1-q_0 \geq \rho(1-\rho q_0)$. Hence $\rho \leq (1-q_0)/(1-\rho q_0) = a_0/\rho$, implying that $a_0 \geq \rho^2$, which is $\overline{G}_1 \geq \overline{G}_0^2$ using (7.2.47). This proves that $\{g_n;\ n=0,1,2,\cdots\}$ is DS-DFR if $H(x)$ is IMRL.

Note also that from (7.2.47),

$$g_n = \overline{G}_{n-1} - \overline{G}_n = a_{n-2} - a_{n-1} = p_{n-1}; \quad n = 2,3,\cdots,$$

and thus from (7.2.6), if $Q(\tau) = 1/\rho$, then $z_0 = \tau$ and

$$g_n \sim \frac{1-\rho}{\rho Q'(z_0)}\left(\frac{1}{z_0}\right)^n, \quad n \to \infty. \tag{7.2.48}$$

Similarly, from (7.2.7) and (7.2.47),

$$\overline{G}_n \sim \frac{1-\rho}{\rho(z_0-1)Q'(z_0)}\left(\frac{1}{z_0}\right)^n, \quad n \to \infty. \tag{7.2.49}$$

Also, from (7.2.47), theorem 7.2.3 and 7.2.4,

$$\beta_{n-1}\left(\frac{1}{z_0}\right)^n \leq \overline{G}_n \leq \gamma_{n-1}\left(\frac{1}{z_0}\right)^n; \quad n = 1,2,3,\cdots, \tag{7.2.50}$$

and from corollary 7.2.1,

$$\overline{G}_n \leq \left(\frac{1}{z_0}\right)^n; \quad n = 1,2,3,\cdots. \tag{7.2.51}$$

If $H(x)$ is 2-NWU, then $H_1(x)$ is NWU, and it follows from Cai and Kalashnikov (2000) or Block and Savits (1980) and (7.2.40) that $\{q_n;\ n =$

$0, 1, 2, \cdots\}$ is DS-NWU. Thus with $\overline{G}_0 = \rho$ and $Q(z_0) = 1/\rho$, it follows from corollary 7.2.2 and (7.2.47) that $\overline{G}_n \leq \overline{G}_0 \left(\frac{1}{z_0}\right)^n$. Similarly, if $H(x)$ is 2-NBU, then $\overline{G}_n \geq \overline{G}_0 \left(\frac{1}{z_0}\right)^n$. Also, $a_n \leq (\geq)\ a_0 \left(\frac{1}{z_0}\right)^n$ implies using (7.2.47) that $\overline{G}_n \leq (\geq)\ \overline{G}_1 \left(\frac{1}{z_0}\right)^{n-1}$ for $n = 1, 2, 3, \cdots$. Other bounds on $\overline{G}_n$ are easily established using (7.2.47) and the results of this section. □

7.3 Application to ruin probabilities

We now consider an application of the results in section 7.1 to insurance ruin theory.

In the classical continuous time risk model, the number of claims from an insurance portfolio is assumed to follow a Poisson process N_t with mean λ. The individual claim sizes $Y_1, Y_2, \cdots$, independent of N_t, are positive, independent and identical random variables with common distribution function (df) $H(y) = Pr(Y \leq y)$ and mean $E(Y) = \int_0^\infty y dH(y)$. The aggregate claims process is $\{S_t;\ t \geq 0\}$ where $S_t = Y_1 + Y_2 + \cdots + Y_{N_t}$ (with $S_t = 0$ if $N_t = 0$). Obviously the aggregate claims process S_t is a compound Poisson process. The insurer's surplus process is $\{U_t;\ t \geq 0\}$ with $U_t = u + ct - S_t$ where $u \geq 0$ is the initial surplus, $c = \lambda E(Y)(1+\theta)$ the premium rate per unit time, and $\theta > 0$ the relative security loading .

Define $T = \inf\{t;\ U(t) < 0\}$ to be the first time that the surplus becomes negative and is called the time of ruin. The probability $\psi(u) = Pr\{T < \infty\}$ is called the probability of (ultimate) ruin. If one conditions on the time and the amount of the first claim, the law of total probability yields

$$\psi(u) = \int_0^\infty \lambda e^{-\lambda t} \int_0^{u+ct} \psi(u+ct-y) dH(y) dt + \int_0^\infty \lambda e^{-\lambda t} \overline{H}(u+ct) dt. \tag{7.3.1}$$

Integration by parts yields

$$\psi(u) = \frac{1}{1+\theta} \int_0^u \psi(u-y) dH_1(y) + \frac{1}{1+\theta} \overline{H}_1(u), \tag{7.3.2}$$

where

$$H_1(y) = \int_0^y \overline{H}(t) dt \Big/ E(Y), \quad y \geq 0, \tag{7.3.3}$$

i.e., $H_1(y)$ is the equilibrium distribution of $H(y)$.

It is easy to see (for example by taking Laplace transforms) from (7.3.2) that $\psi(u)$ is the tail of a compound geometric distribution, namely, $\psi(u) = Pr\{L > u\}$, where L is a compound geometric random variable with

$$p_n = \frac{\theta}{1+\theta} \left(\frac{1}{1+\theta}\right)^n; \quad n = 0, 1, 2, \cdots, \tag{7.3.4}$$

and the claim size distribution given in (7.3.3). The insurance interpretation of $H_1(y)$ is that it is the df of the amount of the drop in the surplus, or equivalently the difference between the initial surplus and the surplus at the time it first falls below the initial surplus. In other words it is the df of $S_t - ct$, given that $S_t - ct > 0$ and $S_x - cx \leq 0,\ 0 \leq x < t$. The random variable L may be interpreted as the maximal aggregate loss, $\max_{t \geq 0}\{S_t - ct\}$, since

$$Pr\{\max_{t\geq 0}\{S_t - ct\} > u\} = Pr\{S_t - ct - u > 0, \text{ for some } t \geq 0\} = \psi(u).$$

See Gerber (1979) or Bowers, et al. (1997) for more details.

Evidently, the results of section 7.1 apply with $\phi = 1/(1+\theta)$, $p_0 = 1 - \phi$, and $F(y) = H_1(y)$. In particular, if $H(y)$ is the df of a mixture of Erlangs, then so is $H_1(y)$, as shown in example 2.2.1, and $\psi(u)$ is easily obtained as in example 7.1.1. Other explicit expressions for $\psi(u)$ result if $H(y)$ is a finite mixture or combination of exponentials (e.g. Gerber, Goovaerts, and Kaas, 1987), or a Coxian-2 df as in example 4.1.3. Abate and Whitt (1999) have obtained an explicit formula for $\psi(u)$ when $H(y)$ is an undamped exponential mixture of inverse Gaussian distributions. Also, the Shiu expansion for $\psi(u)$ is given by (e.g. Willmot 1988b)

$$\psi(u) = 1 - \frac{\theta e^{\tau u}}{1+\theta}\left\{1 + \sum_{k=1}^{\infty} \frac{(-\tau)^k}{k!}\eta_k(u)\right\}, \tag{7.3.5}$$

where $\tau = 1/\{(1+\theta)E(Y)\}$ and

$$\eta_k(u) = \int_0^u (u-t)^k e^{-\tau t} dH^{*k}(t), \tag{7.3.6}$$

with $H^{*k}(t)$ the df of the k-fold convolution of $H(t)$. The expansion (7.3.5) was used by Shiu (1988) in the case with $H(y)$ the df of a discrete distribution, and by Willmot (1988b) with $H(y)$ a gamma df with non-integer shape parameter.

The unique positive solution κ of the equation

$$1 + \kappa(1+\theta)E(Y) = \int_0^{\infty} e^{\kappa t} dH(t) \tag{7.3.7}$$

is termed the adjustment coefficient. Since $H_1(y)$ in (7.3.3) is non-arithmetic, (7.1.9) becomes

$$\psi(u) \sim \frac{\theta E(Y)}{\int_0^{\infty} y e^{\kappa y} dH(y) - (1+\theta)E(Y)} e^{-\kappa u}, \quad u \to \infty, \tag{7.3.8}$$

using (7.3.7) and integration by parts. This asymptotic result is the famous Cramer-Lundberg asymptotic ruin formula. Also, the inequality

$$\psi(u) \leq e^{-\kappa u} \tag{7.3.9}$$

is known as the famous Lundberg inequality.

Integration by parts on (7.1.5) yields (7.3.7) which is a more common (but less insightful) formula than (7.1.5). The right hand side of (7.1.10) is a restatement of the Lundberg inequality. Thus, the results in section 7.1 provide improvements and refinements of the Lundberg inequality in many cases.

For example, Gerber (1979) shows that if $H(y)$ is IFR,

$$\psi(u) \geq \left(1 + \kappa \frac{c}{\lambda}\right)^{-1} e^{-\kappa u}, \tag{7.3.10}$$

and if $H(y)$ is DFR,

$$\psi(u) \leq \left(1 + \kappa \frac{c}{\lambda}\right)^{-1} e^{-\kappa u}. \tag{7.3.11}$$

Since $H(y)$ is DFR implies $H_1(y)$ is IMRL (i.e. $H(y)$ is 3-DFR) and $H(y)$ is IFR implies $H_1(y)$ is DMRL (i.e. $H(y)$ is 3-IFR), corollaries 7.1.3 and 7.1.4 apply. Thus one has

$$\psi(u) \geq \frac{1}{1+\theta} e^{-\kappa u} \tag{7.3.12}$$

if $H(y)$ is IFR, and

$$\psi(u) \leq \frac{1}{1+\theta} e^{-\kappa u} \tag{7.3.13}$$

if $H(y)$ is DFR. Inequalities (7.3.12) and (7.3.13) are tighter than inequalities (7.3.10) and (7.3.11), respectively, since $\psi(0) = 1/(1+\theta)$, and from (7.3.10) ((7.3.11)) with $u = 0$, one obtains $\psi(0) = 1/(1+\theta) \geq (\leq) \; 1/(1 + \kappa\frac{c}{\lambda})$.

The results above are obtained under the assumption that (7.3.7) has a unique positive solution. Other type of exponential bounds can be found based on the first three moments of the claim size distribution.

The following result is due to Kalashnikov (1997a, p. 177).

Theorem 7.3.1 If $q_k = \int_0^\infty x^k dH(x) < \infty$, for $k = 1, 2, 3$, then

$$\frac{1}{(1+\theta)} e^{-\kappa_1 u} - K \leq \psi(u) \leq \frac{1}{(1+\theta)} e^{-\kappa_1 u} + K, \tag{7.3.14}$$

where

$$\kappa_1 = \frac{2q_1\theta}{q_2(1+\theta)} \quad \text{and} \quad K = \frac{4q_1q_3\theta}{3q_2^2(1+\theta)}.$$

Equation (7.3.14) might provide a good estimate for the ruin probabilities when θ is small.

7.4 Compound negative binomial distributions

Compound negative binomial models are of considerable interest in automobile insurance, partly due to the presence of heterogeneous risk among policyholders. Suppose that the number of accidents from each policyholder follows a Poisson distribution with the Poisson mean as the risk parameter characterizing the heterogeneity. The number of accidents arising from the group of policyholders is negative binomial if we assume that the risk parameter follows a gamma distribution. Also, since the negative binomial distribution can be viewed as a compound Poisson distribution with the logarithmic distribution being a secondary distribution (e.g. Panjer and Willmot, 1992, section 6.8), it is of use in modeling the number of accident claims from a policyholder, in the case of multiple claims arising from one accident. In this case, the Poisson distribution is the number of accidents distribution while the logarithmic distribution is the number of claims distribution arising from each accident.

We now discuss bounds on the tail of compound negative binomial distributions. The negative binomial distribution (example 2.5.1) has pf

$$p_n = \binom{\alpha+n-1}{n}(1-\phi)^\alpha \phi^n; \quad n = 0, 1, 2, \cdots \tag{7.4.1}$$

where $\alpha > 0$ and $0 < \phi < 1$. Since $p_n = \phi\{1 + (\alpha-1)/n\}p_{n-1}$, it follows that $\{p_n; \ n = 0, 1, 2, \cdots\}$ is log-convex for $0 < \alpha \leq 1$ and log-concave for $\alpha \geq 1$. Therefore, if $\kappa > 0$ satisfies

$$\int_0^\infty e^{\kappa y} dF(y) = \frac{1}{\phi}, \tag{7.4.2}$$

it follows from corollary 4.2.1 that for $0 < \alpha \leq 1$,

$$\overline{G}(x) \leq \frac{1-(1-\phi)^\alpha}{\phi}\alpha_1(x)e^{-\kappa x}, \quad x \geq 0, \tag{7.4.3}$$

where $\alpha_1(x)$ is given by (4.2.17). Since the radius of convergence of the negative binomial pgf is $z_0 = 1/\phi$, $\kappa > 0$ obtained from (7.4.2) is the best possible exponential coefficient, as discussed in section 2.6. Unfortunately, (7.4.3) does not follow from the previous results for $\alpha > 1$, and a "best" exponential upper bound cannot be obtained using these arguments. While an alternative exponential bound can be constructed (due to log-concavity one could replace $1/\phi$ by a_0/a_1 in (7.4.2)), we will construct a different upper bound using (7.4.3). In what follows it is convenient to slightly weaken (7.4.3). That is, let θ be defined by

$$\frac{1}{\theta} = \inf_{z \geq 0, \overline{F}(z) > 0} \int_z^\infty e^{\kappa y} dF(y) \Big/ \left\{e^{\kappa z}\overline{F}(z)\right\}, \tag{7.4.4}$$

where, again, κ satisfies (7.4.2). That is, $\theta = \lim_{x\to\infty} \alpha_1(x)$ where $\alpha_1(x)$ is given by (4.2.17). Clearly $\alpha_1(x)$ is nondecreasing in x and so $\alpha_1(x) \leq \theta$, and from (7.4.3) one obtains for $0 < \alpha \leq 1$

$$\overline{G}(x) \leq C_\alpha e^{-\kappa x}, \quad x \geq 0 \tag{7.4.5}$$

where

$$C_\alpha = \frac{\theta}{\phi}\left\{1 - (1-\phi)^\alpha\right\}. \tag{7.4.6}$$

If one sets $z = 0$ in the right hand side of (7.4.4), one obtains $1/\theta \leq 1/\phi$, or $\theta \geq \phi$ since (7.4.2) holds. Similarly, $\int_z^\infty e^{\kappa y} dF(y) \geq \int_z^\infty e^{\kappa z} dF(y) = e^{\kappa z}\overline{F}(z)$, and from (7.4.4) one must have $1/\theta \geq 1$, or $\theta \leq 1$. That is, $\phi \leq \theta \leq 1$.

Also

$$\frac{d}{d\alpha} C_\alpha = \frac{\theta}{\phi}\left\{-\ln(1-\phi)\right\}(1-\phi)^\alpha > 0$$

for $0 < \alpha \leq 1$. Therefore, when viewed as a function of α, C_α is an increasing function of α from $C_0 = 0$ to $C_1 = \theta \leq 1$, and $0 < C_\alpha \leq 1$.

In what follows we shall need the fact that stochastic ordering is preserved under convolution. We have the following well-known lemma.

Lemma 7.4.1 Suppose that $X_1, X_2, \cdots, X_m$ are nonnegative and independent random variables, as are $Z_1, Z_2, \cdots, Z_m$, where $m \geq 1$. If, in addition, X_i and Z_j are independent for $1 \leq i, j \leq m$, and $\Pr(X_i > x) \leq \Pr(Z_i > x)$ for $x \geq 0$ and $i = 1, 2, \cdots, m$, then

$$Pr\left(X_1 + X_2 + \cdots + X_m > x\right) \leq Pr\left(Z_1 + Z_2 + \cdots + Z_m > x\right), \quad x \geq 0.$$

Proof: We shall prove the result for $m = 2$, and the result follows for $m > 2$ by applying the argument repeatedly to the sequence of partial sums. By the law of total probability, we have

$$\begin{aligned}
Pr\left(X_1 + X_2 > x\right) &= Pr\left(X_1 > x\right) + \int_0^x Pr\left(X_2 > y - x\right) dPr\left(X_1 \leq y\right) \\
&\leq Pr\left(X_1 > x\right) + \int_0^x Pr\left(Z_2 > y - x\right) dPr\left(X_1 \leq y\right) \\
&= Pr\left(X_1 + Z_2 > x\right) \\
&= Pr\left(Z_2 > x\right) + \int_0^x Pr\left(X_1 > y - x\right) dPr\left(Z_2 \leq y\right) \\
&\leq Pr\left(Z_2 > x\right) + \int_0^x Pr\left(Z_1 > y - x\right) dPr\left(Z_2 \leq y\right) \\
&= Pr\left(Z_1 + Z_2 > x\right).
\end{aligned}$$

□

We now give the main upper bound, from Willmot and Lin (1997b). Cai (1998) has considered lower bounds.

Theorem 7.4.1 Suppose that p_n satisfies (7.4.1) for $n = 0, 1, 2, \cdots$ with $0 < \phi < 1$, and $\kappa > 0$ satisfies (7.4.2). If $m \geq \alpha$ is a positive integer, and $\{\alpha_1, \alpha_2, \cdots, \alpha_m\}$ satisfies $0 < \alpha_j \leq 1$ for $j = 1, 2, \cdots, m$ and $\sum_{j=1}^{m} \alpha_j = \alpha$, then

$$\overline{G}(x) \leq \sum_{j=1}^{m} r_j \overline{H}_j(x), \quad x \geq 0, \tag{7.4.7}$$

where

$$\overline{H}_j(x) = e^{-\kappa x} \sum_{i=0}^{j-1} \frac{(\kappa x)^i}{i!}, \quad x \geq 0, \tag{7.4.8}$$

and $\{r_1, r_2, \cdots, r_m\}$ satisfy

$$\sum_{j=0}^{m} r_j z^j = \prod_{k=1}^{m} (1 - C_{\alpha_k} + C_{\alpha_k} z). \tag{7.4.9}$$

Proof: For notational convenience, let X_j have the compound negative binomial distribution with negative binomial parameters α_j and ϕ. Then

$$\int_0^\infty e^{-sx} dPr\,(X_j \leq x) = \left\{ \frac{1-\phi}{1 - \phi \int_0^\infty e^{-sy} dF(y)} \right\}^{\alpha_j}.$$

Since $\sum_{j=1}^{m} \alpha_j = \alpha$ and the above holds,

$$\prod_{j=1}^{m} \int_0^\infty e^{-sx} dPr\,(X_j \leq x) = \left\{ \frac{1-\phi}{1 - \phi \int_0^\infty e^{-sy} dF(y)} \right\}^{\alpha},$$

and so $\overline{G}(x) = Pr\left(\sum_{j=1}^{m} X_j > x\right)$ where the X_j's are independent. Now, suppose that $Z_1, Z_2, \cdots, Z_m$ are independent (also independent of the X_j's), with $Pr\,(Z_j \leq x) = 1 - C_{\alpha_j} e^{-\kappa x}, \quad x \geq 0$. It follows from (7.4.5) that $Pr\,(X_j > x) \leq Pr\,(Z_j > x)$ for $j = 1, 2, \cdots, m$. Hence, using lemma 7.4.1,

$$\overline{G}(x) = Pr\left(\sum_{j=1}^{m} X_j > x\right) \leq Pr\left(\sum_{j=1}^{m} Z_j > x\right)$$

and we need to identify the distribution of $\sum_{j=1}^{m} Z_j$. Now,

$$\int_0^\infty e^{-sx} dPr\,(Z_j \leq x) = 1 - C_{\alpha_j} + C_{\alpha_j} \left(\frac{\kappa}{\kappa + s} \right)$$

and so

$$\int_0^\infty e^{-sx} dPr\left(\sum_{j=1}^m Z_j \le x\right) = \prod_{j=1}^m \left\{1 - C_{\alpha_j} + C_{\alpha_j}\left(\frac{\kappa}{\kappa+s}\right)\right\}$$
$$= \sum_{j=0}^m r_j \left(\frac{\kappa}{\kappa+s}\right)^j$$

using (7.4.9). But, with

$$H_j(x) = \int_0^x \frac{\kappa(\kappa t)^{j-1}e^{-\kappa t}}{(j-1)!}dt,$$

one has

$$\int_0^\infty e^{-sx} dH_j(x) = \left(\frac{\kappa}{\kappa+s}\right)^j,$$

and it follows that

$$\int_0^\infty e^{-sx} dPr\left(\sum_{j=1}^m Z_j \le x\right) = r_0 + \sum_{j=1}^m r_j \int_0^\infty e^{-sx} dH_j(x).$$

Using (2.1.13) and (7.4.8), one has $\overline{H}_j(x) = 1 - H_j(x)$, and by the uniqueness of the Laplace transform $\overline{G}(x) \le Pr\left(\sum_{j=1}^m Z_j > x\right) = \sum_{j=1}^m r_j \overline{H}_j(x)$. □

Interchanging the order of summation in (7.4.7), we obtain the alternative form

$$\overline{G}(x) \le e^{-\kappa x} \sum_{i=0}^{m-1} \overline{R}_i \frac{(\kappa x)^i}{i!}, \quad x \ge 0, \tag{7.4.10}$$

where $\overline{R}_i = \sum_{j=i+1}^m r_j$ for $i = 0, 1, 2, \cdots, m-1$.

It is worth noting that $\{r_0, r_1, \cdots, r_m\}$ satisfying (7.4.8) are the probabilities corresponding to the sum of independent Bernoulli random variables, and (7.4.7) expresses the bounds in terms of the tail of a mixture of Erlangs. We immediately have the following simpler but weaker bound.

Corollary 7.4.1 Suppose that p_n satisfies (7.4.1) for $n = 0, 1, 2, \cdots$ with $0 < \phi < 1$, and $\kappa > 0$ satisfies (7.4.2). If $m \ge \alpha$ is a positive integer, then

$$\overline{G}(x) \le e^{-\kappa x} \sum_{i=0}^{m-1} \frac{(\kappa x)^i}{i!}, \quad x \ge 0. \tag{7.4.11}$$

Proof: It is clear from the above discussion that $\overline{R}_i \le 1$, and (7.4.11) follows from (7.4.10). □

It is worth noting that if $0 < \alpha \leq 1$ and $m = 1$, then (7.4.7) reduces to (7.4.5). We also have the following corollary.

Corollary 7.4.2 Suppose that p_n satisfies (7.4.1) for $n = 0, 1, 2, \cdots$ with $0 < \phi < 1$ and α a positive integer, and $\kappa > 0$ satisfies (7.4.2). Then

$$\overline{G}(x) \leq e^{-\kappa x} \sum_{i=0}^{\alpha-1} \frac{(\kappa x)^i}{i!} \sum_{j=i+1}^{\alpha} \binom{\alpha}{j} \theta^j (1-\theta)^{\alpha-j}, \quad x \geq 0 \tag{7.4.12}$$

where θ is given by (7.4.4).

Proof: With $m = \alpha$ and $\alpha_j = 1$ for $j = 1, 2, \cdots, m$, one has from (7.4.6) that $C_{\alpha_j} = \theta$, and from (7.4.9)

$$\sum_{j=0}^{m} r_j z^j = \prod_{k=1}^{m} (1 - \theta + \theta z) = (1 - \theta + \theta z)^\alpha,$$

which implies that $r_j = \binom{\alpha}{j} \theta^j (1-\theta)^{\alpha-j}$ for $j = 0, 1, 2, \cdots, \alpha$. The result follows from (7.4.10). □

We now show that the bound is sharp since it is an equality in one special case.

Example 7.4.1. Compound Pascal-exponential

If α is a positive integer then (7.4.12) holds. If $F(y) = 1 - e^{-\beta y}$ for $y > 0$ then (7.4.2) becomes

$$\frac{1}{\phi} = \int_0^\infty e^{\kappa y} \beta e^{-\beta y} dy = \frac{\beta}{\beta - \kappa},$$

from which one obtains $\kappa = \beta(1 - \phi)$. Then for $z \geq 0$

$$\frac{\int_z^\infty e^{\kappa y} dF(y)}{e^{\kappa z} \overline{F}(z)} = \frac{\int_z^\infty e^{\kappa y} \beta e^{-\beta y} dy}{e^{-(\beta-\kappa)z}} = \frac{\beta}{\beta - \kappa} = \frac{1}{\phi},$$

and from (7.4.4) one obtains $\theta = \phi$. Then from (7.4.12) one obtains

$$\overline{G}(x) \leq e^{-\beta(1-\phi)x} \sum_{i=0}^{\alpha-1} \frac{\{\beta(1-\phi)x\}^i}{i!} \sum_{j=i+1}^{\alpha} \binom{\alpha}{j} \phi^j (1-\phi)^{\alpha-j}, \quad x \geq 0. \tag{7.4.13}$$

But comparison with (4.1.11) reveals that (7.4.13) is an equality. □

In corollary 7.4.2 (as well as example 7.4.1), m was chosen to be as small as possible, namely $m = \alpha$. This is intuitively reasonable, and must be done to get an optimal bound for large x. To see this, note from (7.4.8) that for $j > 1$

$$\lim_{x\to\infty}(\kappa x)^{1-j}e^{\kappa x}\overline{H}_j(x) = \lim_{x\to\infty}\sum_{i=0}^{j-1}\frac{(\kappa x)^{i-(j-1)}}{i!} = \frac{1}{(j-1)!},$$

i.e.

$$\overline{H}_j(x) \sim \frac{(\kappa x)^{j-1}e^{-\kappa x}}{(j-1)!}, \quad x\to\infty, \tag{7.4.14}$$

which obviously holds for $j = 1$ as well, i.e. for $j = 1, 2, \cdots$. Therefore, if $j < k$

$$\lim_{x\to\infty}\frac{\overline{H}_j(x)}{\overline{H}_k(x)} = \frac{(k-1)!}{(j-1)!}\lim_{x\to\infty}(\kappa x)^{j-k} = 0.$$

If $m(1)$ and $m(2)$ are integers such that $\alpha \le m(1) < m(2)$ and $\{\alpha_1(1), \alpha_2(1), \cdots, \alpha_{m(1)}(1)\}$ and $\{\alpha_1(2), \alpha_2(2), \cdots, \alpha_{m(2)}(2)\}$ are chosen to satisfy the conditions of theorem 7.4.1, then, in an obvious notation

$$\overline{G}(x) \le \sum_{j=1}^{m(k)} r_j(k)\overline{H}_j(x), \quad x \ge 0$$

holds for $k = 1, 2$. But

$$\lim_{x\to\infty}\frac{\sum_{j=1}^{m(1)} r_j(1)\overline{H}_j(x)}{\sum_{j=1}^{m(2)} r_j(2)\overline{H}_j(x)} = \lim_{x\to\infty}\frac{\sum_{j=1}^{m(1)} r_j(1)\overline{H}_j(x)\Big/\overline{H}_{m(2)}(x)}{\sum_{j=1}^{m(2)} r_j(2)\overline{H}_j(x)\Big/\overline{H}_{m(2)}(x)} = 0.$$

This implies that the inequality

$$\overline{G}(x) \le \sum_{j=1}^{m(1)} r_j(1)\overline{H}_j(x) \le \sum_{j=1}^{m(2)} r_j(2)\overline{H}_j(x)$$

holds for x sufficiently large. Hence m should be chosen as small as possible.

Therefore, we define $m_\alpha = -[-\alpha]$ to be the smallest integer greater than or equal to α (with $[\cdot]$ the greatest integer function). Evidently, $m = m_\alpha$ is the best choice of m in theorem 7.4.1.

If α is a positive integer, then $m_\alpha = \alpha$ and one must have $\alpha_j = 1$ for $j = 1, 2, \cdots, m_\alpha$. Then corollary 7.4.2 applies, and there is no choice in the selection of α_j. If α is not an integer, then the choice of the α_j's is not unique, even with $m = m_\alpha$. For example, if $\alpha = 1.5$, then either of $\alpha_1 = \alpha_2 = .75$ or $\alpha_1 = 1, \alpha_2 = .5$ suffice. One simple solution is to choose all the α_j's equal. That is, $\alpha_j = \alpha/m_\alpha$ for $j = 1, 2, \cdots, m_\alpha$. Then from (7.4.9)

$$\sum_{j=0}^{m_\alpha} r_j z^j = \prod_{k=1}^{m_\alpha}\left(1 - C_{\alpha/m_\alpha} - C_{\alpha/m_\alpha}z\right) = \left(1 - C_{\alpha/m_\alpha} + C_{\alpha/m_\alpha}z\right)^{m_\alpha},$$

and so

$$r_j = \binom{m_\alpha}{j} C_{\alpha/m_\alpha}^j \left(1 - C_{\alpha/m_\alpha}\right)^{m_\alpha - j}, \quad j = 1, 2, \cdots, m_\alpha. \tag{7.4.15}$$

Thus, from (7.4.10),

$$\overline{G}(x) \le e^{-\kappa x} \sum_{i=0}^{m_\alpha - 1} \frac{(\kappa x)^i}{i!} \sum_{j=i+1}^{m_\alpha} \binom{m_\alpha}{j} C_{\alpha/m_\alpha}^j \left(1 - C_{\alpha/m_\alpha}\right)^{m_\alpha - j}, \quad x \ge 0. \tag{7.4.16}$$

It is not hard to see that (7.4.16) generalizes (7.4.12), since $C_1 = \theta$ and $m_\alpha = \alpha$ if α is an integer.

A second alternative is to choose all the α_j's equal to 1 except one. That is, let $\alpha_j = 1$ for $j = 1, 2, \cdots, m_\alpha - 1$ and $\alpha_{m_\alpha} = \alpha - [\alpha]$. Substitution into (7.4.9) yields

$$\sum_{j=0}^{m_\alpha} r_j z^j = (1 - \theta + \theta z)^{m_\alpha - 1} \left(1 - C_{\alpha - [\alpha]} + C_{\alpha - [\alpha]} z\right)$$

from which one obtains

$$r_j = \left(1 - C_{\alpha - [\alpha]}\right) q_j + C_{\alpha - [\alpha]} q_{j-1}; \quad j = 1, 2, \cdots, m_\alpha, \tag{7.4.17}$$

where

$$q_j = \binom{m_\alpha - 1}{j} (1 - \theta)^{m_\alpha - 1 - j} \theta^j. \tag{7.4.18}$$

While there are other choices for r_j besides the first (7.4.15) and the second (7.4.17), one of these two extremes appear to be the best to minimize the right hand side of (7.4.7). The difficulty, however, is that the choice may depend on x.

To illustrate this, suppose that $\alpha \in (1, 2)$ and so $m_\alpha = 2$. Then $\alpha = \alpha_1 + \alpha_2$, i.e. $\alpha_2 = \alpha - \alpha_1$. Therefore, there is one free variable α_1 which may be chosen to minimize the right hand side of (7.4.7), subject to the constraint $\alpha - 1 < \alpha_1 \le 1$. Now, from (7.4.9),

$$\sum_{j=0}^{2} r_j z^j = (1 - C_{\alpha_1} + C_{\alpha_1} z)(1 - C_{\alpha - \alpha_1} + C_{\alpha - \alpha_1} z) \tag{7.4.19}$$

and so using (7.4.6)

$$\begin{aligned} r_2 &= C_{\alpha_1} C_{\alpha - \alpha_1} \\ &= \left(\frac{\theta}{\phi}\right)^2 \{1 - (1 - \phi)^{\alpha_1}\}\{1 - (1 - \phi)^{\alpha - \alpha_1}\} \\ &= \left(\frac{\theta}{\phi}\right)^2 \{1 - (1 - \phi)^{\alpha_1} - (1 - \phi)^{\alpha - \alpha_1} + (1 - \phi)^{\alpha}\}. \end{aligned}$$

With the definition

$$b(\alpha_1) = (1-\phi)^{\alpha_1} + (1-\phi)^{\alpha-\alpha_1}, \tag{7.4.20}$$

it follows that

$$r_2 = K_2 - \left(\frac{\theta}{\phi}\right)^2 b(\alpha_1) \tag{7.4.21}$$

where K_2 does not depend on α_1. Similarly,

$$\begin{aligned} r_1 &= C_{\alpha_1}(1-C_{\alpha-\alpha_1}) + C_{\alpha-\alpha_1}(1-C_{\alpha_1}) \\ &= C_{\alpha_1} + C_{\alpha-\alpha_1} - 2C_{\alpha_1}C_{\alpha-\alpha_1} \\ &= \frac{\theta}{\phi}\{1-(1-\phi)^{\alpha_1} + 1 - (1-\phi)^{\alpha-\alpha_1}\} - 2r_2 \\ &= \frac{\theta}{\phi}\{2-b(\alpha_1)\} - 2K_2 + 2\left(\frac{\theta}{\phi}\right)^2 b(\alpha_1). \end{aligned}$$

That is

$$r_1 = \left\{2\left(\frac{\theta}{\phi}\right)^2 - \frac{\theta}{\phi}\right\} b(\alpha_1) + K_1 \tag{7.4.22}$$

where K_1 does not depend on α_1. Then, from (7.4.10),

$$\begin{aligned} \overline{G}(x) &\leq e^{-\kappa x}\{(r_1+r_2) + r_2(\kappa x)\} \\ &= e^{-\kappa x}\left\{\frac{\theta}{\phi}\left(\frac{\theta}{\phi}-1\right) b(\alpha_1) - (\kappa x)\left(\frac{\theta}{\phi}\right)^2 b(\alpha_1) + K_3(x)\right\} \end{aligned} \tag{7.4.23}$$

where $K_3(x)$ does not depend on α_1.

Now, from (7.4.20),

$$b'(\alpha_1) = \ln(1-\phi)\left\{(1-\phi)^{\alpha_1} - (1-\phi)^{\alpha-\alpha_1}\right\}$$

and

$$b''(\alpha_1) = \{\ln(1-\phi)\}^2\left\{(1-\phi)^{\alpha_1} + (1-\phi)^{\alpha-\alpha_1}\right\}.$$

Therefore, $b'(\alpha/2) = 0$ and $b''(\alpha_1) \geq 0$. This means that $b(\alpha_1)$ is convex with a unique minimum at $\alpha_1 = \alpha/2$. Recalling that $\alpha - 1 < \alpha_1 \leq 1$, this also implies that $b(\alpha_1)$ has a unique maximum on $(\alpha-1, 1]$ at $\alpha_1 = 1$.

If $\theta = \phi$, (7.4.23) becomes

$$\overline{G}(x) \leq e^{-\kappa x}\{K_3(x) - (\kappa x)b(\alpha_1)\},$$

and the right hand side is minimized when $b(\alpha_1)$ is maximized, i.e. when $\alpha_1 = 1$. On the other hand, if $\theta > \phi$ (recall that $\phi \leq \theta \leq 1$), then (7.4.23) may be expressed as

$$\overline{G}(x) \leq e^{-\kappa x}\left(\frac{\theta}{\phi}\left\{\frac{\theta}{\phi}(1-\kappa x) - 1\right\} b(\alpha_1) + K_3(x)\right),$$

which implies that the right hand side is minimized when $b(\alpha_1)$ is minimized (i.e. with $\alpha_1 = \alpha/2$) as long as $\frac{\theta}{\phi}(1 - \kappa x) - 1 \geq 0$, i.e. if $x \leq (\theta - \phi)/(\theta\kappa)$. Conversely, if $x > (\theta - \phi)/(\theta\kappa)$, the minimum is attained when $b(\alpha_1)$ is maximized, i.e. when $\alpha_1 = 1$.

To summarize, when $1 < \alpha < 2$, the minimum bound is obtained when $\theta = \phi$ by the choice $\alpha_1 = 1$ and $\alpha_2 = \alpha - 1$, which is the choice (7.4.17). However, when $\theta > \phi$, the minimum bound is obtained with $\alpha_1 = \alpha_2 = \alpha/2$, i.e. (7.4.15), as long as $x \leq (\theta - \phi)/(\theta\kappa)$, and with $\alpha_1 = 1$ and $\alpha_2 = \alpha - 1$, i.e. (7.4.17), when $x > (\theta - \phi)/(\theta\kappa)$. The minimum bound is thus obtained from the two extreme choices (7.4.15) and (7.4.17) but the choice depends on x.

For general $\alpha > 1$ with α not an integer, it can be shown (Willmot and Lin, 1997b) that (7.4.17) is the best choice for all $x \geq 0$ if $\theta = \phi$, and for x sufficiently large if $\theta > \phi$. Both of these were observed to be true for $1 < \alpha < 2$. When $\theta > \phi$, it should not be expected that there exists a unique minimizing choice for all $x \geq 0$, as demonstrated when $1 < \alpha < 2$. Also, a simple but weaker bound holds when $\theta = \phi$.

Corollary 7.4.3 Suppose that p_n satisfies (7.4.1) for $n = 0, 1, 2, \cdots$, with $0 < \phi < 1$, and $\kappa > 0$ satisfies (7.4.2). If $F(y)$ is NWUC then

$$\overline{G}(x) \leq (1 - p_0)e^{-\kappa x} \sum_{i=0}^{m_\alpha} \frac{(\kappa x)^i}{i!}, \quad x \geq 0, \tag{7.4.24}$$

where m_α is the smallest integer greater than or equal to α.

Proof: It is clear from (7.4.4) that $\theta = \lim_{x\to\infty} \alpha_1(x)$ where $\alpha_1(x)$ is given by (4.2.17). But, from the discussion following proposition 6.1.1, $\alpha_1(x) = \phi$ since $F(y)$ is NWUC and (7.4.2) holds. Hence $\theta = \phi$ and from (7.4.6) one has $1 - C_\alpha = (1 - \phi)^\alpha$. Thus from (7.4.9)

$$r_0 = \prod_{k=1}^{m} (1 - C_{\alpha_k}) = \prod_{k=1}^{m} (1 - \phi)^{\alpha_k} = (1 - \phi)^\alpha$$

and from (7.4.1), $r_0 = p_0$. Hence from theorem 7.4.1 with $m = m_\alpha$,

$$\overline{G}(x) \leq \sum_{j=1}^{m_\alpha} r_j \overline{H}_j(x) \leq \sum_{j=1}^{m_\alpha} r_j \overline{H}_{m_\alpha}(x) = (1 - r_0)\overline{H}_{m_\alpha}(x).$$

□

8
Tijms approximations

A compound distribution is in general complex and hence a closed form is not available for its tail $\overline{G}(x)$ except for a few special cases. In the previous chapters, we have discussed upper and lower bounds on compound distribution tails. In this chapter various approximations are considered. These approximations are motivated by an approximation given in Tijms (1986, p. 61), in which Tijms proposed the use of a combination of two exponentials to approximate a compound geometric tail. The parameters are chosen as such that the probability mass at 0 and the mean of two distributions are matched. We extend this idea by considering a combination of an exponential distribution with the adjustment coefficient as its parameter and a general distribution. Again, the probability mass at 0 and the mean are matched. As will be demonstrated, this approximation provides a very satisfactory result when the number of claims distribution is asymptotically geometric. For a large class of individual claim amount distributions, the approximating distribution has the same asymptotic behaviour as that of the compound distribution.

8.1 The asymptotic geometric case

Suppose that the number of claims distribution p_n; $n = 0, 1, 2, \cdots$, satisfies

$$p_n \sim C_1 \phi^n, \quad n \to \infty, \tag{8.1.1}$$

where $C_1 > 0$, and $0 < \phi < 1$, i.e. $p_n;\ n = 0, 1, 2, \cdots$, is asymptotically geometric. Then if $\kappa > 0$ satisfies the Lundberg adjustment equation

$$\int_0^\infty e^{\kappa y} dF(y) = \frac{1}{\phi}, \tag{8.1.2}$$

it follows from the asymptotic formula (4.1.9) that if $F(y)$ is non-arithmetic,

$$\overline{G}(x) \sim Ce^{-\kappa x}, \quad x \to \infty, \tag{8.1.3}$$

where

$$C = C_1 \left\{ \kappa\phi \int_0^\infty y e^{\kappa y} dF(y) \right\}^{-1}. \tag{8.1.4}$$

In other words, the aggregate claims distribution is asymptotically exponential with the adjustment coefficient as its parameter.

We define the Tijms approximation to the tail $\overline{G}(x)$ by

$$\overline{G}_T(x) = (1 - p_0 - C)\,\overline{H}(x) + Ce^{-\kappa x}, \quad x \geq 0, \tag{8.1.5}$$

where $H(x)$ is a df on $(0, \infty)$ required to preserve the mean, i.e.

$$E(N)E(Y) = (1 - p_0 - C) \int_0^\infty \overline{H}(x)dx + \frac{C}{\kappa}. \tag{8.1.6}$$

The motivation for the approximation is the improvement of the asymptotic formula by the addition of a correction term. The approximation is chosen to match the true value $\overline{G}(0) = 1 - p_0$ of $\overline{G}(x)$ at $x = 0$, the mean $E(N)E(Y)$, and will also match the asymptotic value at $x = \infty$ if $\overline{G}_T(x) \sim Ce^{-\kappa x}, x \to \infty$, or equivalently $\lim_{x\to\infty} e^{\kappa x}\overline{H}(x) = 0$.

A convenient choice of $\overline{H}(x)$ for analytic (as well as numerical) purposes is $\overline{H}(x) = e^{-\mu x}$, i.e. (8.1.5) becomes

$$\overline{G}_T(x) = (1 - p_0 - C)\,e^{-\mu x} + Ce^{-\kappa x}, \quad x \geq 0, \tag{8.1.7}$$

where, from (8.1.6),

$$\mu = (1 - p_0 - C) \left\{ E(N)E(Y) - \frac{C}{\kappa} \right\}^{-1}. \tag{8.1.8}$$

We remark that for the modified geometric-exponential distribution where $\overline{G}(x)$ is itself an exponential tail given by the explicit formula (7.1.15), (8.1.3) is an equality for all $x \geq 0$. Thus, $C = 1 - p_0$, and an additive correction term as in (8.1.5) is inappropriate. In this case, for the exponential df $F(y) = 1 - e^{-y/E(Y)}$, (8.1.2) holds with $\kappa = (1 - \phi)/E(Y)$, and for the modified geometric distribution given in (7.1.1), $E(N) = (1 - p_0)/(1 - \phi)$. Thus we wish to exclude this situation when using the Tijms approximation. In the following we give an asymptotic result.

Theorem 8.1.1 Suppose that p_n is asymptotically geometric satisfying (8.1.1) with $0 < \phi < 1$, $a_n \leq (\geq)\ a_0\phi^n$ for $n = 1, 2, 3, \cdots$, and the adjustment coefficient $\kappa > 0$ satisfies (8.1.2). Suppose also that C given by (8.1.4) satisfies $C \neq 1 - p_0$, and either $\kappa \neq (1-\phi)/E(Y)$ or $E(N) \neq (1-p_0)/(1-\phi)$. Let $H(x)$ be an absolutely continuous df satisfying (8.1.6) with failure rate $\mu_H(x) = -\frac{d}{dx}\ln \overline{H}(x)$ satisfying $\mu_H(\infty) = \lim_{x\to\infty}\mu_H(x) \geq 1/\int_0^\infty \overline{H}(x)dx$. If the non-arithmetic df $F(y)$ is NWUC (NBUC), then $C < (>)\ 1 - p_0$ and $\overline{G}_T(x)$ given by (8.1.5) satisfies

$$\overline{G}_T(x) \sim Ce^{-\kappa x}, \quad x \to \infty. \tag{8.1.9}$$

Proof: From theorem 7.1.2 and corollary 6.1.4, $\overline{G}(x) \leq (\geq)\ (1-p_0)\,e^{-\kappa x}$ and from (8.1.3), $\overline{G}(x)$ also satisfies $\overline{G}(x) \sim Ce^{-\kappa x}, x \to \infty$. Thus, it must be the case that $C \leq (\geq)\ 1 - p_0$, and since $C \neq 1 - p_0$, it follows that $C < (>)\ 1 - p_0$.

Clearly,

$$\frac{1}{\kappa} - \frac{1}{\mu_H(\infty)} \geq \frac{1}{\kappa} - \int_0^\infty \overline{H}(x)dx,$$

and since (8.1.6) holds,

$$\frac{1}{\kappa} - \frac{1}{\mu_H(\infty)} \geq \frac{1}{1-p_0-C}\left\{\frac{1-p_0-C}{\kappa} - E(N)E(Y) + \frac{C}{\kappa}\right\},$$

i.e.

$$\frac{1}{\kappa} - \frac{1}{\mu_H(\infty)} \geq \frac{1-p_0}{1-p_0-C}\left\{\left(\frac{1}{\kappa} - \frac{E(Y)}{1-\phi}\right) + E(Y)\left(\frac{1}{1-\phi} - \frac{E(N)}{1-p_0}\right)\right\}. \tag{8.1.10}$$

Also, $E(N) = \sum_{n=0}^\infty a_n \leq (\geq)\ a_0\sum_{n=0}^\infty \phi^n = (1-p_0)\,/(1-\phi)$, or

$$\frac{1}{1-\phi} - \frac{E(N)}{1-p_0} \geq (\leq)\ 0. \tag{8.1.11}$$

Also, $F(y)$ is NWUC (NBUC) implies that $F(y)$ is NWUE (NBUE), i.e. $\overline{F}(y) \leq (\geq)\ \overline{F}_1(y)$ where $F_1(y) = 1 - \overline{F}_1(y) = \int_y^\infty \overline{F}(t)dt/E(Y)$ is the equilibrium df of $F(y)$. Therefore (e.g. Ross, 1996, p. 405)

$$\frac{1}{\phi} = \int_0^\infty e^{\kappa y}dF(y) \leq (\geq) \int_0^\infty e^{\kappa y}dF_1(y) = \frac{\int_0^\infty e^{\kappa y}dF(y) - 1}{\kappa E(Y)},$$

i.e. $\phi^{-1} \leq (\geq)\ (\phi^{-1} - 1)\,/\,\{\kappa E(Y)\}$, or

$$\frac{1}{\kappa} - \frac{E(Y)}{1-\phi} \geq (\leq)\ 0. \tag{8.1.12}$$

But both of $\kappa = (1-\phi)/E(Y)$ and $E(N) = (1-p_0)/(1-\phi)$ cannot hold simultaneously, and so at least one of (8.1.11) and (8.1.12) must be a strict inequality, implying that

$$\left(\frac{1}{\kappa} - \frac{E(Y)}{1-\phi}\right) + E(Y)\left(\frac{1}{1-\phi} - \frac{E(N)}{1-p_0}\right) > (<)\ 0. \tag{8.1.13}$$

Since $1 - p_0 - C > (<)\ 0$, it follows from (8.1.13) that the right hand side of (8.1.10) is the ratio of two positive (negative) terms and is thus positive. That is $\kappa^{-1} - \{\mu_H(\infty)\}^{-1} > 0$ or $\mu_H(\infty) > \kappa$. Thus, by the definition of the limit, $\mu_H(x) > \kappa$ for all x sufficiently large, implying that $\int_0^\infty \{\mu_H(x) - \kappa\}\,dx = \infty$, or equivalently $\lim_{x\to\infty} e^{\kappa x}\overline{H}(x) = 0$. Finally, it follows from (8.1.5) that $\lim_{x\to\infty} e^{\kappa x}\overline{G}_T(x) = C$, i.e. (8.1.9) holds. □

The approximation (8.1.5) is greater (smaller) than the asymptotic result (8.1.3) since $1 - p_0 - C > (<)\ 0$, and exhibits the correct asymptotic behavior as $x \to \infty$ if $F(y)$ is NWUC (NBUC).

We remark that if $H(x)$ is absolutely continuous with mean residual lifetime $r_H(x)$ satisfying $r_H(\infty) \le r_H(0)$, then by (2.3.11), the conditions of theorem 8.1.1 are satisfied. The UBAE condition implies this, for example, and the result is trivially true if $\overline{H}(x) = e^{-\mu x}$. This special case actually reproduces the true value of $\overline{G}(x)$ for all x in some cases, as shown by the following theorem.

Theorem 8.1.2 Suppose that p_n satisfies (8.1.1) with $0 < \phi < 1, \kappa > 0$ satisfies (8.1.2), and $\int_0^\infty y e^{\kappa y} dF(y) < \infty$. If $F(y)$ is non-arithmetic, and

$$\overline{G}(x) = A_1 e^{-R_1 x} + A_2 e^{-R_2 x}, \quad x \ge 0, \tag{8.1.14}$$

where $A_1 \neq 0, A_2 \neq 0$, and $R_1 \neq R_2$, then $\overline{G}(x) = \overline{G}_T(x)$ where $\overline{G}_T(x)$ is given by (8.1.7) with μ given by (8.1.8).

Proof: Since (8.1.1) and (8.1.2) hold, it follows that (8.1.3) holds with C given by (8.1.4). That is, $\lim_{x\to\infty} e^{\kappa x}\overline{G}(x) = C$. Since (8.1.14) holds with $R_1 \neq R_2$, assume without loss of generality that $R_1 > R_2$. Then

$$\lim_{x\to\infty} e^{R_2 x}\overline{G}(x) = \lim_{x\to\infty}\left\{A_1 e^{-(R_1-R_2)x} + A_2\right\} = A_2.$$

But this implies that $R_2 = \kappa$ since if $R_2 \neq \kappa$,

$$\lim_{x\to\infty} e^{R_2 x}\overline{G}(x) = C \lim_{x\to\infty} e^{(R_2-\kappa)x}$$

which equals 0 if $R_2 < \kappa$ and ∞ if $R_2 > \kappa$. Since $R_2 = \kappa$, it then follows that $A_2 = C$, and (8.1.14) becomes $\overline{G}(x) = A_1 e^{-R_1 x} + C e^{-\kappa x}, x \ge 0$. With $x = 0$ one obtains $1 - p_0 = A_1 + C$, i.e. $A_1 = 1 - p_0 - C$. The mean is

$$E(N)E(Y) = \int_0^\infty \overline{G}(x)dx = \frac{1-p_0-C}{R_1} + \frac{C}{\kappa},$$

and solving for R_1 yields $R_1 = (1 - p_0 - C)\left\{E(N)E(Y) - \frac{C}{K}\right\}^{-1} = \mu$ from (8.1.8). □

As in example 4.1.3 when the distribution of Y is a combination of two exponentials, (8.1.14) holds. Thus the Tijms approximation (8.1.7) is an equality in this case.

There are situations when it is desirable to generalize (8.1.7) (to improve the approximation or to ensure that (8.1.9) holds), and a useful generalization of $\overline{H}(x) = e^{-\mu x}$ is the gamma tail

$$\overline{H}(x) = \int_x^\infty \frac{\mu(\mu t)^{\alpha-1}e^{-\mu t}}{\Gamma(\alpha)}dt, \quad x \geq 0. \tag{8.1.15}$$

Useful properties of (8.1.15) include the fact that $H(x)$ is UBAE if $\alpha \geq 1$, since $H(x)$ is IFR with $\mu_H(\infty) = \mu$, as shown in example 2.1.2, and its two parameters allows for greater flexibility for approximation purposes. Moreover, one can always ensure that (8.1.9) is satisfied by selecting a sufficiently large value of α, as we now demonstrate.

If (8.1.15) holds, the mean preserving condition (8.1.6) becomes

$$\frac{\alpha}{\mu}(1 - p_0 - C) = E(N)E(Y) - \frac{C}{\kappa}. \tag{8.1.16}$$

Applying L'Hopital's rule to (8.1.15), one obtains

$$\begin{aligned}
\lim_{x\to\infty} \frac{x^{\alpha-1}e^{-\mu x}}{\overline{H}(x)} &= \lim_{x\to\infty} \frac{(\alpha-1)x^{\alpha-2}e^{-\mu x} - \mu x^{\alpha-1}e^{-\mu x}}{-\mu^\alpha x^{\alpha-1}e^{-\mu x}/\Gamma(\alpha)} \\
&= \frac{\Gamma(\alpha)}{\mu^\alpha} \lim_{x\to\infty}\left(\mu - \frac{\alpha-1}{x}\right) = \frac{\Gamma(\alpha)}{\mu^{\alpha-1}},
\end{aligned}$$

i.e. $\overline{H}(x) \sim (\mu x)^{\alpha-1}e^{-\mu x}/\Gamma(\alpha)$. It is clear from (8.1.5) that the asymptotic result (8.1.9) is equivalent to $\lim_{x\to\infty} e^{\kappa x}\overline{H}(x) = 0$. But from the above, this is equivalent to $\lim_{x\to\infty} x^{\alpha-1}e^{-(\mu-\kappa)x} = 0$ which holds if $\mu > \kappa$. Clearly, since $\alpha/\mu = k\alpha/(k\mu)$ for any $k \neq 0$, μ satisfying (8.1.16) may be chosen arbitrarily large by simultaneously increasing α.

Also, using (8.1.16),

$$\begin{aligned}
\frac{1}{\kappa} - \frac{1}{\mu} &= \frac{1}{\alpha(1-p_0-C)}\left\{\frac{\alpha(1-p_0-C)}{\kappa} - E(N)E(Y) + \frac{C}{\kappa}\right\} \\
&= \frac{1}{\alpha\kappa(1-p_0-C)}\{\alpha(1-p_0) + C(1-\alpha) - \kappa E(N)E(Y)\} \\
&= \frac{1}{\kappa} - \frac{1}{\alpha\kappa}\left\{1 - \frac{(1-p_0) - \kappa E(Y)E(N)}{1-p_0-C}\right\}.
\end{aligned}$$

Thus, $\mu > \kappa$ is equivalent to $\alpha\kappa\left(\kappa^{-1} - \mu^{-1}\right) > 0$, i.e. if $\alpha > \alpha_0$ where

$$\alpha_0 = 1 - \frac{(1-p_0) - \kappa E(Y)E(N)}{1-p_0-C}. \tag{8.1.17}$$

To summarize, (8.1.9) holds if $\alpha > \alpha_0$ where α_0 is given by (8.1.17), and $1 - \overline{H}(x)$ may be DFR if $\alpha_0 < 1$.

For analytical purposes, it is convenient if α is a positive integer, since $\overline{H}(x) = e^{-\mu x}\sum_{j=0}^{\alpha-1}\frac{(\mu x)^j}{j!}$ in this case. Thus, an obvious approach to use is to choose α to be the smallest positive integer greater than α_0, that is

$$\alpha = \min\{1, 1 + [\alpha_0]\}, \tag{8.1.18}$$

where $[x]$ is the greatest integer less than or equal to x. Then from (8.1.16),

$$\mu = \alpha(1 - p_0 - C)\{E(N)E(Y) - \frac{C}{\kappa}\}^{-1}. \tag{8.1.19}$$

Finally, with α and μ given by (8.1.18) and (8.1.19), (8.1.5) becomes

$$\overline{G}_T(x) = Ce^{-\kappa x} + (1 - p_0 - C)e^{-\mu x}\sum_{j=0}^{\alpha-1}\frac{(\mu x)^j}{j!}, \quad x \geq 0, \tag{8.1.20}$$

which is a direct generalization of (8.1.7). Clearly, (8.1.20) matches the true value $\overline{G}(0) = 1 - p_0$ at $x = 0$, the limiting value at $x = \infty$, and the mean $E(N)E(Y)$.

An example of a distribution $\{p_n; n = 0, 1, 2, \cdots\}$ satisfying the conditions of theorem 8.1.1 is the mixed Poisson

$$p_n = \int_0^\infty \frac{(\lambda x)^n e^{-\lambda x}}{n!}k(x)dx \tag{8.1.21}$$

where $k(x)$ is a UWA (UBA) pdf satisfying

$$k(x) \sim C_2 e^{-\beta x}, x \to \infty. \tag{8.1.22}$$

It follows from theorem 3.4.1 that

$$p_n \sim \frac{C_2}{\lambda + \beta}\left(\frac{\lambda}{\lambda + \beta}\right)^n, \quad n \to \infty, \tag{8.1.23}$$

which is of the form (8.1.1), and if $k(x)$ is a UWA (UBA) pdf then $a_{n+1} \leq (\geq)\ a_n/z_0$ with $z_0 = 1 + \beta/\lambda$, which implies that $a_n \leq (\geq)\ a_0(1/z_0)^n$, as follows from theorem 3.3.1. In particular, the results apply if $k(x)$ is a finite mixture (sum) of exponentials in which case $\{p_n; n = 0, 1, 2, \cdots\}$ is a finite mixture (sum) of geometrics.

Another example of a distribution $\{p_n;\ n = 0, 1, 2, \cdots\}$ satisfying the conditions of theorem 8.1.1 is the compound geometric distribution with

$$P(z) = \frac{1 - a_0}{1 - a_0 Q(z)}, \tag{8.1.24}$$

$Q(z) = \sum_{n=1}^{\infty} q_n z^n$, and the distribution $\{q_n;\ n = 1,2,3,\cdots\}$ satisfies $Q(1/\phi) = 1/a_0$ and (7.2.22). See theorem 7.2.5 for details.

In the next section we consider the modified geometric distribution of section 7.1.

8.2 The modified geometric distribution

We consider the results of the previous section where p_n is given by (7.1.1), i.e. $p_n = (1-p_0)(1-\phi)\phi^{n-1}; n = 1,2,3,\cdots$, and thus from (7.1.2), $a_n = (1-p_0)\,\phi^n; n = 0,1,2,\cdots$. In this case (8.1.1) is an equality with $C_1 = (1-p_0)(1-\phi)/\phi$, and $E(N) = \sum_{n=0}^{\infty} a_n = (1-p_0)/(1-\phi)$. The Tijms approximation is given by (8.1.5) and the mean preserving condition (8.1.6) becomes

$$\frac{(1-p_0)E(Y)}{1-\phi} = (1-p_0-C)\int_0^{\infty} \overline{H}(x)dx + \frac{C}{\kappa}. \tag{8.2.1}$$

In this special case the correct limiting behavior is exhibited for a larger class of distributions, as is now stated (e.g. Willmot, 1998).

Corollary 8.2.1 Suppose p_n satisfies (7.1.1) for $n = 1,2,3,\cdots$ with $0 < \phi < 1$, and $\kappa > 0$ satisfies (8.1.2). Suppose also that C given by (8.1.4) satisfies $C \neq 1-p_0$, and $\kappa \neq (1-\phi)/E(Y)$. Let $1-\overline{H}(x)$ be an absolutely continuous df satisfying (8.2.1) with failure rate $\mu_H(x) = -\frac{d}{dx}\ln \overline{H}(x)$ satisfying $\mu_H(\infty) = \lim_{x\to\infty} \mu_H(x) \geq 1/\int_0^{\infty} \overline{H}(x)dx$. If the non-arithmetic df $F(y)$ is NWUC (NBUC), then $C < (>)\ 1-p_0$ and $\overline{G}_T(x)$ given (8.1.5) satisfies $\overline{G}_T(x) \sim Ce^{-\kappa x},\quad x \to \infty$.

Proof: Since $a_n = a_0\phi^n$, both versions of theorem 8.1.1 apply. □

In this case the correct asymptotic behavior is exhibited by the Tijms approximation for both the NWUC and NBUC classes. We omit the details, but mention the fact that $\overline{G}(x)$ is of the form (8.1.14) when $F(y)$ has a Coxian-2 density (e.g. Tijms, 1986). This includes, for example, the density of the sum and mixtures of two exponentials (example 4.1.3) and the mixture of an exponential and a gamma with shape parameter 2. Thus, $\overline{G}_T(x)$ given by (8.1.7) equals the true value $\overline{G}(x)$ in this case, as follows from theorem 8.1.2.

The following numerical example demonstrates that if $F(y)$ is neither NBUC or NWUC, the asymptotic right tail behavior may not be reproduced by (8.1.7).

Example 8.2.1 Suppose that $\phi = .4, p_0 = .6$, and $F(y)$ has density

$$F'(y) = \sum_{j=1}^{3} q_j \frac{\beta(\beta y)^{j-1} e^{-\beta y}}{(j-1)!}$$

where $q_1 = 5/6, q_2 = q_3 = 1/12$, and $\beta = 5/4$. Then $E(Y) = 1$, and $\kappa = 0.5994$ satisfies (8.1.2). This in turn implies from (8.1.4) that $C = .4080$ and from (8.1.8) $\mu = .5729 < \kappa$. Thus $\overline{G}_T(x)$ given by (8.1.7) does not satisfy the asymptotic condition (8.1.9). One has $\alpha_0 = 1.0463$ from (8.1.17), i.e. the correct asymptotic behavior is exhibited if $\alpha > 1.0463$, which excludes the exponential where $\alpha = 1$. The estimates from (8.1.18) and (8.1.19) are $\alpha = 2$ and $\mu = 1.1458$, respectively, which may be substituted into (8.1.20).

□

8.3 Transform derivation of the approximation

There are situations where it can be verified that a particular distribution is of compound geometric form based on various criteria. For instance, Van Harn (1978) considers the relationship with log-convexity, complete monotonicity, and related notions. Also, it is well known that the ruin probability in the case when the number of claims process is a renewal process is a compound geometric tail (e.g. Embrechts et al., 1997, pp. 26-7). In some cases the associated Laplace transform may be evaluated using Wiener-Hopf factorization (e.g. De Smit, 1995, or Cohen, 1982). The distribution itself may be difficult to obtain, however, and so we consider a Tijms approximation.

Consequently, we shall assume that the Laplace transform

$$\tilde{g}(s) = \int_0^\infty e^{-sx} dG(x) \tag{8.3.1}$$

is known, and we wish to construct the Tijms approximation (8.1.7) using (8.3.1). Thus, we assume that (8.3.1) may be expressed as

$$\tilde{g}(s) = \frac{1-\phi}{1-\phi\tilde{f}(s)}, \tag{8.3.2}$$

with $0 < \phi < 1$ and $\tilde{f}(s) = \int_0^\infty e^{-sy} dF(y)$. By a Tauberian result,

$$\phi = 1 - \tilde{g}(\infty) \tag{8.3.3}$$

where $\tilde{g}(\infty) = \lim_{s\to\infty} \tilde{g}(s)$. Then substitution of (8.3.3) into (8.3.2) and solving for $\tilde{f}(s)$ yields

$$\tilde{f}(s) = \frac{\tilde{g}(s) - \tilde{g}(\infty)}{\{1-\tilde{g}(\infty)\}\tilde{g}(s)}. \tag{8.3.4}$$

The adjustment coefficient (assuming its existence) is the smallest positive solution κ to the equation

$$\tilde{f}(-\kappa) = 1/\{1 - \tilde{g}(\infty)\}, \tag{8.3.5}$$

which follows from (8.3.3) and (8.1.2). Then (7.1.9) with $p_0 = 1 - \phi = \tilde{g}(\infty)$ may be expressed as $\overline{G}(x) \sim Ce^{-\kappa x}$, $x \to \infty$, where

$$C = -\frac{\tilde{g}(\infty)}{\kappa\{1 - \tilde{g}(\infty)\}\tilde{f}'(-\kappa)}. \tag{8.3.6}$$

Then (8.1.8) becomes

$$\mu = -\frac{1 - \tilde{g}(\infty) - C}{\tilde{g}'(0) + \frac{C}{\kappa}}. \tag{8.3.7}$$

Finally, the Tijms approximation is, from (8.1.7)

$$\overline{G}_T(x) = \{1 - \tilde{g}(\infty) - C\}e^{-\mu x} + Ce^{-\kappa x}, \quad x \geq 0. \tag{8.3.8}$$

9

Defective renewal equations

Defective renewal equations arise in many different areas of application of applied probability. For example, the equilibrium waiting time distribution in many queueing models and the number of offspring distribution in branching processes satisfy certain defective renewal equations. For a detailed discussion of these applications, see Feller (1971) or Resnick (1992). Defective renewal equations also play an important role in insurance risk theory. Many functionals of interest associated with the time of ruin in the classical risk model are in the form of the solution of a defective renewal equation, as will be seen later. On the other hand, compound geometric distributions are appealing for the reasons that (i) the analytic expression of a compound geometric tail is obtainable for some individual claim amount distributions, (ii) tight upper and lower bounds based on reliability classifications are available for compound geometric tails, and (iii) there exists a satisfactory approximation to compound geometric tails. These desirable properties are discussed in chapters 7 and 8. Motivated by the fact that a compound geometric distribution may be viewed as the solution of a special defective renewal equation, we explore a close connection between defective renewal equations and compound geometric distributions in this chapter. The solution of a defective renewal equation is expressed in terms of a compound geometric distribution. This connection allows for the use of analytic properties of a compound geometric distribution and for the application of exact and approximate results which have been developed for the compound geometric tail in chapters 7 and 8 to the solution of the renewal equation itself. This approach is discussed in section 9.1. Applications to various situations are then presented. In particular, the expected discounted penalty function related to the classical risk model is considered. Special cases of this function include the moments of the deficit at the time of ruin and the Laplace transform of the time of ruin. As will be seen in section 9.2, the expected discounted penalty may be expressed as the solution of a defective renewal equation. We further identify the components of this defective renewal equation and investigate their relations to the underlying risk model. The results in the early chapters and section 9.1 thus apply. Finally, we extend the application to a risk model with diffusion, and a queueing model.

9.1 Some properties of defective renewal equations

In this section, we identify the close connection between defective renewal equations and compound geometric distributions. Suppose that $0 < \phi < 1$ and $F(y)$ is a df on $[0, \infty)$ with $F(0) = 0$, referred to again as the individual claim amount distribution. Let $m(x)$ satisfy the defective renewal equation

$$m(x) = \phi \int_0^x m(x-y)dF(y) + v(x), \quad x \geq 0, \tag{9.1.1}$$

where $v(x)$ is a continuous function on $[0, \infty)$. Equations of the form (9.1.1) arise repeatedly in various areas of application of applied probability such as insurance risk theory, branching processes, queueing theory, and inventory theory. See, for example, Ross (1996), Karlin and Taylor (1975), and Tijms (1994). See Feller (1971) and Resnick (1992) for a detailed discussion of the solution of (9.1.1).

In the present context we shall demonstrate how some of the ideas presented in earlier chapters may be applied to the solution $m(x)$ of (9.1.1). Central to the discussion is the compound geometric df $G(x)$ defined by

$$\overline{G}(x) = \sum_{n=1}^{\infty} (1-\phi)\phi^n \overline{F}^{*n}(x), \quad x \geq 0. \tag{9.1.2}$$

Note that $G(x)$ has a mass point $G(0) = 1 - \phi$ at 0 and it does not involve the function $v(x)$. The following theorem expresses the solution to (9.1.1) in terms of $G(x)$.

Theorem 9.1.1 The solution $m(x)$ to (9.1.1) may be expressed as

$$m(x) = \frac{1}{1-\phi} \int_0^x v(x-y)dG(y) + v(x). \tag{9.1.3}$$

Proof: It follows from (4.1.3) and (9.1.2) that

$$\tilde{g}(s) = \frac{1-\phi}{1-\phi\tilde{f}(s)} \tag{9.1.4}$$

where

$$\tilde{g}(s) = \int_0^\infty e^{-sx} dG(x) + G(0) \text{ and } \tilde{f}(s) = \int_0^\infty e^{-sx} dF(x). \tag{9.1.5}$$

Thus, from (9.1.1),

$$\int_0^\infty e^{-sx} m(x)dx = \int_0^\infty e^{-sx} v(x)dx / \{1 - \phi \tilde{f}(s)\} = \frac{\tilde{g}(s)}{1-\phi} \int_0^\infty e^{-sx} v(x)dx.$$

Inversion of this Laplace transform equation yields (9.1.3). □

We remark that in what follows, integrals of the type in (9.1.3) are assumed to be of Stieltjes type, and exclude the mass point at 0. Also, with $F^{*n}(x) = 1 - \overline{F}^{*n}(x)$, (9.1.3) may also be expressed as

$$m(x) = \sum_{n=1}^{\infty} \phi^n \int_0^x v(x-y) dF^{*n}(y) + v(x), \tag{9.1.6}$$

as follows from (9.1.2). If the df $F^{*n}(y)$ of the n-fold convolution may be identified, (9.1.6) may be of use. Simpler solutions may be available, however, as in the following example.

Example 9.1.1 Mixture of Erlangs

As in example 4.1.1, if $F(y)$ has density

$$F'(y) = \sum_{k=1}^{r} q_k \frac{\beta(\beta y)^{k-1} e^{-\beta y}}{(k-1)!},$$

then (4.1.7) holds, and differentiating yields

$$G'(x) = \sum_{n=1}^{\infty} c_n \frac{\beta(\beta x)^{n-1} e^{-\beta x}}{(n-1)!}, \quad x > 0, \tag{9.1.7}$$

where

$$C(z) = \sum_{n=0}^{\infty} c_n z^n = \frac{1-\phi}{1 - \phi \sum_{k=1}^{r} q_k z^k}. \tag{9.1.8}$$

As discussed in section 7.2, the probability distribution $\{c_0, c_1, c_2, \cdots\}$ may sometimes be obtained by expansion of (9.1.8) in powers of z (e.g., if $r = 1$ or 2). Alternatively, since (9.1.8) may be reexpressed as $C(z) = \phi C(z) \sum_{k=1}^{n} q_k z^k + (1-\phi)$, it follows that the c_n may be computed recursively from

$$c_n = \phi \sum_{k=1}^{\min(n,r)} q_k c_{n-k}, \quad n = 1, 2, 3, \cdots, \tag{9.1.9}$$

beginning with $c_0 = 1 - \phi$, a result which also follows from (7.2.2). Then from (9.1.7) and (9.1.3),

$$m(x) = \frac{1}{1-\phi}\sum_{n=1}^{\infty} c_n \frac{\beta^n}{(n-1)!}\int_0^x v(x-y)y^{n-1}e^{-\beta y}dy + v(x), \quad x \geq 0. \tag{9.1.10}$$

Further simplification is possible in the exponential case with $F(y) = 1 - e^{-\beta y}$. Then $q_1 = 1$, and from (9.1.8) it follows that $c_n = (1-\phi)\phi^n$. Substitution into (9.1.12) yields

$$\begin{aligned} m(x) &= \sum_{n=1}^{\infty}\frac{(\phi\beta)^n}{(n-1)!}\int_0^x v(x-y)y^{n-1}e^{-\beta y}dy + v(x) \\ &= \phi\beta\int_0^x v(x-y)e^{-\beta y}\left\{\sum_{n=1}^{\infty}\frac{(\phi\beta y)^{n-1}}{(n-1)!}\right\}dy + v(x) \\ &= \phi\beta\int_0^x v(x-y)e^{-\beta(1-\phi)y}dy + v(x), \end{aligned}$$

a result which may also be obtained from (4.1.12) and (9.1.3). □

Since many of the properties of $\overline{G}(x)$ developed in the previous chapters such as bounds and approximations, are not readily applicable to the density $dG(y)$, it would be useful to express $m(x)$ as a function of $\overline{G}(x)$ instead. We have the following alternative expression for $m(x)$.

Theorem 9.1.2 If $v(x)$ is differentiable, then the solution $m(x)$ to (9.1.1) may be expressed as

$$m(x) = \frac{1}{1-\phi}v(x) - \frac{v(0)}{1-\phi}\overline{G}(x) - \frac{1}{1-\phi}\int_0^x \overline{G}(x-y)v'(y)dy. \tag{9.1.11}$$

Proof: Integration by parts yields

$$\begin{aligned} \int_0^x v(x-y)dG(y) &= -v(x-y)\overline{G}(y)\,|_{y=0}^{x} - \int_0^x v'(x-y)\overline{G}(y)dy \\ &= v(x)\overline{G}(0) - v(0)\overline{G}(x) - \int_0^x v'(x-y)\overline{G}(y)dy. \end{aligned}$$

Since $\overline{G}(0) = \phi$, it follows from (9.1.3) that

$$m(x) = \frac{1}{1-\phi}\left\{\phi v(x) - v(0)\overline{G}(x) - \int_0^x v'(x-y)\overline{G}(y)dy\right\} + v(x)$$

which is (9.1.11). □

Bounds on $m(x)$ may be obtained from (9.1.11) by substitution of bounds for $\overline{G}(x)$ into (9.1.11). Exponential bounds involve the adjustment coefficient $\kappa > 0$ which satisfies the Lundberg adjustment equation

$$\int_0^\infty e^{\kappa y} dF(y) = \frac{1}{\phi}. \tag{9.1.12}$$

The following corollary gives but one of many bounds which may be obtained along these lines. It is obtained from the general compound geometric bound of corollary 7.1.1, and special cases based on various reliability based classifications of $F(y)$ are easily established. Moreover, it is assumed that $v(x)$ has a nonpositive derivative, as is a common occurrence in applications.

Corollary 9.1.1 Suppose that $\kappa > 0$ satisfies (9.1.12), and that $v(x)$ is differentiable with $v'(x) \le 0$. Then

$$\begin{aligned} m(x) \le\ & \frac{1-\alpha_1(\infty)}{1-\phi} v(x) + \frac{v(0)}{1-\phi}\{\alpha_1(\infty)e^{-\kappa x} - \overline{G}(x)\} \\ & + \frac{\kappa \alpha_1(\infty)}{1-\phi} \int_0^x e^{-\kappa y} v(x-y) dy, \quad x \ge 0, \end{aligned} \tag{9.1.13}$$

and

$$\begin{aligned} m(x) \ge\ & \frac{1-\alpha_2(\infty)}{1-\phi} v(x) + \frac{v(0)}{1-\phi}\{\alpha_2(\infty)e^{-\kappa x} - \overline{G}(x)\} \\ & + \frac{\kappa \alpha_2(\infty)}{1-\phi} \int_0^x e^{-\kappa y} v(x-y) dy, \quad x \ge 0, \end{aligned} \tag{9.1.14}$$

where $\alpha_1(x)$ and $\alpha_2(x)$ satisfy (7.1.6) and (7.1.7) respectively.

Proof: It follows from (7.1.8) that $\overline{G}(x) \le \alpha_1(x)e^{-\kappa x} \le \alpha_1(\infty)e^{-\kappa x}$ since $\alpha_1(x)$ is nondecreasing. Since $v'(x) \le 0$,

$$\int_0^x \overline{G}(x-y) v'(y) dy \ge \alpha_1(\infty) \int_0^x e^{-\kappa(x-y)} v'(y) dy.$$

Integration by parts yields

$$\begin{aligned} \int_0^x e^{-\kappa(x-y)} v'(y) dy &= e^{-\kappa(x-y)} v(y) \,\Big|_{y=0}^x - \kappa \int_0^x e^{-\kappa(x-y)} v(y) dy \\ &= v(x) - e^{-\kappa x} v(0) - \kappa \int_0^x e^{-\kappa y} v(x-y) dy. \end{aligned}$$

Thus,

$$\int_0^x \overline{G}(x-y) v'(y) dy \ge \alpha_1(\infty) \left\{ v(x) - e^{-\kappa x} v(0) - \kappa \int_0^x e^{-\kappa y} v(x-y) dy \right\}.$$

Substitution into (9.1.11) yields

$$\begin{aligned} m(x) &\leq \frac{v(x)}{1-\phi} - \frac{v(0)}{1-\phi}\overline{G}(x) \\ &- \frac{\alpha_1(\infty)}{1-\phi}\left\{v(x) - e^{-\kappa x}v(0) - \kappa\int_0^x e^{-\kappa y}v(x-y)dy\right\} \\ &= \frac{1-\alpha_1(\infty)}{1-\phi}v(x) + \frac{v(0)}{1-\phi}\left\{\alpha_1(\infty)e^{-\kappa x} - \overline{G}(x)\right\} \\ &+ \frac{\kappa\alpha_1(\infty)}{1-\phi}\int_0^x e^{-\kappa y}v(x-y)dy \end{aligned}$$

and (9.1.13) follows. The derivation of (9.1.14) follows by reversing the inequalities. □

There is one choice of $v(x)$ in (9.1.1) for which a relatively simple solution for $m(x)$ exists. It follows from (9.1.4) and Feller (1971, p. 435) that

$$\int_0^\infty e^{-sx}\overline{G}(x)dx = \frac{1}{s}\left\{1 - \frac{1-\phi}{1-\phi\tilde{f}(s)}\right\} = \frac{\phi}{1-\phi\tilde{f}(s)}\left\{\frac{1-\tilde{f}(s)}{s}\right\}$$

which may be restated as

$$\int_0^\infty e^{-sx}\overline{G}(x)dx = \phi\tilde{f}(s)\int_0^\infty e^{-sx}\overline{G}(x)dx + \phi\frac{1-\tilde{f}(s)}{s},$$

which yields upon inversion

$$\overline{G}(x) = \phi\int_0^x \overline{G}(x-y)dF(y) + \phi\overline{F}(x), \quad x \geq 0. \tag{9.1.15}$$

Clearly, (9.1.15) is of the form (9.1.1) with $v(x) = \phi\overline{F}(x)$, and thus by uniqueness of the solution to (9.1.1), it follows that with this choice of $v(x)$ one obtains the solution $m(x) = \overline{G}(x)$. Thus, from (9.1.3), one obtains immediately that

$$\overline{G}(x) = \frac{\phi}{1-\phi}\int_0^x \overline{F}(x-y)dG(y) + \phi\overline{F}(x), \quad x \geq 0. \tag{9.1.16}$$

A bound based on properties of $v(x)$ follows from the above relationships.

Corollary 9.1.2 Suppose that $v(x) \leq (\geq)\ k\overline{F}(x)$, $x \geq 0$. Then $m(x) \leq (\geq)\ k\overline{G}(x)/\phi$, $x \geq 0$.

Proof: It follows from (9.1.3) and (9.1.16) that

$$m(x) \leq (\geq)\ \frac{k}{1-\phi}\int_0^x \overline{F}(x-y)dG(y) + k\overline{F}(x) = \frac{k}{\phi}\overline{G}(x).$$

□

Recall from (7.1.9) that if (9.1.12) holds, then $\overline{G}(x) \sim Ce^{-\kappa x}$, $x \to \infty$. In light of the close connection between $\overline{G}(x)$ and $m(x)$, it is not surprising that a similar asymptotic result holds for $m(x)$ under fairly general conditions.

Theorem 9.1.3 Suppose that $F(y)$ is the df of a non-arithmetic distribution, $\kappa > 0$ satisfies (9.1.12), and $e^{\kappa x}v(x)$ is directly Riemann integrable on $(0, \infty)$. Then

$$m(x) \sim \frac{\int_0^\infty e^{\kappa y}v(y)dy}{\phi \int_0^\infty ye^{\kappa y}dF(y)} e^{-\kappa x}, \quad x \to \infty. \tag{9.1.17}$$

Proof: Multiplication of (9.1.1) by $e^{\kappa x}$ results in

$$m_*(x) = \int_0^x m_*(x-y)dF_*(y) + e^{\kappa x}v(x)$$

where $m_*(x) = e^{\kappa x}m(x)$ and $dF_*(y) = \phi e^{\kappa y}dF(y)$. By (9.1.12), $F_*(y)$ is a proper df and the above is a proper renewal equation. Thus, by the renewal theorem (e.g. Karlin and Taylor, 1975, p. 191)

$$\lim_{x\to\infty} m_*(x) = \int_0^\infty e^{\kappa y}v(y)dy \Big/ \int_0^\infty ydF_*(y)$$

which is (9.1.17). □

The concept of direct Riemann integrability is discussed by Feller (1971, pp. 362-33) and Resnick (1992, section 3.10). Sufficient conditions for a function $h(x)$ to be directly Riemann integrable include any one of the following:

a) $h(x)$ is nonnegative, nonincreasing, and Riemann integrable.

b) $h(x)$ is monotone and absolutely integrable.

c) $h(x)$ is bounded by a directly Riemann integrable function.

For the present application, it is often the case that $v(x) = k\overline{A}(x)$ where $1-\overline{A}(x)$ is a df on $(0, \infty)$. In this case a sufficient condition for $e^{\kappa x}v(x)$ to be directly Riemann integrable is that $\int_0^\infty e^{(\kappa+\epsilon)x}\overline{A}(x)dx < \infty$ for some $\epsilon > 0$, hardly a strong assumption in light of the numerator term on the right hand side of (9.1.17). To see this, note that $e^{(\kappa+\epsilon)x}\overline{A}(x)$ is locally bounded on $(0, \infty)$ and $\lim_{x\to\infty} e^{(\kappa+\epsilon)x}\overline{A}(x) = 0$. Therefore, there exists $K < \infty$ such that $e^{(\kappa+\epsilon)x}\overline{A}(x) \leq K$. That is, $e^{\kappa x}\overline{A}(x) \leq Ke^{-\epsilon x}$, $x \geq 0$. By a), $Ke^{-\epsilon x}$ is directly Riemann integrable, and by c), $e^{\kappa x}\overline{A}(x)$ is as well.

Next, we consider approximation of the solution $m(x)$ to the renewal equation (9.1.1). Deligonul (1985, and references therein) has considered this problem. Here we shall concern ourselves with Tijms type approximations. If the adjustment equation (9.1.12) holds and $F(y)$ is non-arithmetic, then from (7.1.9), $\overline{G}(x) \sim Ce^{-\kappa x}$, $x \to \infty$ where

$$C = \frac{1-\phi}{\phi\kappa \int_0^\infty ye^{\kappa y}dF(y)}. \tag{9.1.18}$$

The Tijms approximation (8.1.7) to $\overline{G}(x)$ is, in this case, $\overline{G}_T(x)$ where

$$\overline{G}_T(x) = (\phi - C)e^{-\mu x} + Ce^{-\kappa x}, \quad x \geq 0, \tag{9.1.19}$$

where, from (8.1.8),

$$\mu = (\phi - C)\left\{\frac{\phi}{1-\phi}\int_0^\infty ydF(y) - \frac{C}{\kappa}\right\}^{-1}. \tag{9.1.20}$$

We define a Tijms approximation to $m(x)$ by

$$m_T(x) = \frac{1}{1-\phi}\int_0^x v(x-y)dG_T(y) + v(x), \quad x \geq 0, \tag{9.1.21}$$

based on (9.1.3). It follows from (9.1.19) that (9.1.21) may be expressed as

$$m_T(x) = \frac{\mu(\phi - C)}{1-\phi}e^{-\mu x}\int_0^x e^{\mu y}v(y)dy + \frac{\kappa C}{1-\phi}e^{-\kappa x}\int_0^x e^{\kappa y}v(y)dy + v(x). \tag{9.1.22}$$

It is not hard to see by using integration by parts that (9.1.21) may also be expressed as

$$m_T(x) = \frac{1}{1-\phi}v(x) - \frac{v(0)}{1-\phi}\overline{G}_T(x) - \frac{1}{1-\phi}\int_0^x \overline{G}_T(x-y)v'(y)dy, \tag{9.1.23}$$

if $v(y)$ is differentiable. Evidently, $m_T(x)$ may also be motivated by replacement of $\overline{G}(x)$ by $\overline{G}_T(x)$ in (9.1.11). A slight variation involves only replacing $\overline{G}(x-y)$ by $\overline{G}_T(x-y)$ in the integral term in (9.1.11), and this idea was used by Willmot and Lin (1998).

It is not hard to see that the approximation retains many of the properties of $\overline{G}_T(x)$, such as matching the true value at 0, reproducing the true value when $\overline{G}_T(x)$ does (as in theorem 8.1.2), and matching the limiting value (9.1.17) under certain conditions.

We end this section by considering an application in branching processes.

Example 9.1.2 Age-dependent branching processes

An individual produces a random number of offspring with probability distribution $\{q_0, q_1, \cdots\}$. All offspring produce offspring independently of all others, the number of which also has probability distribution $\{q_0, q_1, \cdots\}$. The lifetimes of all individuals are independent with common distribution $F(t)$. Then $\{X(t); t \geq 0\}$ is a stochastic process with $X(t)$ the number of organisms above at t. Let $m(t) = E\{X(t)\}$. If $X(0) = 1$ (i.e. initially a single individual), and $\phi = \sum_{j=1}^{\infty} jq_j$, then $m(t)$ satisfies the renewal equation

$$m(t) = \phi \int_0^t m(t-y)dF(y) + \overline{F}(t). \tag{9.1.24}$$

See Ross (1996, pp. 121-2), or Resnick (1992, pp. 198-9), for example. Clearly, (9.1.24) is of the form (9.1.1), and so if $\phi < 1$ one obtains immediately from theorem 9.1.1 that

$$m(t) = \frac{1}{1-\phi} \int_0^t \overline{F}(t-y)dG(y) + \overline{F}(t), \tag{9.1.25}$$

where $G(x) = 1 - \overline{G}(x)$ is the compound geometric df given by (9.1.2). Furthermore, from (9.1.25) and (9.1.16), it follows that

$$m(t) = \overline{G}(t)/\phi, \quad t \geq 0. \tag{9.1.26}$$

Thus, if $\kappa > 0$ satisfies

$$\int_0^\infty e^{\kappa y} dF(y) = \frac{1}{\phi},$$

it follows from corollary 7.1.1 that

$$\frac{\alpha_2(t)}{\phi} e^{-\kappa t} \leq m(t) \leq \frac{\alpha_1(t)}{\phi} e^{-\kappa t}, \quad t \geq 0, \tag{9.1.27}$$

where $\alpha_1(t)$ and $\alpha_2(t)$ satisfy (7.1.6) and (7.1.7) respectively. In addition, if $F(t)$ is non-arithmetic, then from (7.1.9),

$$m(t) \sim \frac{1-\phi}{\kappa \phi^2 \int_0^\infty y e^{\kappa y} dF(y)} e^{-\kappa t}, \quad t \to \infty. \tag{9.1.28}$$

Other properties of $m(t)$ follow directly from properties of $\overline{G}(t)$, discussed in section 7.1.

9.2 The time of ruin and related quantities

In this section we consider a detailed example involving the time of ruin. We shall use the notation and model of section 7.3. Furthermore, define the deficit (negative surplus) at the time of ruin to be $|U_T|$, where T is the time of ruin, and we are interested in a nonnegative function $w(|U_T|)$ of the deficit. The function $w(x)$ may be viewed as a penalty when the deficit $|U_T| = x$. See Gerber and Shiu (1998). Let $I(A) = 1$ if the event A occurs and $I(A) = 0$ otherwise (i.e. $I(\cdot)$ is an indicator function). Then for $\delta \geq 0$, define the expected discounted penalty

$$m(u) = E\{e^{-\delta T} w(|U_T|) I(T < \infty)\} \tag{9.2.1}$$

where u is the initial surplus and δ is interpreted as a force of interest. The function $m(u)$ contains many important quantities in the classical risk model as special cases. For example, if $\delta = 0$ and $w(x) = 1$, one reproduces the probability of ruin $\psi(u)$ given in section 7.3. For $\delta = 0$ and $w(x) = x^k$, one obtains the k-th moment of the deficit at the time of ruin. An explicit expression for this k-th moment is given by Lin and Willmot (2000). The function $m(u)$ may also be viewed in terms of Laplace transforms with δ the argument and this approach will be used to derive the moments of the time of ruin.

We now introduce a quantity $\rho = \rho(\delta)$ which plays an important role in the analysis of the function $m(u)$, as will be seen later in this section. Thus, let $\tilde{h}(s) = \int_0^\infty e^{-sx} dH(x)$ be the Laplace-Stieltjes transform of the claim amount distribution, and consider the two equations $y_1(x) = \lambda \tilde{h}(x)$ and $y_2(x) = \lambda + \delta - cx$, shown in figure 9.1.

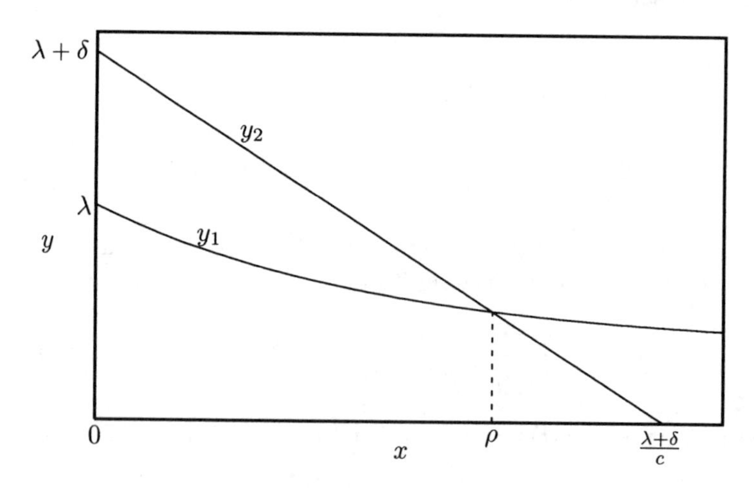

FIGURE 9.1. The graph of $y_1(x) = \lambda \tilde{h}(x)$ and $y_2(x) = \lambda + \delta - cx$

If $\delta > 0$ then $y_2(0) = \lambda + \delta > \lambda = y_1(0)$. But $y_1'(x) < 0$ and $y_1''(x) > 0$ for $x > 0$ whereas $y_2(x)$ is a decreasing straight line. Thus, there is a unique positive root $\rho = \rho(\delta)$ in $(0, \frac{\lambda+\delta}{c})$ to the equation $y_1(x) = y_2(x)$, i.e. ρ satisfies

$$\lambda \tilde{h}(\rho) = \lambda + \delta - c\rho. \tag{9.2.2}$$

If $\delta = 0$ then $y_1(0) = y_2(0)$ and $y_1'(0) = -\lambda E(Y) > -c = y_2'(0)$, and thus $\rho = 0$. Therefore there is a unique nonnegative root $\rho = \rho(\delta)$ to (9.2.2) satisfying $\rho(0) = 0$.

Gerber and Shiu (1998) showed that $m(u)$ in (9.2.1) satisfies the relation

$$m(u) = \frac{\lambda}{c} \int_0^u m(u-y) \int_y^\infty e^{-\rho(t-y)} dH(t) dy + v(u) \tag{9.2.3}$$

where

$$v(u) = \frac{\lambda}{c} e^{\rho u} \int_u^\infty e^{-\rho y} \int_y^\infty w(t-y) dH(t) dy. \tag{9.2.4}$$

We shall now demonstrate that (9.2.3) is a defective renewal equation. Recall that $1 - \overline{H}(y+t)/\overline{H}(t)$ is the residual lifetime df of $H(y)$ as discussed in section 2.3. Consequently, $F(y)$ is also a df, where $\overline{F}(y) = 1 - F(y)$ is defined by

$$\overline{F}(y) = \frac{\int_0^\infty e^{-\rho t} \overline{H}(y+t) dt}{\int_0^\infty e^{-\rho t} \overline{H}(t) dt}, \tag{9.2.5}$$

since it may be viewed as a mixture over t of the df $1 - \overline{H}(y+t)/\overline{H}(t)$ with mixing density proportional to $e^{-\rho t}\overline{H}(t)$. Note that (9.2.5) may be expressed as

$$\overline{F}(y) = \frac{e^{\rho y} \int_y^\infty e^{-\rho t} \overline{H}(t) dt}{\int_0^\infty e^{-\rho t} \overline{H}(t) dt}, \tag{9.2.6}$$

which guarantees that $F(y)$ is absolutely continuous. Recall that the equilibrium density of $H(t)$ is $H_1'(t) = \overline{H}(t)/E(Y)$. Define the Laplace-Stieltjes transform $\tilde{h}_1(s) = \int_0^\infty e^{-st} dH_1(t)$, and integration by parts yields $\tilde{h}_1(s) = \{1 - \tilde{h}(s)\}/\{sE(Y)\}$. See also Feller (1971, p. 435). Thus, $\int_0^\infty e^{-\rho t}\overline{H}(t)dt = E(Y)\tilde{h}_1(\rho)$, and differentiating yields the density

$$F'(y) = -\overline{F}'(y) = \frac{\overline{H}(y) - \rho e^{\rho y} \int_y^\infty e^{-\rho t} \overline{H}(t) dt}{E(Y)\tilde{h}_1(\rho)}.$$

Integration by parts yields

$$\rho \int_y^\infty e^{-\rho t} \overline{H}(t) dt = e^{-\rho y} \overline{H}(y) - \int_y^\infty e^{-\rho t} dH(t)$$

which implies that

$$F'(y) = \frac{e^{\rho y} \int_y^\infty e^{-\rho t} dH(t)}{E(Y)\tilde{h}_1(\rho)}. \tag{9.2.7}$$

Therefore, $\int_y^\infty e^{-\rho(t-y)} dH(t) = E(Y)\tilde{h}_1(\rho)F'(y)$ and substitution into (9.2.3) yields

$$m(u) = \frac{\lambda E(Y)\tilde{h}_1(\rho)}{c} \int_0^u m(u-y) F'(y) dy + v(u).$$

Since $c = \lambda E(Y)(1+\theta)$, this means that (9.2.3) may be expressed as

$$m(u) = \phi \int_0^u m(u-y)dF(y) + v(u) \tag{9.2.8}$$

where

$$\phi = \frac{\tilde{h}_1(\rho)}{1+\theta}. \tag{9.2.9}$$

Clearly, $0 < \phi < 1$, and thus (9.2.3) is equivalent to (9.2.8), a defective renewal equation. Also, since

$$\frac{\lambda}{c} = \frac{1}{E(Y)(1+\theta)} = \frac{\phi}{E(Y)\tilde{h}_1(\rho)},$$

it follows that (9.2.4) may be expressed as

$$v(u) = \frac{\phi}{E(Y)\tilde{h}_1(\rho)} e^{\rho u} \int_u^\infty e^{-\rho y} \int_y^\infty w(t-y)dH(t)dy. \tag{9.2.10}$$

In the following theorem, we identify the connection between the associated compound geometric tail $\overline{G}(u)$ given in (9.1.2) and a special expected discounted penalty. This connection is useful in calculating the moments of the time of ruin as shown later in this section.

Theorem 9.2.1 Define for $\delta \geq 0$ the function

$$\overline{G}(u) = E\{e^{-\delta T} I(T < \infty)\}, \tag{9.2.11}$$

i.e. $\overline{G}(u)$ is the expected discounted penalty with $w(x) = 1$. Then $\overline{G}(u)$ is the compound geometric tail given by

$$\overline{G}(u) = \sum_{n=1}^{\infty} (1-\phi)\phi^n \overline{F}^{*n}(u) \tag{9.2.12}$$

where ϕ satisfies (9.2.9) and $F(u)$ has density (9.2.7). In other words, $\overline{G}(u)$ is the associated compound geometric tail to the renewal equation (9.2.3).

Proof: It follows from (9.2.10) with $w(x) = 1$ that $v(u) = \phi\overline{F}(y)$ using (9.2.6). The result follows from the fact that from (9.2.1) one has $m(u) = \overline{G}(u)$ and (9.2.8) reduces to (9.1.15). □

When $\delta = 0$, $\psi(u) = E\{I(T < \infty)\}$, $\phi = 1/(1+\theta)$ from (9.2.9), and $F(y) = H_1(y)$ from (9.2.6). Thus, theorem 9.2.1 is a direct generalization of the compound geometric tail representation of the probability of ruin.

The distribution $F(u)$ is quite relevant for analysis of the compound geometric tail $\overline{G}(u)$, as is evident from section 7.1. Moreover, $F(u)$ retains many of the properties of the claim amount distribution. For more details, see Lin and Willmot (1999). An important example is now given.

Example 9.2.1 Mixture of Erlangs

Suppose that

$$H'(y) = \sum_{k=1}^{r} q_k \frac{\beta(\beta y)^{k-1} e^{-\beta y}}{(k-1)!}, \tag{9.2.13}$$

where $\{q_1, q_2, \cdots, q_r\}$ is a probability distribution. Then

$$\begin{aligned}
\int_y^\infty e^{-\rho t} dH(t) &= \sum_{j=1}^{r} q_j \left(\frac{\beta}{\beta+\rho}\right)^j \int_y^\infty \frac{(\beta+\rho)^j t^{j-1} e^{-(\beta+\rho)t}}{(j-1)!} dt \\
&= \sum_{j=1}^{r} q_j \left(\frac{\beta}{\beta+\rho}\right)^j \sum_{k=1}^{j} \frac{\{(\beta+\rho)y\}^{k-1} e^{-(\beta+\rho)y}}{(k-1)!} \\
&= \frac{e^{-\rho y}}{\beta+\rho} \sum_{k=1}^{r} \left\{ \sum_{j=k}^{r} q_j \left(\frac{\beta}{\beta+\rho}\right)^{j-k} \right\} \frac{\beta(\beta y)^{k-1} e^{-\beta y}}{(k-1)!}.
\end{aligned}$$

Also, from (4.1.5), if $\rho > 0$

$$\begin{aligned}
E(Y)\tilde{h}_1(\rho) &= \int_0^\infty e^{-\rho t} \overline{H}(t) dt \\
&= \frac{1}{\rho} \left\{ 1 - \sum_{j=1}^{r} q_j \left(\frac{\beta}{\beta+\rho}\right)^j \right\} \\
&= \frac{1}{\beta+\rho} \sum_{j=1}^{r} q_j \left\{ \frac{1 - (\frac{\beta}{\beta+\rho})^j}{1 - \frac{\beta}{\beta+\rho}} \right\} \\
&= \frac{1}{\beta+\rho} \sum_{j=1}^{r} q_j \sum_{i=0}^{j-1} \left(\frac{\beta}{\beta+\rho}\right)^i.
\end{aligned}$$

Substitution of these relations into (9.2.7) yields

$$F'(y) = \sum_{k=1}^{r} q_k^* \frac{\beta(\beta y)^{k-1} e^{-\beta y}}{(k-1)!}$$

where

$$q_k^* = \frac{\sum_{j=k}^{r} q_j (\frac{\beta}{\beta+\rho})^{j-k}}{\sum_{j=1}^{r} q_j \sum_{i=0}^{j-1} (\frac{\beta}{\beta+\rho})^i}; \quad k = 1, 2, \cdots, r.$$

Now, $\{q_1^*, q_2^*, \cdots, q_r^*\}$ is a probability distribution and the result holds with $\rho = 0$ since $F(y)$ reduces to $H_1(y)$ in this case (see example 2.2.1). Then

$\overline{G}(u)$ may be evaluated as in example 7.1.1. □

As we have seen in section 9.1, asymptotic relations and exponential bounds for $\overline{G}(u)$ depend on the adjustment coefficient $\kappa > 0$ satisfying the Lundberg adjustment equation

$$\frac{1}{\phi} = \tilde{f}(-\kappa) = \int_0^\infty e^{\kappa y} dF(y). \tag{9.2.14}$$

As in Gerber and Shiu (1998), (9.2.14) may be expressed in a simple manner. Integration by parts yields

$$\int_0^\infty e^{-(s-\rho)y} \int_y^\infty e^{-\rho t} dH(t) dy = \frac{\tilde{h}(s) - \tilde{h}(\rho)}{\rho - s}$$

and thus from (9.2.7),

$$\tilde{f}(s) = \frac{\rho}{\rho - s} \left\{ \frac{\tilde{h}(s) - \tilde{h}(\rho)}{1 - \tilde{h}(\rho)} \right\}. \tag{9.2.15}$$

Substitution into (9.2.14) yields

$$\frac{\rho}{\rho + \kappa} \left\{ \frac{\tilde{h}(-\kappa) - \tilde{h}(\rho)}{1 - \tilde{h}(\rho)} \right\} = \frac{(1+\theta)\rho E(Y)}{1 - \tilde{h}(\rho)}$$

using (9.2.9) and $\tilde{h}_1(\rho) = \{1 - \tilde{h}(\rho)\}/\{\rho E(Y)\}$. Therefore

$$\frac{\tilde{h}(-\kappa) - \tilde{h}(\rho)}{\rho + \kappa} = (1+\theta)E(Y) = \frac{c}{\lambda},$$

in turn,

$$\lambda \tilde{h}(-\kappa) - \lambda \tilde{h}(\rho) = c(\rho + \kappa)$$

and using (9.2.2), it follows that the Lundberg adjustment equation (9.2.14) is equivalent to

$$\lambda \tilde{h}(-\kappa) = \lambda + \delta + c\kappa. \tag{9.2.16}$$

Clearly, (9.2.16) implies that $-\kappa$ is a negative solution of (9.2.2).

We now examine the failure rate, the equilibrium df, and the mean residual lifetime of $F(y)$ and establish the connection to the claim amount distribution $H(y)$. It follows from (9.2.6) that the failure rate of $F(y)$ is given by

$$\mu(y) = -\frac{d}{dy} \ln \overline{F}(y) = \rho \left\{ \frac{1}{\gamma(y)} - 1 \right\} \tag{9.2.17}$$

where

$$\gamma(y) = \rho \int_0^\infty e^{-\rho t} \frac{\overline{H}(y+t)}{\overline{H}(y)} dt. \tag{9.2.18}$$

Note that from (9.2.5),

$$\begin{aligned}\int_y^\infty \overline{F}(x)dx &= \frac{\int_0^\infty e^{-\rho t}\int_y^\infty \overline{H}(x+t)dx\,dt}{\int_0^\infty e^{-\rho t}\overline{H}(t)dt}\\ &= \frac{E(Y)}{\int_0^\infty e^{-\rho t}\overline{H}(t)dt}\int_0^\infty e^{-\rho t}\overline{H}_1(y+t)dt.\end{aligned}$$

Thus with $y=0$, we obtain the mean

$$\int_0^\infty \overline{F}(x)dx = E(Y)\frac{\int_0^\infty e^{-\rho t}\overline{H}_1(t)dt}{\int_0^\infty e^{-\rho t}\overline{H}(t)dt}. \tag{9.2.19}$$

Also, the tail of the equilibrium distribution of $F(y)$ is, using the above result,

$$\overline{F}_1(y) = \frac{\int_y^\infty \overline{F}(x)dx}{\int_0^\infty \overline{F}(x)dx} = \frac{\int_0^\infty e^{-\rho t}\overline{H}_1(y+t)dt}{\int_0^\infty e^{-\rho t}\overline{H}_1(t)dt}. \tag{9.2.20}$$

Evidently, (9.2.20) is of the same form as (9.2.5), but with H replaced by H_1. Therefore, if $r(y)$ is the mean residual lifetime of $F(y)$, it follows from (2.3.5), (9.2.17), and (9.2.18) that

$$\frac{1}{r(y)} = \rho\left\{\frac{1}{\gamma_1(y)} - 1\right\}$$

where

$$\gamma_1(y) = \rho\int_0^\infty e^{-\rho t}\frac{\overline{H}_1(y+t)}{\overline{H}_1(y)}dt. \tag{9.2.21}$$

That is,

$$r(y) = \frac{1}{\rho\left\{\frac{1}{\gamma_1(y)} - 1\right\}}. \tag{9.2.22}$$

An alternative representation for $r(y)$ is also possible. It follows from (9.2.20) that

$$\overline{F}_1(y) = \frac{e^{\rho y}\int_y^\infty e^{-\rho t}\overline{H}_1(t)dt}{\int_0^\infty e^{-\rho t}\overline{H}_1(t)dt}. \tag{9.2.23}$$

Therefore, using (2.3.4) and (9.2.19), one obtains together with (9.2.6) and (9.2.23),

$$\begin{aligned}r(y) &= \frac{\overline{F}_1(y)}{\overline{F}(y)}\int_0^\infty \overline{F}(x)dx = E(Y)\frac{\overline{F}_1(y)\int_0^\infty e^{-\rho t}\overline{H}_1(t)dt}{\overline{F}(y)\int_0^\infty e^{-\rho t}\overline{H}(t)dt}\\ &= E(Y)\frac{\int_y^\infty e^{-\rho t}\overline{H}_1(t)dt}{\int_y^\infty e^{-\rho t}\overline{H}(t)dt}.\end{aligned}$$

Let $r_H(y) = E(Y)\overline{H}_1(y)/\overline{H}(y)$ be the mean residual lifetime associated with the df $H(y)$, and it follows that

$$r(y) = \frac{\int_y^\infty e^{-\rho t}\overline{H}(t)r_H(t)dt}{\int_y^\infty e^{-\rho t}\overline{H}(t)dt}. \tag{9.2.24}$$

The representation (9.2.24) expresses $r(y)$ as an average of values of $r_H(y)$, with weights proportional to $e^{-\rho y}\overline{H}(y)$.

The following theorem demonstrates that reliability classifications for $F(y)$ depend heavily on those for $H(y)$.

Theorem 9.2.2 Suppose that $F(y)$ satisfies (9.2.5) with $\rho > 0$. Then the following class implications hold.

a) If $H(y)$ is DFR (IFR) then $F(y)$ is DFR (IFR).

b) If $H(y)$ is IMRL (DMRL) then $F(y)$ is IMRL (DMRL).

c) If $H(y)$ is 3-DFR (3-IFR) then $F(y)$ is 3-DFR (3-IFR).

d) If $H(y)$ is UWA (UBA) then $F(y)$ is UWA (UBA).

e) If $H(y)$ is UWAE (UBAE) then $F(y)$ is UWAE (UBAE).

f) If $H(y)$ is 2-NWU (2-NBU) then $F(y)$ is NWUE (NBUE).

Proof: If $H(y)$ is DFR (IFR), then $\overline{H}(y+t)/\overline{H}(y)$ is nondecreasing (nonincreasing) in y for fixed t, implying the same for $\gamma(y)$ from (9.2.18), and a) follows from (9.2.17). Similarly, if $H(y)$ is IMRL (DMRL), then $H_1(y)$ is DFR (IFR) and $\gamma_1(y)$ in (9.2.21) is nondecreasing (nonincreasing). Then b) follows from (9.2.22). Since (9.2.20) is of the same form as (9.2.5), b) implies that if $H_1(y)$ is IMRL (DMRL) then $F_1(y)$ is IMRL (DMRL), proving c). To prove d), it follows from (9.2.24) and L'Hopital's rule that

$$\lim_{y\to\infty} r(y) = \lim_{y\to\infty} \frac{-e^{-\rho y}\overline{H}(y)r_H(y)}{-e^{-\rho y}\overline{H}(y)} = \lim_{y\to\infty} r_H(y),$$

and thus

$$r(\infty) = r_H(\infty), \tag{9.2.25}$$

a result which is valid as long as $r_H(\infty)$ is well defined. Thus, from (9.2.5),

$$\begin{aligned}\overline{F}(x+y) &= \frac{\int_0^\infty e^{-\rho t}\overline{H}(x+y+t)dt}{\int_0^\infty e^{-\rho t}\overline{H}(t)dt}\\ &\leq (\geq) \frac{\int_0^\infty e^{-\rho t}\overline{H}(y+t)e^{-x/r_H(\infty)}dt}{\int_0^\infty e^{-\rho t}\overline{H}(t)dt} = \overline{F}(y)e^{-x/r(\infty)}\end{aligned}$$

and d) is proved. If $r_H(t) \leq (\geq)\ r_H(\infty)$ then from (9.2.24) it follows that $r(y) \leq (\geq)\ r_H(\infty) = r(\infty)$ using (9.2.25). Thus e) is proved. Finally, if $H(y)$ is 2-NWU (2-NBU), then $\overline{H}_1(y+t) \geq (\leq)\ \overline{H}_1(y)\overline{H}_1(t)$, implying from (9.2.21) that $\gamma_1(y) \geq (\leq)\ \gamma_1(0)$, and from (9.2.22) that $r(y) \geq (\leq)\ r(0)$. We have thus proved f). □

The importance of theorem 9.2.2 lies in the fact that bounds for the compound geometric tail $\overline{G}(u)$ depend on reliability classifications of $F(y)$, as is clear from section 7.1. In a similar vein, bounds on the failure rate $\mu(y)$ in (9.2.17) and the mean residual lifetime $r(y)$ are of use. A more detailed discussion of failure rate and mean residual lifetime bounds and orders (e.g. Shaked and Shanthikumar, 1994, chapter 1) may be found in Lin and Willmot (1999).

In what follows, we explore further the relationship between $F(x)$ and $H_1(x)$, and the mean residual lifetime plays a central role here as well. To begin, let $\tilde{h}_1(s) = \int_0^\infty e^{-sy} dH_1(y)$. Then since (9.2.20) is of the same form as (9.2.5), but with $H(y)$ replaced by $H_1(y)$, it follows from (9.2.15) that

$$\tilde{f}_1(s) = \int_0^\infty e^{-sy} dF_1(y) = \frac{\rho}{\rho - s}\left\{\frac{\tilde{h}_1(s) - \tilde{h}_1(\rho)}{1 - \tilde{h}_1(\rho)}\right\}, \tag{9.2.26}$$

and $\tilde{h}_1(s) = \{1 - \tilde{h}(s)\}/\{sE(Y)\}$. Therefore, using (9.2.15)

$$\begin{aligned}
&\{1 - \tilde{h}_1(\rho)\}\tilde{f}_1(s) + \tilde{h}_1(\rho)\tilde{f}(s) \\
= &\ \frac{\rho}{\rho - s}\{\tilde{h}_1(s) - \tilde{h}_1(\rho)\} + \frac{1}{\rho - s}\left\{\frac{\tilde{h}(s) - \tilde{h}(\rho)}{E(Y)}\right\} \\
= &\ \frac{\rho E(Y)\{\tilde{h}_1(s) - \tilde{h}_1(\rho)\} + \tilde{h}(s) - \tilde{h}(\rho)}{E(Y)(\rho - s)} \\
= &\ \frac{\frac{\rho}{s}\{1 - \tilde{h}(s)\} - \{1 - \tilde{h}(\rho)\} + \tilde{h}(s) - \tilde{h}(\rho)}{E(Y)(\rho - s)} \\
= &\ \frac{\rho\{1 - \tilde{h}(s)\} - s\{1 - \tilde{h}(s)\}}{sE(Y)(\rho - s)}.
\end{aligned}$$

That is

$$\tilde{h}_1(s) = \{1 - \tilde{h}_1(\rho)\}\tilde{f}_1(s) + \tilde{h}_1(\rho)\tilde{f}(s), \tag{9.2.27}$$

which demonstrates that $H_1(y)$ is a mixture of $F(y)$ and $F_1(y)$ with weights $\tilde{h}_1(\rho)$ and $1 - \tilde{h}_1(\rho)$ respectively. Thus,

$$\overline{H}_1(y) = \{1 - \tilde{h}_1(\rho)\}\overline{F}_1(y) + \tilde{h}_1(\rho)\overline{F}(y). \tag{9.2.28}$$

Now, (9.2.15) may be written as

$$(\rho - s)\tilde{f}(s) = \frac{\tilde{h}(s) - \tilde{h}(\rho)}{E(Y)\tilde{h}_1(\rho)}$$

and differentiation yields

$$(\rho - s)\tilde{f}'(s) - \tilde{f}(s) = \frac{\tilde{h}'(s)}{E(Y)\tilde{h}_1(\rho)},$$

which with $s = 0$ yields

$$\rho \int_0^\infty \overline{F}(y)dy + 1 = \frac{1}{\tilde{h}_1(\rho)}.$$

That is, (9.2.19) may be restated as

$$r(0) = \int_0^\infty \overline{F}(y)dy = \frac{1 - \tilde{h}_1(\rho)}{\rho \tilde{h}_1(\rho)}. \tag{9.2.29}$$

Higher moments are given by Lin and Willmot (1999). It then follows that (9.2.28) may be expressed as

$$\begin{aligned}\overline{H}_1(y) &= \rho r(0)\tilde{h}_1(\rho)\overline{F}_1(y) + \tilde{h}_1(\rho)\overline{F}(y) \\ &= \tilde{h}_1(\rho)\overline{F}(y)\left\{\frac{\rho r(0)\overline{F}_1(y)}{\overline{F}(y)} + 1\right\}.\end{aligned}$$

Thus, using (2.3.4), one obtains the useful identity

$$\overline{F}(y) = \frac{\overline{H}_1(y)}{\tilde{h}_1(\rho)\{1 + \rho r(y)\}}, \quad y \geq 0. \tag{9.2.30}$$

It follows immediately from (9.2.30) and (9.2.25) that

$$\overline{F}(y) \sim \frac{\overline{H}_1(y)}{\tilde{h}_1(\rho)\{1 + \rho r_H(\infty)\}}, \quad y \to \infty, \tag{9.2.31}$$

as long as $r_H(\infty)$ is well defined. We now consider stochastic orders.

Theorem 9.2.3 Suppose that $F(y)$ satisfies (9.2.5) with $\rho > 0$. Then the following stochastic implications hold.

a) $\overline{F}(y) \leq \overline{H}_1(y)/\tilde{h}_1(\rho)$.

b) If the df $A(y) = 1 - \overline{A}(y)$ satisfies $\overline{H}_1(y+t)/\overline{H}_1(t) \geq (\leq)\ \overline{A}(y)$, then

$$\overline{F}(y) \leq (\geq)\ \frac{\tilde{a}(\rho)}{\tilde{h}_1(\rho)}\overline{H}_1(y), \tag{9.2.32}$$

where $\tilde{a}(\rho) = \int_0^\infty e^{-\rho t} dA(t)$. In particular,

i) if $H(y)$ is 2-NWU (2-NBU) then $\overline{F}(y) \leq (\geq)\ \overline{H}_1(y)$.

ii) if $H(y)$ is NWUC (NBUC) then $\overline{F}(y) \leq (\geq) \{\tilde{h}(\rho)/\tilde{h}_1(\rho)\}\overline{H}_1(y)$.

iii) if $H(y)$ is NWUE (NBUE) then

$$\overline{F}(y) \leq (\geq) \frac{\overline{H}_1(y)}{\tilde{h}_1(\rho)\{1+\rho E(Y)\}}.$$

iv) if $H(y)$ is UBAE (UWAE) then

$$\overline{F}(y) \leq (\geq) \frac{\overline{H}_1(y)}{\tilde{h}_1(\rho)\{1+\rho r_H(\infty)\}}.$$

c) If the df $A(y) = 1 - \overline{A}(y)$ satisfies $\overline{H}(y+t)/\overline{H}(t) \geq (\leq) \overline{A}(y)$, then

$$\overline{A}(y) \leq (\geq) \overline{F}(y) \leq (\geq) \frac{\tilde{a}(\rho)}{\tilde{h}_1(\rho)}\overline{H}_1(y), \tag{9.2.33}$$

where $\tilde{a}(\rho) = \int_0^\infty e^{-\rho t} dA(t)$. In particular,

i) if $H(y)$ is NWU (NBU) then

$$\overline{H}(y) \leq (\geq) \overline{F}(y) \leq (\geq) \frac{\tilde{h}(\rho)}{\tilde{h}_1(\rho)}\overline{H}_1(y).$$

ii) if $H(y)$ is UBA (UWA) then

$$e^{-y/r_H(\infty)} \leq (\geq) \overline{F}(y) \leq (\geq) \frac{\overline{H}_1(y)}{\tilde{h}_1(\rho)\{1+\rho r_H(\infty)\}}.$$

Proof: Since $r(y) \geq 0$, part a) follows from (9.2.30). To prove (9.2.32), note that from (9.2.21),

$$\gamma_1(y) \geq (\leq) \rho \int_0^\infty e^{-\rho t}\overline{A}(t)dt = 1 - \tilde{a}(\rho).$$

Thus from (9.2.22),

$$r(y) \geq (\leq) \rho^{-1}\left\{\frac{1}{1-\tilde{a}(\rho)} - 1\right\}^{-1} = \frac{1-\tilde{a}(\rho)}{\rho\tilde{a}(\rho)},$$

i.e. $1 + \rho r(y) \geq (\leq) 1/\tilde{a}(\rho)$, and (9.2.32) follows from (9.2.30). Then b)i), b)ii), b)iii), and b)iv) are special cases with $A(y) = H_1(y)$, $A(y) = H(y)$, $A(y) = 1 - e^{-y/E(Y)}$, and $A(y) = 1 - e^{-y/r_H(\infty)}$, respectively (the latter two with the help of theorem 2.4.3). Thus b) is proved. The left-hand inequality in (9.2.33) follows easily from (9.2.5). For the right-hand side,

$$\begin{aligned} \overline{H}_1(y+t) &= \int_{y+t}^\infty \frac{\overline{H}(x)}{E(Y)}dx = \int_t^\infty \frac{\overline{H}(y+x)}{E(Y)}dx \\ &\geq (\leq)\ \overline{A}(y)\int_t^\infty \frac{\overline{H}(x)}{E(Y)}dx = \overline{A}(y)\overline{H}_1(t), \end{aligned}$$

and part b) applies. □

Since $H(y)$ is DFR implies that $H(y)$ is both NWU and 2-NWU, it follows from theorem 9.2.3 that

$$\overline{H}(y) \leq \overline{F}(y) \leq \overline{H}_1(y), \tag{9.2.34}$$

whereas if $H(y)$ is IFR

$$\overline{H}_1(y) \leq \overline{F}(y) \leq \overline{H}(y). \tag{9.2.35}$$

Clearly, (9.2.34) and (9.2.35) give fairly precise information about $F(y)$.

We remark that if $r_H(x) \geq (\leq)\ r$, then using theorem 2.4.3 it follows easily that (9.2.32) holds with $\tilde{a}(\rho) = 1/(1+\rho r)$.

We now turn to an interpretation of $\overline{G}(u)$ as a Laplace transform with argument δ rather than δ as a discount factor. Hence we wish to differentiate $\overline{G}(u)$ with respect to δ. The following lemma is of use in this regard.

Lemma 9.2.1 The function $\overline{G}(u) = E\{e^{-\delta T} I(T < \infty)\}$ satisfies

$$\overline{G}(u) = \frac{\lambda}{c}\int_0^u \overline{G}(u-y)\tau(y,\rho)dy + \frac{\lambda}{c}\int_u^\infty \tau(x,\rho)dx \tag{9.2.36}$$

where

$$\tau(y,\rho) = \int_y^\infty e^{-\rho(t-y)}dH(t). \tag{9.2.37}$$

Proof: Clearly, (9.2.3) may be expressed using (9.2.37) as

$$\overline{G}(u) = \frac{\lambda}{c}\int_0^u \overline{G}(u-y)\tau(y,\rho)dy + v(u)$$

and we need only show that $v(u) = \frac{\lambda}{c}\int_u^\infty \tau(x,\rho)dx$ in this case. But from theorem 9.2.1, $v(u) = \phi\overline{F}(u) = \phi\int_u^\infty F'(x)dx$. Thus, from (9.2.7) and (9.2.9),

$$\phi F'(x) = \frac{e^{\rho x}\int_x^\infty e^{-\rho t}dH(t)}{E(Y)(1+\theta)} = \frac{\lambda}{c}\tau(x,\rho).$$

□

With the help of the above lemma, we may now give the mean of the time of ruin (e.g. Rolski et al, 1999, section 11.3.5).

Theorem 9.2.4 The mean time to ruin is given by

$$E(T|T<\infty) = \frac{\psi_1(u)}{\psi(u)} \tag{9.2.38}$$

where $\psi_1(u)$ satisfies the defective renewal equation

$$\psi_1(u) = \frac{1}{1+\theta}\int_0^u \psi_1(u-x)dH_1(x) + \frac{1}{c}\int_u^\infty \psi(x)dx \tag{9.2.39}$$

and is given by

$$\psi_1(u) = \frac{1}{\lambda\theta E(Y)}\left\{\int_0^u \psi(u-x)\psi(x)dx + \int_u^\infty \psi(x)dx - \frac{E(Y^2)}{2\theta E(Y)}\psi(u)\right\}. \tag{9.2.40}$$

Proof: We shall use lemma 9.2.1 and differentiate (9.2.36) with respect to δ. This yields

$$\begin{aligned}\frac{c}{\lambda}\frac{\partial \overline{G}(u)}{\partial \delta} &= \int_0^u \left\{\frac{\partial}{\partial \delta}\overline{G}(u-y)\right\}\tau(y,\rho)dy \\ &+ \rho'(\delta)\int_0^u \overline{G}(u-y)\frac{\partial}{\partial \rho}\tau(y,\rho)dy + \rho'(\delta)\int_u^\infty \frac{\partial}{\partial \rho}\tau(y,\rho)dy.\end{aligned} \tag{9.2.41}$$

Now, from (9.2.2), $\lambda\tilde{h}'\{\rho(\delta)\}\rho'(\delta) = 1 - c\rho'(\delta)$, and since $\rho(0) = 0$, it follows that $-\lambda E(Y)\rho'(0) = 1 - c\rho'(0)$, i.e. $\rho'(0) = \{c - \lambda E(Y)\}^{-1} = \{\lambda\theta E(Y)\}^{-1}$. Also, from (9.2.37),

$$\frac{\partial}{\partial \rho}\tau(y,\rho)\,|_{\rho=0} = -\int_y^\infty (t-y)dH(t) = -E(Y)\overline{H}_1(y)$$

using integration by parts. Now, $\psi_1(u) = E\{TI(T<\infty)\} = -\frac{\partial}{\partial\delta}\overline{G}(u)\,|_{\delta=0}$. Thus, (9.2.41) with $\delta = 0$ becomes

$$\begin{aligned}-\frac{c}{\lambda}\psi_1(u) &= \int_0^u \{-\psi_1(u-y)\}\,\overline{H}(y)dy \\ &+ \frac{1}{\lambda\theta E(Y)}\int_0^u \psi(u-y)\{-E(Y)\overline{H}_1(y)\}dy \\ &+ \frac{1}{\lambda\theta E(Y)}\int_u^\infty \{-E(Y)\overline{H}_1(y)\}dy.\end{aligned}$$

That is

$$\begin{aligned}\psi_1(u) &= \frac{1}{1+\theta}\int_0^u \psi_1(u-y)dH_1(y) \\ &+ \frac{1}{c\theta}\int_0^u \psi(u-y)\overline{H}_1(y)dy + \frac{1}{c\theta}\int_u^\infty \overline{H}_1(y)dy,\end{aligned}$$

and (9.2.39) follows if we can show that

$$\theta\int_u^\infty \psi(y)dy = \int_0^u \psi(u-y)\overline{H}_1(y)dy + \int_u^\infty \overline{H}_1(y)dy. \tag{9.2.42}$$

To prove (9.2.42), note that when $\delta = 0$, (9.1.15) becomes

$$(1+\theta)\psi(y) = \int_0^y \psi(y-x)dH_1(x) + \overline{H}_1(y).$$

Integration from u to ∞ yields

$$(1+\theta)\int_u^\infty \psi(y)dy = \int_u^\infty \int_0^y \psi(y-x)dH_1(x)dy + \int_u^\infty \overline{H}_1(y)dy.$$

Reverse the order of integration to obtain

$$\begin{aligned}
&\int_u^\infty \int_0^y \psi(y-x)dH_1(x)dy \\
=& \int_0^u \int_u^\infty \psi(y-x)dydH_1(x) + \int_u^\infty \int_x^\infty \psi(y-x)dydH_1(x) \\
=& \int_0^u \int_{u-x}^\infty \psi(y)dydH_1(x) + \overline{H}_1(u)\int_0^\infty \psi(y)dy.
\end{aligned}$$

Thus

$$\begin{aligned}
&(1+\theta)\int_u^\infty \psi(y)dy \\
=& \int_0^u \int_{u-x}^\infty \psi(y)dydH_1(x) + \overline{H}_1(u)\int_0^\infty \psi(y)dy + \int_u^\infty \overline{H}_1(y)dy.
\end{aligned}$$

Now, integration by parts yields

$$\begin{aligned}
&\int_0^u \int_{u-x}^\infty \psi(y)dydH_1(x) \\
=& -\overline{H}_1(x)\int_{u-x}^\infty \psi(y)dy \,|_{x=0}^u + \int_0^u \psi(u-x)\overline{H}_1(x)dx \\
=& \int_u^\infty \psi(y)dy - \overline{H}_1(u)\int_0^\infty \psi(y)dy + \int_0^u \psi(u-x)\overline{H}_1(x)dx.
\end{aligned}$$

Therefore,

$$(1+\theta)\int_u^\infty \psi(y)dy = \int_u^\infty \psi(y)dy + \int_0^u \psi(u-x)\overline{H}_1(x)dx + \int_u^\infty \overline{H}_1(y)dy$$

from which (9.2.42) follows. Hence (9.2.39) holds, and is a defective renewal equation of the form (9.1.1) with $m(x) = \psi_1(x)$, $\phi = 1/(1+\theta)$, and $v(x) = \int_x^\infty \psi(t)dt/c$. The solution may thus be obtained from (9.1.2), yielding

$$\begin{aligned}
\psi_1(u) &= \frac{1+\theta}{\theta c}\int_u^\infty \psi(x)dx - \frac{1+\theta}{\theta c}\left\{\int_0^\infty \psi(x)dx\right\}\psi(u) \\
&+ \frac{1+\theta}{\theta c}\int_0^u \psi(u-x)\psi(x)dx.
\end{aligned}$$

But this is (9.2.40) since $(1+\theta)/(\theta c) = 1/\{\lambda\theta E(Y)\}$ and $\int_0^\infty \psi(x)dx$ is the compound geometric mean, namely $E(Y^2)/\{2\theta E(Y)\}$ using (2.2.3). □

We remark that $\psi_1(u)$ is also given in Rolski et al (1999, section 11.3). It is a relatively simple matter to evaluate $\psi_1(u)$ in (9.2.40) when the claim amount distribution is exponential, a mixture of exponentials, or a mixture of Erlang distributions. Moreover, an approximation to the mean time to ruin is easily established.

Example 9.2.2 Tijms approximation to the mean time to ruin

It follows from (9.1.19) that the ruin probability may be approximated by

$$\psi_T(u) = C_1 e^{-\kappa_1 u} + Ce^{-\kappa u},$$

where, from (9.1.20), $C_1 = \left(\frac{1}{1+\theta} - C\right)$ and $\kappa_1 = C_1\left\{\frac{E(Y^2)}{2\theta E(Y)} - \frac{C}{\kappa}\right\}^{-1}$. Clearly, if we replace $\psi(u)$ by $\psi_T(u)$ in (9.2.38) and (9.2.40) we obtain an approximation to the mean time to ruin, namely,

$$E(T|T<\infty) \approx \frac{\int_0^u \psi_T(u-x)\psi_T(x)dx + \int_u^\infty \psi_T(x)dx}{\lambda\theta E(Y)\psi_T(u)} - \frac{E(Y^2)}{2\lambda\theta^2\{E(Y)\}^2}.$$

After a little algebra, one obtains

$$E(T|T<\infty) \approx \frac{C_1(C_1 u + \frac{1}{\kappa_1} - \frac{2C}{\kappa_1-\kappa})e^{-\kappa_1 u} + C(Cu + \frac{1}{\kappa} + \frac{2C_1}{\kappa_1-\kappa})e^{-\kappa u}}{2\theta E(Y)(C_1 e^{-\kappa_1 u} + Ce^{-\kappa u})} - \frac{E(Y^2)}{2\lambda\theta^2\{E(Y)\}^2}. \tag{9.2.43}$$

□

Clearly, (9.2.43) is an equality for certain choices of $F(y)$ when $\psi_T(u) = \psi(u)$. Also, in the exponential case, the exact answer is obtained with $C_1 = 0$ in (9.2.43) as in Gerber (1979, p. 138).

Finally, we have the following simple upper bound for the mean time to ruin.

Corollary 9.2.1

$$E(T|T<\infty) \leq \frac{1}{\lambda\theta E(Y)}\left\{\frac{\{2\theta E(Y)\}u - E(Y^2)}{2\theta E(Y)} + \frac{1}{\psi(u)}\int_u^\infty \psi(x)dx\right\}. \tag{9.2.44}$$

Proof: Brown (1990) has shown that compound geometric df's are NWU, i.e. $\psi(u-x)\psi(x) \leq \psi(u)$. Thus, substitution into (9.2.40) yields (9.2.44). □

Further bounds may be obtained in the usual way based on bounds for the compound geometric tail $\psi(x)$, as in section 7.1.

Higher moments of the time to ruin T are obtainable by differentiating the Laplace transform $\overline{G}(u)$ given in (9.2.11) with respect to δ. Following this line of argument, Lin and Willmot (2000) have derived a recursive set of defective renewal equations for higher moments of T, i.e.

$$E(T^k \mid T < \infty) = \frac{\psi_k(u)}{\psi(u)}, \tag{9.2.45}$$

where

$$\psi_k(u) = \frac{1}{1+\theta}\int_0^u \psi_k(u-x)dH_1(x) + \frac{k}{c}\int_u^\infty \psi_{k-1}(x)dx, \tag{9.2.46}$$

for $k = 2, 3, 4, \cdots$. Thus, from (9.1.2),

$$\begin{aligned} \psi_k(u) &= \frac{k}{\lambda\theta E(Y)}\left\{\int_0^u \psi(u-x)\psi_{k-1}(x)dx\right. \\ &+ \left.\int_u^\infty \psi_{k-1}(x)dx - \psi(u)\int_0^\infty \psi_{k-1}(x)dx\right\}. \end{aligned} \tag{9.2.47}$$

An obvious advantage of the above formula is that it is an explicit expression for the k-th moment and thus solving a renewal equation is unnecessary.

9.3 Convolutions involving compound geometric distributions

As in many previous sections, let $\overline{G}(x) = 1 - G(x) = \sum_{n=1}^{\infty}(1-\phi)\phi^n\overline{F}^{*n}(x)$, $x \geq 0$, be a compound geometric tail. In this section we consider the tail of the convolution with a df $A(x) = 1 - \overline{A}(x)$, $x \geq 0$ with $A(0) = 0$. That is, let $W(x) = 1 - \overline{W}(x)$ be the df satisfying

$$\overline{W}(x) = \int_x^\infty dA{*}G(y) = \overline{A}(x) + \int_0^x \overline{G}(x-y)dA(y), \quad x \geq 0. \tag{9.3.1}$$

Convolutions of this type arise repeatedly in many different areas of applied probability such as ruin theory and queueing theory. See Neuts (1986), Dufresne and Gerber (1991), and references therein for details.

The following result shows that $\overline{W}(x)$ satisfies a defective renewal equation.

Theorem 9.3.1 The tail $\overline{W}(x)$ satisfies

$$\overline{W}(x) = \phi\int_0^x \overline{W}(x-y)dF(y) + \phi\overline{F}(x) + (1-\phi)\overline{A}(x). \tag{9.3.2}$$

Proof: By the law of total probability, one may interchange the role of A and G in (9.3.1), yielding

$$\overline{W}(x) = \int_0^x \overline{A}(x-y)dG(y) + \overline{G}(x) + (1-\phi)\overline{A}(x).$$

Let $m(x)$ be defined by $(1-\phi)m(x) = \overline{W}(x) - \overline{G}(x)$, and thus

$$m(x) = \frac{1}{1-\phi}\int_0^x \overline{A}(x-y)dG(y) + \overline{A}(x),$$

and from theorem 9.1.1 it follows that $m(x)$ satisfies the defective renewal equation

$$m(x) = \phi \int_0^x m(x-y)dF(y) + \overline{A}(x),$$

that is

$$\overline{W}(x) - \overline{G}(x) = \phi \int_0^x \{\overline{W}(x-y) - \overline{G}(x-y)\}dF(y) + (1-\phi)\overline{A}(x).$$

But (9.1.15) yields $\overline{G}(x) - \phi\int_0^x \overline{G}(x-y)dF(y) = \phi\overline{F}(x)$, and (9.3.2) follows. □

Theorem 9.3.1 implies that the results of section 9.1 may be applied to $\overline{W}(x)$.

Corollary 9.3.1 If $\overline{A}(x) \leq (\geq)\ \overline{F}(x)$ then $\overline{W}(x) \leq (\geq)\ \overline{G}(x)/\phi$.

Proof: Clearly, $\phi\overline{F}(x) + (1-\phi)\overline{A}(x) \leq (\geq)\ \overline{F}(x)$, and the result follows from corollary 9.1.2 with $k = 1$ and theorem 9.1.1. □

Similarly, one obtains an asymptotic formula for $\overline{W}(x)$, in terms of $\kappa > 0$ satisfying

$$\int_0^\infty e^{\kappa y} dF(y) = \frac{1}{\phi}. \tag{9.3.3}$$

Corollary 9.3.2 Suppose that $F(y)$ is the df of a non-arithmetic distribution, $\kappa > 0$ satisfies (9.3.3), and for some $\epsilon > 0$

$$\int_0^\infty e^{(\kappa+\epsilon)y} dA(y) < \infty \quad \text{and} \quad \int_0^\infty e^{(\kappa+\epsilon)y} dF(y) < \infty.$$

Then

$$\overline{W}(x) \sim \frac{(1-\phi)\displaystyle\int_0^\infty e^{\kappa y} dA(y)}{\phi\kappa \displaystyle\int_0^\infty y e^{\kappa y} dF(y)} e^{-\kappa x}, \quad x \to \infty. \tag{9.3.4}$$

Proof: Integration by parts yields

$$\begin{aligned}
&\phi\int_0^\infty e^{\kappa y}\overline{F}(y)dy+(1-\phi)\int_0^\infty e^{\kappa y}\overline{A}(y)dy \\
&= \phi\left\{\frac{\int_0^\infty e^{\kappa y}dF(y)-1}{\kappa}\right\}+(1-\phi)\left\{\frac{\int_0^\infty e^{\kappa y}dA(y)-1}{\kappa}\right\} \\
&= \phi\left\{\frac{\phi^{-1}-1}{\kappa}\right\}+\frac{1-\phi}{\kappa}\left\{\int_0^\infty e^{\kappa y}dA(y)-1\right\} \\
&= \frac{1-\phi}{\kappa}\int_0^\infty e^{\kappa y}dA(y).
\end{aligned}$$

The result follows from theorems 9.1.3 and 9.3.1. □

Bounds may be obtained from the corresponding bounds for $\overline{G}(x)$ and (9.3.1). The following is a special case of a result obtained by Willmot and Lin (1996).

Theorem 9.3.2 Suppose that $\kappa > 0$ satisfies (9.3.3), the quantity ϕ_A^{-1} satisfies

$$\frac{1}{\phi_A}=\int_0^\infty e^{\kappa y}dA(y), \tag{9.3.5}$$

and the functions $\eta_1(x)$ and $\eta_2(x)$ satisfy

$$\frac{1}{\eta_1(x)}\le\frac{\int_x^\infty e^{\kappa y}dA(y)}{e^{\kappa x}\overline{A}(x)}\le\frac{1}{\eta_2(x)}. \tag{9.3.6}$$

Then

$$\overline{W}(x)\le\left\{1-\frac{\alpha_1(x)}{\eta_1(x)}\right\}\overline{A}(x)+\left\{\frac{\alpha_1(x)}{\phi_A}\right\}e^{-\kappa x},\quad x\ge 0, \tag{9.3.7}$$

and

$$\overline{W}(x)\ge\left\{1-\frac{\alpha_2(x)}{\eta_2(x)}\right\}\overline{A}(x)+\left\{\frac{\alpha_2(x)}{\phi_A}\right\}e^{-\kappa x},\quad x\ge 0, \tag{9.3.8}$$

where $\alpha_1(x)$ and $\alpha_2(x)$ satisfy (7.1.6) and (7.1.7) respectively.

Proof: It follows from (9.3.1) and corollary 7.1.1 that

$$\begin{aligned}
\overline{W}(x) &\le \overline{A}(x)+\int_0^x \alpha_1(x-y)e^{-\kappa(x-y)}dA(y) \\
&\le \overline{A}(x)+\alpha_1(x)e^{-\kappa x}\int_0^x e^{\kappa y}dA(y)
\end{aligned}$$

$$= \overline{A}(x) + \alpha_1(x)e^{-\kappa x}\left\{\frac{1}{\phi_A} - \int_x^\infty e^{\kappa y} dA(y)\right\}$$

$$\leq \overline{A}(x) + \frac{\alpha_1(x)}{\phi_A}e^{-\kappa x} - \alpha_1(x)\left\{\frac{\overline{A}(x)}{\eta_1(x)}\right\}$$

which is (9.3.7). Similarly, (9.3.8) follows by reversing the inequalities. □

We remark that if $A(x)$ is NWUC (NBUC), then as in the proof of corollary 7.4.3, we may choose $\eta_1(x)$ ($\eta_2(x)$) equal to ϕ_A in theorem 9.3.2, and the resulting bound is a weighted average of $\overline{A}(x)$ and $e^{-\kappa x}$. We now consider two examples.

Example 9.3.1 Ruin with diffusion

Dufresne and Gerber (1991) considered the model

$$U_t = x + ct - S_t + W_t, \quad t \geq 0,$$

where U_t is the surplus at time t, $x \geq 0$ is the initial surplus, S_t is the aggregate claims process which is assumed to be compound Poisson with Poisson parameter λ and individual claim amount distribution with df $H(y)$ and mean $E(Y) = \int_0^\infty y dH(y)$, and W_t is a Wiener process with infinitesimal drift 0 and infinitesimal variance $2D$. The premium rate is $c > \lambda E(Y)$. Let $W(x)$ be the probability of ultimate survival, i.e. $W(x) = Pr\{U_t \geq 0 \text{ for all } t \geq 0\}$. Let $A(x) = 1 - e^{-\frac{c}{D}x}$, $x \geq 0$, i.e. $A(x)$ is an exponential df. Also, let $H_1(y) = \int_0^x \overline{H}(y)dy/E(Y)$ be the equilibrium df of $H(y)$, and $F(x) = A * H_1(x)$ the convolution df. Then Dufresne and Gerber (1991) have shown that, if

$$\overline{G}(x) = \sum_{n=1}^{\infty}\left[1 - \frac{\lambda E(Y)}{c}\right]\left[\frac{\lambda E(Y)}{c}\right]^n \overline{F}^{*n}(x), \quad x \geq 0, \tag{9.3.9}$$

then $W(x) = A * G(x)$, and the results of this section apply with

$$\phi = \frac{\lambda E(Y)}{c}. \tag{9.3.10}$$

In particular, since $\overline{F}(x) = \overline{A}(x) + \int_0^x \overline{H}_1(x-y)dA(y) \geq \overline{A}(x)$, corollary 9.3.1 yields

$$\overline{W}(x) \leq \frac{c}{\lambda E(Y)}\overline{G}(x), \quad x \geq 0. \tag{9.3.11}$$

Let $\tilde{a}(s) = \int_0^\infty e^{-sx} dA(x) = c/(c + Ds)$, and (9.3.5) becomes $\tilde{a}(-\kappa) = 1/\phi_A$, that is

$$\phi_A = \frac{c - D\kappa}{c}. \tag{9.3.12}$$

Similarly, let $\tilde{h}(s) = \int_0^\infty e^{-sx} dH(x)$ and $\tilde{h}_1(s) = \int_0^\infty e^{-sx} dH_1(x)$. Then integration by parts yields $\tilde{h}_1(s) = \{1 - \tilde{h}(s)\}/\{sE(Y)\}$ (see also Feller, 1971, p. 435). Also let $\tilde{f}(s) = \int_0^\infty e^{-sx} dF(x)$, and thus $\tilde{f}(s) = \tilde{h}_1(s)\tilde{a}(s)$. Then (9.3.3) becomes $\tilde{f}(-\kappa) = 1/\phi$, i.e.

$$\tilde{h}_1(-\kappa) = \frac{\phi_A}{\phi}. \tag{9.3.13}$$

Alternatively, (9.3.13) may be expressed as

$$\frac{\tilde{h}(-\kappa) - 1}{\kappa E(Y)} = \frac{c - D\kappa}{\lambda E(Y)},$$

i.e. $\kappa > 0$ satisfies $\lambda\{\tilde{h}(-\kappa) - 1\} = \kappa(c - D\kappa)$, that is

$$\lambda \int_0^\infty e^{\kappa y} dH(y) = \lambda + \kappa(c - D\kappa). \tag{9.3.14}$$

Next, consider the asymptotic formula (9.3.4). Since

$$\tilde{f}(s) = \frac{1 - \tilde{h}(s)}{s} \cdot \frac{1}{c + Ds} \cdot \frac{c}{E(Y)},$$

(9.3.10) implies that

$$\phi \tilde{f}(s)(cs + Ds^2) = \lambda\{1 - \tilde{h}(s)\}.$$

Differentiation yields

$$\phi \tilde{f}'(s)(cs + Ds^2) + \phi \tilde{f}(s)(c + 2Ds) = -\lambda \tilde{h}'(s).$$

Since $c + Ds = c/\tilde{a}(s)$, this implies that

$$\frac{c\phi s \tilde{f}'(s)}{\tilde{a}(s)} = -\lambda \tilde{h}'(s) - \phi \tilde{f}(s)(c + 2Ds).$$

Consequently,

$$\frac{(1 - \phi)\tilde{a}(s)}{\phi s \tilde{f}'(s)} = \frac{c(1 - \phi)}{-\lambda \tilde{h}'(s) - \phi \tilde{f}(s)(c + 2Ds)}.$$

Since $c(1 - \phi) = c - \lambda E(Y)$ and $\tilde{f}(-\kappa) = 1/\phi$, it follows by substitution of $s = -\kappa$ into the above that (9.3.4) becomes

$$\overline{W}(x) \sim \frac{c - \lambda E(Y)}{\lambda \int_0^\infty y e^{\kappa y} dH(y) - (c - 2D\kappa)} e^{-\kappa x}, \quad x \to \infty. \tag{9.3.15}$$

Next, consider the bounds in theorem 9.3.1. Since $A(x) = 1 - e^{-\frac{c}{D}x}$, an easy calculation yields

$$\frac{\int_x^\infty e^{\kappa y} dA(y)}{e^{\kappa x}\overline{A}(x)} = \frac{1}{\phi_A}$$

using (9.3.12). Thus, with $\eta_1(x) = \phi_A$, (9.3.7) becomes

$$\overline{W}(x) \le \left\{1 - \frac{\alpha_1(x)}{\phi_A}\right\} e^{-\frac{c}{D}x} + \left\{\frac{\alpha_1(x)}{\phi_A}\right\} e^{-\kappa x}, \quad x \ge 0. \tag{9.3.16}$$

Similarly, with $\eta_2(x) = \phi_A$, (9.3.8) becomes

$$\overline{W}(x) \ge \left\{1 - \frac{\alpha_2(x)}{\phi_A}\right\} e^{-\frac{c}{D}x} + \left\{\frac{\alpha_2(x)}{\phi_A}\right\} e^{-\kappa x}, \quad x \ge 0. \tag{9.3.17}$$

It is worth noting that if $H(y)$ is DMRL, then $H_1(y)$ is IFR, and the convolution $F(x) = A * H_1(x)$ is also IFR (since IFR is preserved under convolution) and hence NBUC. Since (9.3.3) holds, it follows from proposition 6.1.1 and (7.1.7) that $\alpha_2(x) = \phi$. Thus, using (9.3.13), (9.3.17) becomes

$$\overline{W}(x) \ge \left\{1 - \frac{1}{\tilde{h}_1(-\kappa)}\right\} e^{-\frac{c}{D}x} + \left\{\frac{1}{\tilde{h}_1(-\kappa)}\right\} e^{-\kappa x}, \quad x \ge 0. \tag{9.3.18}$$

Evidently, since $\tilde{h}_1(-\kappa) > 1$, the bound in (9.3.18) is a simple mixture of two exponential tails. □

Example 9.3.2 The M/G/c queue waiting time

Van Hoorn (1984) has proposed an approximation to the equilibrium waiting time distribution in the M/G/c queue. The arrivals process to the queue is a Poisson process with rate λ, and service is in order of arrival by one of c servers with service time distribution $H(y)$ and mean $E(Y)$. The traffic intensity is $\phi = \lambda E(Y)/c < 1$. Let $H_1(y) = 1 - \overline{H}_1(y) = \int_0^x \overline{H}(y)dy/E(Y)$ be the equilibrium df of $H(y)$, and let $F(y) = H_1(cy)$. Also, let $A(y) = 1 - \{\overline{H}_1(y)\}^c$. The approximate equilibrium waiting time distribution (given that there is a wait) is then given by $W(t) = 1 - \overline{W}(t)$ where $W(t)$ is defined in this section (9.3) with the above parameters. Since the probability of having to wait is approximated by

$$P_W = \frac{\dfrac{\phi}{1-\phi}\dfrac{\{\lambda E(Y)\}^{c-1}}{(c-1)!}}{\displaystyle\sum_{n=0}^{c-1}\frac{\{\lambda E(Y)\}^n}{n!} + \frac{\{\lambda E(Y)\}^c}{c!(1-\phi)}}, \tag{9.3.19}$$

the unconditional equilibrium waiting time distribution is approximated by $W_*(t) = 1 - \overline{W}_*(t)$ where $\overline{W}_*(t) = P_W\overline{W}(t)$. Equation (9.3.3) for κ

becomes, in this case

$$\int_0^\infty e^{\frac{\kappa}{c}y} dH_1(y) = \frac{1}{\phi},$$

or equivalently,

$$\int_0^\infty e^{\frac{\kappa}{c}y} dH(y) = 1 + \frac{\kappa E(Y)}{c\phi} = 1 + \frac{\kappa}{\lambda}.$$

That is, $\kappa > 0$ satisfies

$$\lambda \int_0^\infty e^{\frac{\kappa}{c}y} dH(y) = \lambda + \kappa, \tag{9.3.20}$$

a quantity which is useful in determining the asymptotic formula $\overline{W}_*(t) \sim Ce^{-\kappa t}$, $t \to \infty$, from (9.3.4), as well as the bounds in (9.3.7) and (9.3.8).

Simpler bounds are often possible. If $H(y)$ is 2-NWU, then $H_1(y)$ is NWU. Thus

$$\overline{F}(x+y) = \overline{H}_1(cx+cy) \geq \overline{H}_1(cx)\overline{H}_1(cy) = \overline{F}(x)\overline{F}(y),$$

and $F(y)$ is NWU. Similarly,

$$\begin{aligned} \overline{A}(x+y) &= \{\overline{H}_1(x+y)\}^c \geq \{\overline{H}_1(x)\overline{H}_1(y)\}^c \\ &= \{\overline{H}_1(x)\}^c\{\overline{H}_1(y)\}^c = \overline{A}(x)\overline{A}(y), \end{aligned}$$

and $A(y)$ is also NWU. Therefore, if $H(y)$ is 2-NWU, it follows that both $A(y)$ and $F(y)$ are NWU and hence NWUC. Thus, from proposition 6.1.1, $\alpha_1(x) = \phi$ where $\alpha_1(x)$ is given by (7.1.6), and by an identical argument $\eta_1(x) = \phi_A$ satisfies (9.3.6) since (9.3.5) holds. Consequently, if $H(y)$ is 2-NWU, (9.3.7) becomes

$$\overline{W}(x) \leq (1 - \frac{\phi}{\phi_A})\{\overline{H}_1(x)\}^c + (\frac{\phi}{\phi_A})e^{-\kappa x}, \quad x \geq 0. \tag{9.3.21}$$

Furthermore, if $H(y)$ is 2-NWU, then $H_1(y)$ is NWU, and since c is an integer, it follows that $\overline{H}_1(cy) \geq \{\overline{H}_1(y)\}^c$, i.e. $\overline{F}(y) \geq \overline{A}(y)$. But this implies using (9.3.3), (9.3.5), and Ross (1996, p. 405), that

$$\frac{1}{\phi} = \int_0^\infty e^{\kappa y} dF(y) \geq \int_0^\infty e^{\kappa y} dA(y) = \frac{1}{\phi_A}.$$

That is $\phi/\phi_A \leq 1$, and the bound in (9.3.21) is a mixture.

In the case when $H(y)$ is 2-NBU, similar reasoning yields, from (9.3.8),

$$\overline{W}(x) \geq (1 - \frac{\phi}{\phi_A})\{\overline{H}_1(x)\}^c + (\frac{\phi}{\phi_A})e^{-\kappa x}, \quad x \geq 0, \tag{9.3.22}$$

but $\phi/\phi_A \geq 1$, and the bound in (9.3.22) is not a mixture.

Other bounds involving $G(x)$ are possible when $H(y)$ is 2-NWU or 2-NBU. If $H(y)$ is 2-NWU, then $\overline{F}(y) \geq \overline{A}(y)$ and since $\phi = \lambda E(Y)/c$, it follows from corollary 9.3.1 that

$$\overline{W}(x) \leq \frac{c\overline{G}(x)}{\lambda E(Y)}, \quad x \geq 0. \tag{9.3.23}$$

Similarly, if $H(x)$ is 2-NBU,

$$\overline{W}(x) \geq \frac{c\overline{G}(x)}{\lambda E(Y)}, \quad x \geq 0. \tag{9.3.24}$$

Further results may be obtained by using bounds for $\overline{G}(x)$ in (9.3.23) and (9.3.24).

□

10
The severity of ruin

The conditional distribution of the deficit at the time of ruin, given that ruin has occurred, is the subject matter of this chapter. This quantity may be viewed as an expected discounted penalty introduced in section 9.2, where the penalty function $w(x)$ takes a special form. As discussed in section 9.2, we are thus able to express the conditional distribution of the deficit as the solution of a defective renewal equation, which conveniently leads to the use of many of the techniques developed in earlier chapters. As will be demonstrated in what follows, the defective renewal equation associated with the conditional distribution of the deficit is simple and has a desirable structure. This allows for a mixture representation for the conditional distribution in terms of the probability of ruin and thus for construction of bounds and approximations for it. The advantage of the use of a mixture representation becomes evident when an explicit expression for the conditional distribution is derived in the special case when the claim amount distribution is an Erlang mixture or a general Erlang mixture. The approach of this chapter follows that of Willmot and Lin (1998) and Willmot (2000). We shall continue to use the notation and model of sections 7.3 and 9.2.

10.1 The associated defective renewal equation

As in section 9.2, consider a penalty function $w(|U_T|)$ of the deficit at the time of ruin but let $\delta = 0$. Then, from (9.2.2), $\rho = 0$, and from (9.2.1),

(9.2.3), and (9.2.4), it follows that

$$m(u) = E\{w(|U_T|)I(T < \infty)\} \tag{10.1.1}$$

satisfies the defective renewal equation

$$m(u) = \frac{1}{1+\theta}\int_0^u m(u-x)dH_1(x) + v(u) \tag{10.1.2}$$

where

$$v(u) = \frac{\lambda}{c}\int_u^\infty \int_x^\infty w(t-x)dH(t)dx. \tag{10.1.3}$$

Rather than appeal to the properties of defective renewal equations to analyze $m(u)$ for arbitrary $w(\cdot)$, we wish to consider the distribution of $|U_T|$, which is of interest both in its own right but also in connection with the surplus immediately before ruin (Dickson, 1992), and the duration of negative surplus (Dickson and Egídio dos Reis, 1996). Thus, define

$$G(u,y) = Pr(T < \infty, |U_T| \leq y), \quad y \geq 0. \tag{10.1.4}$$

In words, $G(u,y)$ is the probability that ruin occurs beginning with initial surplus u, and the deficit at the time of ruin is at most y. Thus $G(u,y) = E\{I(T < \infty)I(|U_T| \leq y)\}$ and so choose $w(t) = 1$, $t \leq y$, and $w(t) = 0$ otherwise. Then $w(t-x) = 1$ if $t - x \leq y$, i.e. if $t \leq x + y$, and (10.1.3) yields

$$\begin{aligned} v(u) &= \frac{\lambda}{c}\int_u^\infty \int_x^{x+y} dH(t)dx \\ &= \frac{\lambda}{c}\int_u^\infty \{\overline{H}(x) - \overline{H}(x+y)\}dx. \end{aligned}$$

In other words, in terms of the equilibrium df $H_1(y)$,

$$v(u) = \frac{1}{1+\theta}\{\overline{H}_1(u) - \overline{H}_1(u+y)\}. \tag{10.1.5}$$

Thus, $G(u,y)$ satisfies

$$G(u,y) = \frac{1}{1+\theta}\int_0^u G(u-x,y)dH_1(x) + \frac{1}{1+\theta}\{\overline{H}_1(u) - \overline{H}_1(u+y)\}. \tag{10.1.6}$$

It is instructive to note that (10.1.6) may be obtained directly using probabilistic reasoning. Recall from the discussion following (7.3.3) that $H_1(y)$ may be interpreted as the df of the amount of the drop in the surplus below its previous low, given that a drop occurs. Also, it follows from (7.3.2) that $\psi(0) = 1/(1+\theta)$, which is also the probability that a drop does occur (obviously, this drop must take place when a claim occurs). Since the process 'starts over' probabilistically at this time from the new surplus level, we can obtain $G(u, y)$ by conditioning on the amount of the first drop in surplus. By the law of total probability, if the amount of the drop is x where $0 < x \leq u$, then ruin does not occur but the process 'starts over' probabilistically from the new surplus level of $u - x$. If the drop is between u and $u + y$, then ruin occurs and the deficit at ruin is between 0 and y (if the drop exceeds $u + y$ then the deficit at ruin exceeds y). Thus (10.1.6) follows. Clearly, the ruin probability $\psi(u)$ satisfies $\psi(u) = \lim_{y \to \infty} G(u, y)$, and (10.1.6) reduces to

$$\psi(u) = \frac{1}{1+\theta} \int_0^u \psi(u-x) dH_1(x) + \frac{1}{1+\theta} \overline{H}_1(u), \tag{10.1.7}$$

which is well known. Since

$$\psi(u) = \sum_{n=1}^{\infty} \frac{\theta}{1+\theta} (\frac{1}{1+\theta})^n \overline{H}_1^{*n}(u), \quad u \geq 0, \tag{10.1.8}$$

as discussed in section 7.3, it is clear from the results of section 9.1 that $1 - \psi(u)$ is the relevant compound geometric df associated with the defective renewal equation (10.1.2) and hence (10.1.6). Also note that the derivative of $\psi(u)$ for $u > 0$ always exists because the derivative of $H_1(x)$ exists. The following theorem is the starting point for the analysis of the distribution $G(u, y)$.

Theorem 10.1.1

$$G(u, y) = \psi(u) - \frac{1}{\theta} \int_0^u \overline{H}_1(y+u-x)\{-\psi'(x)\}dx - \frac{1}{1+\theta} \overline{H}_1(u+y). \tag{10.1.9}$$

Proof: It is convenient for notational purposes to introduce $\overline{G}(u, y) = \psi(u) - G(u, y) = Pr(T < \infty, |U_T| > y)$. Then subtraction of (10.1.6) from (10.1.7) yields a defective renewal equation for $\overline{G}(u, y)$, namely

$$\overline{G}(u, y) = \frac{1}{1+\theta} \int_0^u \overline{G}(u-x, y) dH_1(x) + \frac{1}{1+\theta} \overline{H}_1(u+y). \tag{10.1.10}$$

Then from theorem 9.1.1 with $\phi = 1/(1+\theta)$, $v(x) = \overline{H}_1(x+y)/(1+\theta)$, and $G(x)$ replaced by $1 - \psi(x)$, one obtains

$$\overline{G}(u, y) = \frac{1}{\theta} \int_0^u \overline{H}_1(y+u-x)\{-\psi'(x)\}dx + \frac{1}{1+\theta} \overline{H}_1(u+y) \tag{10.1.11}$$

which is (10.1.9). □

We remark that (10.1.10) may be obtained probabilistically using the same arguments as those following (10.1.6).

An explicit solution for $G(u,y)$ follows from the above results.

Corollary 10.1.1

$$\begin{aligned} G(u,y) &= H_1(y)\psi(u) + \frac{1+\theta}{\theta}\{\overline{H}_1(y)\psi(u) - \psi(u+y)\} \\ &+ \frac{1}{\theta}\int_0^y \psi(u+y-x)dH_1(x). \end{aligned} \tag{10.1.12}$$

Proof: It follows from theorem 9.1.2 with $\phi = 1/(1+\theta)$, $v(x) = \overline{H}_1(x+y)/(1+\theta)$, and $G(x)$ replaced by $1-\psi(x)$ that

$$\begin{aligned} \overline{G}(u,y) &= \frac{\overline{H}_1(u+y) - \overline{H}_1(y)\psi(u)}{\theta} + \frac{1}{\theta}\int_0^u \psi(u-t)\frac{\overline{H}(t+y)}{E(Y)}dt \\ &= \frac{\overline{H}_1(u+y) - \overline{H}_1(y)\psi(u)}{\theta} + \frac{1}{\theta}\int_y^{u+y} \psi(u+y-t)dH_1(t). \end{aligned}$$

But from (10.1.7) with u replaced by $u+y$, we obtain

$$\begin{aligned} \overline{G}(u,y) &= \frac{\overline{H}_1(u+y) - \overline{H}_1(y)\psi(u)}{\theta} + \frac{1}{\theta}\{(1+\theta)\psi(u+y) - \overline{H}_1(u+y)\} \\ &- \frac{1}{\theta}\int_0^y \psi(u+y-t)dH_1(t), \end{aligned}$$

i.e.

$$\overline{G}(u,y) = \frac{1+\theta}{\theta}\psi(u+y) - \frac{\overline{H}_1(y)\psi(u)}{\theta} - \frac{1}{\theta}\int_0^y \psi(u+y-t)dH_1(t). \tag{10.1.13}$$

Thus, since $G(u,y) = \psi(u) - \overline{G}(u,y)$, we obtain

$$G(u,y) = \{1 + \frac{\overline{H}_1(y)}{\theta}\}\psi(u) - \frac{1+\theta}{\theta}\psi(u+y) + \frac{1}{\theta}\int_0^y \psi(u+y-t)dH_1(t)$$

from which (10.1.12) follows. □

In the next section we consider the conditional distribution of the deficit given that ruin occurs.

10.2 A mixture representation for the conditional distribution

Let $G_u(y) = 1 - \overline{G}_u(y) = Pr(|U_T| \le y \mid T < \infty)$, $y \ge 0$, be the conditional distribution of the deficit given that ruin occurs. Then $G_u(y) =$

$G(u,y)/\psi(u)$ and $\overline{G}_u(y) = \overline{G}(u,y)/\psi(u)$. It follows immediately from (10.1.6) and (10.1.7) with $u = 0$ that

$$G_0(y) = H_1(y), \quad y \geq 0. \tag{10.2.1}$$

In what follows it is convenient to introduce the df $H_{1,x}(y) = 1 - \overline{H}_{1,x}(y)$ defined for $x \geq 0$ by

$$\overline{H}_{1,x}(y) = \frac{\overline{H}_1(x+y)}{\overline{H}_1(x)}, \quad y \geq 0. \tag{10.2.2}$$

Evidently, from sections 2.2 and 2.3, $H_{1,x}(y)$ is the residual lifetime distribution associated with the equilibrium df $H_1(y)$. The following theorem expresses $G_u(y)$ as a mixture over x of df's of the form (10.2.2).

Theorem 10.2.1

$$G_u(y) = \frac{\int_0^u H_{1,u-x}(y)\overline{H}_1(u-x)\{-\psi'(x)\}dx + \frac{\theta}{1+\theta}H_{1,u}(y)\overline{H}_1(u)}{\int_0^u \overline{H}_1(u-x)\{-\psi'(x)\}dx + \frac{\theta}{1+\theta}\overline{H}_1(u)}, \quad y \geq 0. \tag{10.2.3}$$

Proof: Clearly, from (10.1.7) and (10.1.10), $\psi(u) = \overline{G}(u,0)$, and from (10.1.11),

$$\theta\psi(u) = \int_0^u \overline{H}_1(u-x)\{-\psi'(x)\}dx + \frac{\theta}{1+\theta}\overline{H}_1(u). \tag{10.2.4}$$

Thus, again from (10.1.11),

$$\overline{G}_u(y) = \frac{\overline{G}(u,y)}{\psi(u)} = \frac{\int_0^u \overline{H}_1(y+u-x)\{-\psi'(x)\}dx + \frac{\theta}{1+\theta}\overline{H}_1(y+u)}{\int_0^u \overline{H}_1(u-x)\{-\psi'(x)\}dx + \frac{\theta}{1+\theta}\overline{H}_1(u)}.$$

That is,

$$\overline{G}_u(y) = \frac{\int_0^u \overline{H}_{1,u-x}(y)\overline{H}_1(u-x)\{-\psi'(x)\}dx + \frac{\theta}{1+\theta}\overline{H}_{1,u}(y)\overline{H}_1(u)}{\int_0^u \overline{H}_1(u-x)\{-\psi'(x)\}dx + \frac{\theta}{1+\theta}\overline{H}_1(u)}, \tag{10.2.5}$$

from which (10.2.3) follows. □

Theorem 10.2.1 is an important result since it enables us to apply results for mixed distributions to $G_u(y)$. As will be seen in what follows and also sections 10.3 and 10.4, the mixture representation (10.2.3) for $G_u(y)$ allows for identification of class properties, for derivation of stochastic bounds, and for an explicit expression for $G_u(y)$ when the claim amount distribution $H(x)$ is an Erlang mixture. The following example gives an explicit expression for $G_u(y)$ when the claim amount distribution $H(x)$ is exponential.

Example 10.2.1 Exponential claim amounts
If $H(y) = 1 - e^{-y/E(Y)}$, $y \geq 0$, then $H_1(y) = H_{1,x}(y) = H(y)$, as is easily verified. Since $H_{1,x}(y)$ is independent of x, it follows from (10.2.3) $G_u(y) = 1 - e^{-y/E(Y)} = H(y)$, for all $u \geq 0$. □

Willmot (2000) considered a subset of the beta family as claim amounts, including the constant and uniform distributions as special cases.

The following class properties of $G_u(y)$ follow the mixture representation.

Corollary 10.2.1 The following class implications hold.

a) If $H(y)$ is IMRL then $G_u(y)$ is DFR.

b) If $H(y)$ is 3-DFR then $G_u(y)$ is IMRL.

Proof: It follows from (2.3.5) that if $r_H(y)$ is the mean residual lifetime associated with the df $H(y)$, then $H_1(y)$ has failure rate $1/r_H(y)$. Thus

$$\overline{H}_{1,x}(y) = \frac{\overline{H}_1(x+y)}{\overline{H}_1(x)} = e^{-\int_0^y \{1/r_H(x+t)\}dt}. \tag{10.2.6}$$

That is, $H_{1,x}(y)$ has failure rate $1/r_H(x+y)$ as a function of y, which is obviously nonincreasing if $r_H(y)$ is nondecreasing. Thus, if $H(y)$ is IMRL then $H_{1,x}(y)$ is DFR for $x \geq 0$, and since the DFR property is preserved under mixing, $G_u(y)$ is DFR, proving a). To prove b), let $r_1(y) = \int_y^\infty \overline{H}_1(t)dt/\overline{H}_1(y)$ be the mean residual lifetime of $H_1(y)$. Then, let $r_{1,x}(y)$ be the mean residual lifetime of $H_{1,x}(y)$, and from (10.2.2),

$$r_{1,x}(y) = \frac{\int_y^\infty \overline{H}_{1,x}(t)dt}{\overline{H}_{1,x}(y)} = \frac{\int_y^\infty \overline{H}_1(x+t)dt}{\overline{H}_1(x+y)} = \frac{\int_{x+y}^\infty \overline{H}_1(t)dt}{\overline{H}_1(x+y)} = r_1(x+y).$$

Thus, if $H(y)$ is 3-DFR then $H_1(y)$ is IMRL and $H_{1,x}(y)$ is also IMRL. As discussed in section 2.3, the IMRL property is preserved under mixing, and by theorem 10.2.1, $G_u(y)$ is IMRL. □

The mixture representation also allows easily for construction of bounds on $G_u(y)$.

Corollary 10.2.2 Suppose $\overline{A}_u(y)$ satisfies

$$\overline{H}_1(x+y) \geq (\leq)\, \overline{A}_u(y)\overline{H}_1(x), \quad 0 \leq x \leq u, \quad y \geq 0. \tag{10.2.7}$$

Then $\overline{G}_u(y) \geq (\leq)\, \overline{A}_u(y)$, $y \geq 0$. In particular,

a) if $H(y)$ is DMRL(IMRL) then $\overline{G}_u(y) \geq (\leq)\, \overline{H}_1(u+y)/\overline{H}_1(u)$;

b) if $H(y)$ is 2-NWU (2-NBU) then $\overline{G}_u(y) \geq (\leq)\, \overline{H}_1(y)$;

c) if $H(y)$ is NWUC (NBUC) then $\overline{G}_u(y) \geq (\leq)\ \overline{H}(y)$;

d) if $H(y)$ is NWUE (NBUE) then $\overline{G}_u(y) \geq (\leq)\ e^{-y/E(Y)}$.

e) if $H(y)$ is UBAE (UWAE) then $\overline{G}_u(y) \geq (\leq)\ e^{-y/r_H(\infty)}$.

Proof: Clearly, (10.2.7) implies that $\overline{H}_{1,x}(y) \geq (\leq)\ \overline{A}_u(y)$, and $\overline{G}_u(y) \geq (\leq)\ \overline{A}_u(y)$ follows from (10.2.3). To prove a), note that $H(y)$ is DMRL(IMRL) may be restated as $H_1(y)$ is IFR(DFR), i.e. $\overline{H}_{1,x}(y)$ is nonincreasing (nondecreasing) in x for fixed $y \geq 0$. Thus, $\overline{H}_{1,x}(y) \geq (\leq)\ \overline{H}_{1,u}(y)$, and (10.2.7) holds with $\overline{A}_u(y) = \overline{H}_{1,u}(y)$. Then, b) and c) follow from the definitions of 2-NWU (2-NBU) and NWUC (NBUC). To prove d) and e), note that $r_H(x) \geq (\leq)\ r$ implies that $\overline{H}_{1,x}(y) \geq (\leq)\ e^{-y/r}$ using (10.2.6). Thus, in this case $\overline{A}_u(y) = e^{-y/r}$ and the results follow immediately. □

We remark that corollary 10.2.2.b) is a refinement of a bound given by Dickson and Egídio dos Reis (1996).

Since DMRL implies 2-NBU and IMRL implies 2-NWU, it follows from corollary 10.2.2 a) and b) that if $H(y)$ is DMRL then

$$\frac{\overline{H}_1(u+y)}{\overline{H}_1(u)} \leq \overline{G}_u(y) \leq \overline{H}_1(y), \tag{10.2.8}$$

whereas if $H(y)$ is IMRL then

$$\overline{H}_1(y) \leq \overline{G}_u(y) \leq \frac{\overline{H}_1(u+y)}{\overline{H}_1(u)}. \tag{10.2.9}$$

The mixing distribution is simpler when $u \to \infty$, as is now demonstrated. Suppose that the adjustment coefficient $\kappa > 0$ satisfies the Lundberg adjustment equation

$$1+\theta = \int_0^\infty e^{\kappa y} dH_1(y). \tag{10.2.10}$$

We have the following result for $G_\infty(y) = 1 - \overline{G}_\infty(y) = \lim_{u\to\infty} G_u(y)$.

Theorem 10.2.2 If $\kappa > 0$ satisfies (10.2.10) and $\int_0^\infty e^{(\kappa+\epsilon)y} dH(y) < \infty$ for some $\epsilon > 0$, then

$$G_\infty(y) = \frac{\int_0^\infty H_{1,x}(y) e^{\kappa x} \overline{H}_1(x)dx}{\int_0^\infty e^{\kappa x}\overline{H}_1(x)dx}, \quad y \geq 0. \tag{10.2.11}$$

Proof: Integration by parts yields $\int_0^\infty e^{(\kappa+\epsilon)y} dH_1(y) < \infty$, and theorem 9.1.3 yields, with (10.1.7),

$$\psi(u) \sim \frac{\int_0^\infty e^{\kappa x} \overline{H}_1(x) dx}{\int_0^\infty x e^{\kappa x} dH_1(x)} e^{-\kappa u}, \quad u \to \infty.$$

Similarly, since $\overline{H}_1(u+y) \leq \overline{H}_1(u)$, (10.1.10) yields

$$\overline{G}(u, y) \sim \frac{\int_0^\infty e^{\kappa x} \overline{H}_1(x+y) dx}{\int_0^\infty x e^{\kappa x} dH_1(x)} e^{-\kappa u}, \quad u \to \infty,$$

and therefore

$$\overline{G}_\infty(y) = \lim_{u\to\infty} \overline{G}_u(y) = \lim_{u\to\infty} \frac{e^{\kappa u} \overline{G}(u, y)}{e^{\kappa u} \psi(u)} = \frac{\int_0^\infty e^{\kappa x} \overline{H}_1(x+y) dx}{\int_0^\infty e^{\kappa x} \overline{H}_1(x) dx}, \tag{10.2.12}$$

from which (10.2.11) follows. □

Class properties are somewhat easier for $G_\infty(y)$ than for $G_u(y)$. In particular, (10.2.12) is of the same form as (9.2.20), but with ρ replaced by $-\kappa$. The following result is similar to theorem 9.2.2.

Corollary 10.2.3 Suppose that $G_\infty(y) = 1 - \overline{G}_\infty(y)$ satisfies (10.2.11). Then the following class implications hold.

a) If $H(y)$ is 3-DFR (3-IFR) then $G_\infty(y)$ is DFR (IFR).

b) If $H(y)$ is UWAE (UBAE) then $G_\infty(y)$ is UWA (UBA).

c) If $H_1(y)$ is 2-NWU (2-NBU) then $G_\infty(y)$ is NWUE (NBUE).

Proof: First note that (10.2.12) may be expressed as

$$\overline{G}_\infty(y) = \frac{e^{-\kappa y} \int_y^\infty e^{\kappa x} \overline{H}_1(x) dx}{\int_0^\infty e^{\kappa x} \overline{H}_1(x) dx},$$

from which it follows as with (9.2.17) and (9.2.18) that the failure rate is

$$\mu_\infty(y) = -\frac{d}{dy} \ln \overline{G}_\infty(y) = \kappa + \frac{1}{\eta_1(y)}, \tag{10.2.13}$$

where

$$\eta_1(y) = \int_0^\infty e^{\kappa x} \frac{\overline{H}_1(x+y)}{\overline{H}_1(y)} dx. \tag{10.2.14}$$

Now let $H_2(y) = 1 - \overline{H}_2(y)$ be the equilibrium df of $H_1(y)$, and it follows from the analogy between (9.2.5) and (9.2.20) that the equilibrium tail of $G_\infty(y)$ is

$$\frac{\int_y^\infty \overline{G}_\infty(x)dx}{\int_0^\infty \overline{G}_\infty(x)dx} = \frac{\int_0^\infty e^{\kappa x}\overline{H}_2(x+y)dx}{\int_0^\infty e^{\kappa x}\overline{H}_2(x)dx}.$$

Therefore, since the equilibrium failure rate is $1/r_\infty(y)$ where $r_\infty(y)$ is the mean residual lifetime of $G_\infty(y)$ from (2.3.5), it follows from (10.2.13) and (10.2.14) that

$$r_\infty(y) = \frac{1}{\kappa + \frac{1}{\eta_2(y)}}, \tag{10.2.15}$$

where

$$\eta_2(y) = \int_0^\infty e^{\kappa x} \frac{\overline{H}_2(x+y)}{\overline{H}_2(y)} dx. \tag{10.2.16}$$

Thus, if $H_1(y)$ is 2-NWU(2-NBU), it follows that $\overline{H}_2(x+y)/\overline{H}_2(y) \geq (\leq)\ \overline{H}_2(x)$ and therefore $\eta_2(y) \geq (\leq)\ \eta_2(0)$, in turn $r_\infty(y) \geq (\leq)\ r_\infty(0)$, proving c). To prove a), it follows from (2.2.3) and integration by parts that

$$\int_y^\infty e^{\kappa x}\overline{H}_1(x)dx = \frac{E(Y^2)}{2E(Y)}\left\{e^{\kappa y}\overline{H}_2(y) + \kappa\int_y^\infty e^{\kappa x}\overline{H}_2(x)dx\right\}. \tag{10.2.17}$$

If $r_1(y)$ is the mean residual lifetime associated with the df $H_1(y)$, then it follows from (2.2.3) and (2.3.4) that

$$r_1(y) = \frac{E(Y^2)}{2E(Y)}\frac{\overline{H}_2(y)}{\overline{H}_1(y)}. \tag{10.2.18}$$

Therefore, from (10.2.14), (10.2.17), and (10.2.18),

$$\begin{aligned}
\eta_1(y) &= \frac{e^{-\kappa y}}{\overline{H}_1(y)}\int_y^\infty e^{\kappa x}\overline{H}_1(x)dx \\
&= \frac{E(Y^2)}{2E(Y)}\frac{e^{-\kappa y}}{\overline{H}_1(y)}\left\{e^{\kappa y}\overline{H}_2(y) + \kappa\int_y^\infty e^{\kappa x}\overline{H}_2(x)dx\right\} \\
&= r_1(y)\left\{1 + \kappa\int_0^\infty e^{\kappa x}\frac{\overline{H}_2(x+y)}{\overline{H}_2(y)}dx\right\}.
\end{aligned}$$

That is, using (10.2.16),

$$\eta_1(y) = r_1(y)\{1 + \kappa\eta_2(y)\}. \tag{10.2.19}$$

If $H(y)$ is 3-DFR (3-IFR), then $H_1(y)$ is IMRL (DMRL). Hence, $r_1(y)$ is nondecreasing(nonincreasing) in y and the same is true of $\eta_2(y)$ from (10.2.16) since $H_2(y)$ is DFR(IFR). Thus, (10.2.19) implies that $\eta_1(y)$ is nondecreasing(nonincreasing), and from (10.2.13), $\mu_\infty(y)$ is nonincreasing (nondecreasing). Therefore, a) is proved. To prove b), note that as discussed following theorem 2.4.1, if $H(y)$ is UWAE(UBAE) then $H_1(y)$ is UWA(UBA). Now, (9.2.24) may be restated in the present context as

$$r_\infty(y) = \frac{\int_y^\infty e^{\kappa x}\overline{H}_1(x) r_1(x)dx}{\int_y^\infty e^{\kappa x}\overline{H}_1(x)dx}. \tag{10.2.20}$$

The proof of b) is then identical to the proof of theorem 9.2.2.d) using (10.2.12) and (10.2.20). $\square$

We remark that the class for which $H_1(y)$ is 2-NWU (2-NBU) is also referred to as the class for which $H(y)$ is 3-NWU(3-NBU). See Fagiuoli and Pellerey (1994).

In the next two sections we use the mixture representation to evaluate $G_u(y)$ when $H(y)$ is the df of a mixture of Erlangs.

10.3 Erlang mixtures with the same scale parameter

In this situation we demonstrate that evaluation of the distribution of the conditional severity is relatively straightforward. Let $g_u(y) = \frac{d}{dy}G_u(y)$ be the associated probability density function (differentiability is guaranteed, for example, by (10.2.3)). Suppose that the claim amount density is given by

$$H'(y) = \sum_{k=1}^{r} q_k \frac{\beta(\beta y)^{k-1}e^{-\beta y}}{(k-1)!}, \quad y \geq 0, \tag{10.3.1}$$

a mixture of Erlangs density. Then from example 2.2.1, the equilibrium density of $H(y)$ is

$$H_1'(y) = \frac{\overline{H}(y)}{E(Y)} = \sum_{k=1}^{r} q_k^* \frac{\beta(\beta y)^{k-1}e^{-\beta y}}{(k-1)!}, \quad y \geq 0 \tag{10.3.2}$$

where

$$q_k^* = \frac{\sum_{j=k}^{r} q_j}{\sum_{j=1}^{r} j q_j}, \quad k = 1, 2, \cdots, r. \tag{10.3.3}$$

In what follows, the discrete compound geometric distribution $\{c_0, c_1, c_2, \cdots\}$ with probability generating function

$$C(z) = \sum_{n=0}^{\infty} c_n z^n = \frac{\frac{\theta}{1+\theta}}{1 - \frac{1}{1+\theta}\sum_{k=1}^{r} q_k^* z^k} \tag{10.3.4}$$

will be needed. As mentioned in example 9.1.1 and section 7.2, the coefficients $\{c_0, c_1, c_2, \cdots\}$ may sometimes be obtained analytically by expansion of (10.3.4) as a power series in z. This is the case when $r = 2$, and this special case will be examined further later. Alternatively, recursive calculation is possible using, from (9.1.8),

$$c_n = \frac{1}{1+\theta} \sum_{k=1}^{\min(n,r)} q_k^* c_{n-k}, \quad n = 1, 2, \cdots, \tag{10.3.5}$$

beginning with $c_0 = \theta/(1+\theta)$.

The following theorem which shows that the conditional distribution of the deficit is a mixture of Erlangs may now be proved.

Theorem 10.3.1 If the claim amount distribution is a mixture of Erlangs whose density is given by (10.3.1), then

$$g_u(y) = \sum_{k=1}^{r} \overline{q}_k(u) \frac{\beta(\beta y)^{k-1} e^{-\beta y}}{(k-1)!}, \quad y \geq 0, \tag{10.3.6}$$

where $\{\overline{q}_1(u), \overline{q}_2(u), \cdots, \overline{q}_r(u)\}$ is a probability distribution satisfying

$$\overline{q}_k(u) = \frac{\sum_{j=k}^{r} q_j^* \tau_{j-k}(\beta u)}{\sum_{j=1}^{r} q_j^* \sum_{m=0}^{j-1} \tau_m(\beta u)}, \quad k = 1, 2, \cdots, r, \tag{10.3.7}$$

where

$$\tau_m(x) = \sum_{n=0}^{\infty} c_n \frac{x^{n+m}}{(n+m)!}, \quad m = 0, 1, 2, \cdots, r-1. \tag{10.3.8}$$

Proof: First, differentiation of the mixture representation (10.2.3) yields

$$g_u(y) = \frac{\int_0^u h_{1,u-x}(y)\overline{H}_1(u-x)\{-\psi'(x)\}dx + \frac{\theta}{1+\theta} h_{1,u}(y)\overline{H}_1(u)}{\int_0^u \overline{H}_1(u-x)\{-\psi'(x)\}dx + \frac{\theta}{1+\theta}\overline{H}_1(u)} \tag{10.3.9}$$

where

$$h_{1,x}(y) = \frac{d}{dy}H_{1,x}(y) = \frac{H_1'(x+y)}{\overline{H}_1(x)} \tag{10.3.10}$$

is the density associated with $H_{1,x}(y)$. Since $h_{1,x}(y)$ is the residual lifetime distribution associated with the equilibrium distribution $H_1(y)$, it follows immediately from (10.3.2) and example 2.3.1 that

$$h_{1,x}(y) = \sum_{k=1}^{r} q_k^*(x) \frac{\beta(\beta y)^{k-1} e^{-\beta y}}{(k-1)!}, \quad y \geq 0 \tag{10.3.11}$$

where

$$q_k^*(x) = \frac{\sum_{j=k}^{r} q_j^* \frac{(\beta x)^{j-k}}{(j-k)!}}{\sum_{j=0}^{r-1} \overline{Q}_j^* \frac{(\beta x)^j}{j!}}, \quad k = 1, 2, \cdots, r, \tag{10.3.12}$$

with $\overline{Q}_j^* = \sum_{m=j+1}^{r} q_m^*$. Thus, substitution of (10.3.11) into (10.3.9) yields

$$g_u(y) = \sum_{k=1}^{r} \overline{q}_k(u) \frac{\beta(\beta y)^{k-1} e^{-\beta y}}{(k-1)!}$$

where

$$\overline{q}_k(u) = \frac{\int_0^u q_k^*(u-x)\overline{H}_1(u-x)\{-\psi'(x)\}dx + \frac{\theta}{1+\theta} q_k^*(u)\overline{H}_1(u)}{\int_0^u \overline{H}_1(u-x)\{-\psi'(x)\}dx + \frac{\theta}{1+\theta}\overline{H}_1(u)}. \tag{10.3.13}$$

Thus, (10.3.6) holds, and it remains to demonstrate that $\{\overline{q}_1(u), \overline{q}_2(u), \cdots, \overline{q}_r(u)\}$ is a probability distribution satisfying (10.3.7). Clearly, it follows from (10.3.13) that $\overline{q}_k(u) \geq 0$, and from (10.3.6),

$$1 = \int_0^\infty g_u(y)dy = \sum_{k=1}^{r} \overline{q}_k(u) \int_0^\infty \frac{\beta(\beta y)^{k-1} e^{-\beta y}}{(k-1)!} dy = \sum_{k=1}^{r} \overline{q}_k(u),$$

which implies that $\{\overline{q}_1(u), \overline{q}_2(u), \cdots, \overline{q}_r(u)\}$ is a probability distribution. It follows from (2.1.13) that $\overline{H}_1(x) = e^{-\beta x} \sum_{k=0}^{r-1} \overline{Q}_k^* \frac{(\beta x)^k}{k!}$, and thus, using (10.3.12),

$$q_k^*(x)\overline{H}_1(x) = e^{-\beta x} \sum_{j=k}^{r} q_j^* \frac{(\beta x)^{j-k}}{(j-k)!}. \tag{10.3.14}$$

Now, from (10.1.8) and example 7.1.1,

$$\psi(x) = e^{-\beta x} \sum_{j=0}^{\infty} \overline{C}_j \frac{(\beta x)^j}{j!} \tag{10.3.15}$$

where $\overline{C}_j = \sum_{k=j+1}^{\infty} c_j$. Thus, differentiation of (10.3.15) yields

$$-\psi'(x) = \sum_{i=1}^{\infty} c_i \frac{\beta(\beta x)^{i-1} e^{-\beta x}}{(i-1)!}. \tag{10.3.16}$$

Therefore,

$$\begin{aligned}
&\int_0^u q_k^*(u-x)\overline{H}_1(u-x)\{-\psi'(x)\}dx + \frac{\theta}{1+\theta} q_k^*(u)\overline{H}_1(u) \\
= \;& e^{-\beta u} \int_0^u \left\{ \sum_{j=k}^{r} q_j^* \frac{\beta^{j-k}}{(j-k)!}(u-x)^{j-k} \right\} \left\{ \sum_{i=1}^{\infty} c_i \frac{\beta^i x^{i-1}}{(i-1)!} \right\} dx \\
+ \;& c_0 e^{-\beta u} \sum_{j=k}^{r} q_j^* \frac{(\beta u)^{j-k}}{(j-k)!} \\
= \;& e^{-\beta u} \sum_{i=1}^{\infty} \sum_{j=k}^{r} c_i q_j^* \frac{\beta^{i+j-k}}{(j-k)!(i-1)!} \int_0^u (u-x)^{j-k} x^{i-1} dx \\
+ \;& c_0 e^{-\beta u} \sum_{j=k}^{r} q_j^* \frac{(\beta u)^{j-k}}{(j-k)!}.
\end{aligned}$$

Since

$$\int_0^u (u-x)^{j-k} x^{i-1} dx = u^{i+j-k} \int_0^1 (1-t)^{j-k} t^{i-1} dt = u^{i+j-k} \frac{(j-k)!(i-1)!}{(i+j-k)!},$$

one has

$$\begin{aligned}
&\int_0^u q_k^*(u-x)\overline{H}_1(u-x)\{-\psi'(x)\}dx + \frac{\theta}{1+\theta} q_k^*(u)\overline{H}_1(u) \\
= \;& e^{-\beta u} \sum_{i=1}^{\infty} c_i \sum_{j=k}^{r} q_j^* \frac{(\beta u)^{i+j-k}}{(i+j-k)!} + c_0 e^{-\beta u} \sum_{j=k}^{r} q_j^* \frac{(\beta u)^{j-k}}{(j-k)!} \\
= \;& e^{-\beta u} \sum_{i=0}^{\infty} c_i \sum_{j=k}^{r} q_j^* \frac{(\beta u)^{i+j-k}}{(i+j-k)!}.
\end{aligned}$$

That is, using (10.3.8),

$$\int_0^u q_k^*(u-x)\overline{H}_1(u-x)\{-\psi'(x)\}dx = e^{-\beta u} \sum_{j=k}^{r} q_j^* \tau_{j-k}(\beta u).$$

Consequently, it follows from (10.3.13) that

$$\overline{q}_k(u) = Me^{-\beta u}\sum_{j=k}^{r} q_j^* \tau_{j-k}(\beta u),$$

where M is a normalizing constant satisfying

$$1 = \sum_{k=1}^{r} \overline{q}_k(u) = Me^{-\beta u}\sum_{k=1}^{r}\sum_{j=k}^{r} q_j^* \tau_{j-k}(\beta u) = Me^{-\beta u}\sum_{j=1}^{r} q_j^* \sum_{k=1}^{j} \tau_{j-k}(\beta u).$$

Thus, $Me^{-\beta u} = \{\sum_{j=1}^{r} q_j^* \sum_{m=0}^{j-1} \tau_m(\beta u)\}^{-1}$, and the theorem is proved. □

Evidently, theorem 10.3.1 expresses the conditional distribution of the deficit as a different mixture of the same Erlangs as the claim amount distribution. Also, it is clear from (10.3.8) that $\tau_0(0) = c_0$, and $\tau_m(0) = 0$ for $m = 1, 2, \cdots, r-1$. Thus, from (10.3.7) it follows that $\overline{q}_k(0) = q_k^*$ and thus (10.3.6) reduces to $g_0(y) = H_1'(y)$ from (10.3.2), consistent with (10.2.1). Turning to the case $u = \infty$, note that (10.2.12) is of the same form as (9.2.5), but with ρ replaced by $-\kappa$ and H replaced by H_1. Thus, by analogy with example 9.2.1, it is straightforward to show that, if the claim amount density is given by (10.3.1), then the density corresponding to (10.2.11) is

$$g_\infty(y) = G_\infty'(y) = \sum_{k=1}^{r} \overline{q}_k(\infty)\frac{\beta(\beta y)^{k-1}e^{-\beta y}}{(k-1)!}, \quad y \geq 0, \tag{10.3.17}$$

where

$$\overline{q}_k(\infty) = \frac{\sum_{j=k}^{r} q_j^* (\frac{\beta}{\beta-\kappa})^{j-k}}{\sum_{j=1}^{r} q_j^* \sum_{i=0}^{j-1} (\frac{\beta}{\beta-\kappa})^{i}}, \quad k = 1, 2, \cdots, r. \tag{10.3.18}$$

We remark that little simplification results for the E_r (Erlang) distribution with $q_r = 1$, since $q_k^* = 1/r$ in this case.

When $r = 2$, closed form expressions for the weights $\overline{q}_1(u)$ and $\overline{q}_2(u) = 1 - \overline{q}_1(u)$ are obtainable, as shown in the following example.

Example 10.3.1 The case with r=2

When $r = 2$, (10.3.4) becomes

$$C(z) = \frac{\frac{\theta}{1+\theta}}{1 - \frac{1}{1+\theta}(q_1^* z + q_2^* z^2)} = \frac{-\frac{\theta}{q_2^*}}{z^2 + \frac{q_1^*}{q_2^*} z - \frac{1+\theta}{q_2^*}}.$$

Let

$$r_1 = -\frac{q_1^* + \sqrt{(q_1^*)^2 + 4q_2^*(1+\theta)}}{2q_2^*}$$

and

$$r_2 = \frac{-q_1^* + \sqrt{(q_1^*)^2 + 4q_2^*(1+\theta)}}{2q_2^*}.$$

Since $q_2^* = 1 - q_1^*$, it follows that $\sqrt{(q_1^*)^2 + 4q_2^*(1+\theta)} = \sqrt{(2-q_1^*)^2 + 4q_2^*\theta} > 2 - q_1^*$, from which it is clear that the two roots satisfy $r_1 < -1/q_2^*$ and $r_2 > 1$. Then

$$C(z) = \frac{-\theta/q_2^*}{(z-r_1)(z-r_2)}$$

and a partial fraction expansion yields

$$C(z) = \frac{\theta}{q_2^*(r_2 - r_1)}\left\{\frac{1}{r_2 - z} - \frac{1}{r_1 - z}\right\}.$$

Expansion as a power series in z yields

$$C(z) = \frac{\theta}{q_2^*(r_2 - r_1)}\left\{\sum_{n=0}^{\infty}(\frac{1}{r_2})^{n+1}z^n - \sum_{n=0}^{\infty}(\frac{1}{r_1})^{n+1}z^n\right\}$$

from which it follows that

$$c_n = \frac{\theta}{q_2^*(r_2 - r_1)}\left\{(\frac{1}{r_2})^{n+1} - (\frac{1}{r_1})^{n+1}\right\}, \quad n = 0, 1, 2, \cdots.$$

Substitution into (10.3.8) yields, with $m = 0$,

$$\tau_0(x) = \frac{\theta}{q_2^*(r_2 - r_1)}\left\{\frac{1}{r_2}e^{\frac{x}{r_2}} - \frac{1}{r_1}e^{\frac{x}{r_1}}\right\},$$

and with $m = 1$,

$$\tau_1(x) = \frac{\theta}{q_2^*(r_2 - r_1)}\left\{e^{\frac{x}{r_2}} - e^{\frac{x}{r_1}}\right\}.$$

Thus, from (10.3.7), after a little algebra

$$\overline{q}_2(u) = \frac{q_2^*\tau_0(\beta u)}{\tau_0(\beta u) + q_2^*\tau_1(\beta u)} = \frac{r_2 q_2^* e^{\frac{\beta u}{r_1}} - r_1 q_2^* e^{\frac{\beta u}{r_2}}}{r_2(1 + r_1 q_2^*)e^{\frac{\beta u}{r_1}} - r_1(1 + r_2 q_2^*)e^{\frac{\beta u}{r_2}}},$$

and obviously, $\overline{q}_1(u) = 1 - \overline{q}_2(u)$. □

Egídio dos Reis (1993) considered the E_2 case with $q_2 = 1$ which implies that $q_2^* = 1/2$.

10.4 General Erlang mixtures

The representation of the conditional distribution of the severity as a different mixture of the same Erlangs as the claim amount distribution is not restricted to those with the same scale parameter. The difficulty is that the mixing weights are more complex due to the fact that the ruin probability $\psi(u)$ does not admit a simple analytic form in general. Consider the claim amount density

$$\frac{d}{dy}H(y) = \sum_{i=1}^{n}\sum_{k=1}^{r} q_{ik}\frac{\beta_i^k y^{k-1}e^{-\beta_i y}}{(k-1)!}, \quad y \geq 0, \tag{10.4.1}$$

where $\sum_{i=1}^{n}\sum_{k=1}^{r} q_{ik} = 1$. It is not always necessary that q_{ik} be nonnegative for all i and k, as in the following example (e.g., Gerber, Goovaerts, and Kaas, 1987).

Example 10.4.1 Combinations of exponentials
Suppose that $r = 1$ in (10.4.1), which therefore becomes

$$\frac{d}{dy}H(y) = \sum_{i=1}^{n} q_i\beta_i e^{-\beta_i y}, \quad y \geq 0, \tag{10.4.2}$$

where $\sum_{i=1}^{n} q_i = 1$, and we assume without loss of generality that the β_i are all distinct. As in example 2.2.3, one has $E(Y) = \sum_{i=1}^{n} q_i/\beta_i$ and

$$\overline{H}_1(y) = \sum_{i=1}^{n} q_i^* e^{-\beta_i y}, \quad y \geq 0, \tag{10.4.3}$$

where

$$q_i^* = \frac{q_i}{\beta_i E(Y)}; \quad i = 1, 2, \cdots, n. \tag{10.4.4}$$

The residual density (10.3.10) becomes

$$\begin{aligned} h_{1,x}(y) &= \frac{\overline{H}(x+y)}{E(Y)\overline{H}_1(x)} = \frac{\sum_{i=1}^{n} q_i e^{-\beta_i(x+y)}}{\sum_{k=1}^{n} \frac{q_k}{\beta_k} e^{-\beta_k x}} \\ &= \sum_{i=1}^{n} q_i^*(x)\beta_i e^{-\beta_i y}, \end{aligned} \tag{10.4.5}$$

where

$$q_i^*(x) = \frac{\frac{q_i}{\beta_i}e^{-\beta_i x}}{\sum_{k=1}^{n} \frac{q_k}{\beta_k} e^{-\beta_k x}}; \quad i = 1, 2, \cdots, n. \tag{10.4.6}$$

Thus, (10.4.5) is also a combination of exponentials. Substitution of (10.4.5) into (10.3.9) yields the severity density, namely

$$g_u(y) = \sum_{i=1}^{n} \bar{q}_i(u)\beta_i e^{-\beta_i y}, \quad y \geq 0, \tag{10.4.7}$$

where

$$\bar{q}_i(u) = \frac{\frac{\theta}{1+\theta} q_i^*(u)\overline{H}_1(u) + \int_0^u q_i^*(u-x)\overline{H}_1(u-x)\{-\psi'(x)\}dx}{\theta\psi(u)}, \tag{10.4.8}$$

using (10.2.4). Integration of (10.4.7) with respect to y over $(0, \infty)$ reveals that $\sum_{i=1}^{n} \bar{q}_i(u) = 1$, and so the severity density (10.4.7) is still a combination of exponentials. It remains to simplify $\bar{q}_i(u)$. First note that (10.4.3) and (10.4.6) imply that

$$q_i^*(x)\overline{H}_1(x) = q_i^* e^{-\beta_i x}, \quad x \geq 0. \tag{10.4.9}$$

Also, one has (e.g. Gerber, Goovaerts, and Kaas, 1987)

$$\psi(x) = \sum_{k=1}^{n} C_k e^{-R_k x}, \quad x \geq 0, \tag{10.4.10}$$

where $\{C_k; k = 1, 2, \cdots, n\}$ are constants, and $\{R_k; k = 1, 2, \cdots, n\}$ are the n distinct roots of the Lundberg equation $1 + \theta = \int_0^\infty e^{R_k y} dy$, that is

$$\sum_{i=1}^{n} \frac{q_i^* \beta_i}{\beta_i - R_k} = 1 + \theta.$$

Since $\sum_{i=1}^{n} q_i^* = 1$, this implies that the R_k satisfy

$$R_k \sum_{i=1}^{n} \frac{q_i^*}{\beta_i - R_k} = \theta; \quad k = 1, 2, \cdots, n. \tag{10.4.11}$$

It is clear from (10.4.9) and (10.4.10) that (10.4.8) is straightforward to evaluate. However, to obtain a simplified expression, we need to derive an identity. Substitution of (10.4.3) and (10.4.10) into (10.2.4) yields

$$\begin{aligned}
&\theta\sum_{k=1}^{n} C_k e^{-R_k u} \\
&= \frac{\theta}{1+\theta}\sum_{i=1}^{n} q_i^* e^{-\beta_i u} + \int_0^u \left\{\sum_{k=1}^{n} C_k R_k e^{-R_k x}\right\}\left\{\sum_{i=1}^{n} q_i^* e^{-\beta_i(u-x)}\right\} dx \\
&= \frac{\theta}{1+\theta}\sum_{i=1}^{n} q_i^* e^{-\beta_i u} + \sum_{k=1}^{n} C_k R_k \sum_{i=1}^{n} q_i^* \frac{e^{-R_k u} - e^{-\beta_i u}}{\beta_i - R_k}.
\end{aligned}$$

This yields, upon rearranging,

$$\sum_{k=1}^{n} C_k e^{-R_k u}\left\{\theta - R_k \sum_{i=1}^{n} \frac{q_i^*}{\beta_i - R_k}\right\} = \sum_{i=1}^{n} q_i^* e^{-\beta_i u}\left\{\frac{\theta}{1+\theta} - \sum_{k=1}^{n} \frac{C_k R_k}{\beta_i - R_k}\right\}$$

and since the left hand side is 0, one must

$$\sum_{i=1}^{n} q_i^* \left\{\frac{\theta}{1+\theta} - \sum_{k=1}^{n} \frac{C_k R_k}{\beta_i - R_k}\right\} e^{-\beta_i u} = 0. \tag{10.4.12}$$

Since (10.4.12) holds for all u and the β_i are distinct, the coefficients on the left hand side of (10.4.12) must be 0, i.e.

$$\sum_{k=1}^{n} \frac{C_k R_k}{\beta_i - R_k} = \frac{\theta}{1+\theta}; \quad i = 1, 2, \cdots, n. \tag{10.4.13}$$

The numerator of (10.4.8) may be evaluated using (10.4.9), (10.4.10), and simplified using (10.4.13). One has

$$\begin{aligned}
&\frac{\theta}{1+\theta} q_i^*(u)\overline{H}_1(u) + \int_0^u q_i^*(u-x)\overline{H}_1(u-x)\{-\psi'(x)\}dx \\
&= \frac{\theta}{1+\theta} q_i^* e^{-\beta_i u} + \int_0^u q_i^* e^{-\beta_i(u-x)} \left\{\sum_{k=1}^{n} C_k R_k e^{-R_k x}\right\} dx \\
&= \frac{\theta}{1+\theta} q_i^* e^{-\beta_i u} + \sum_{k=1}^{n} C_k R_k \frac{e^{-R_k u} - e^{-\beta_i u}}{\beta_i - R_k} \\
&= q_i^* \left\{\left(\frac{\theta}{1+\theta} - \sum_{k=1}^{n} \frac{C_k R_k}{\beta_i - R_k}\right) e^{-\beta_i u} + \sum_{k=1}^{n} \frac{C_k R_k}{\beta_i - R_k} e^{-R_k u}\right\} \\
&= q_i^* \sum_{k=1}^{n} \frac{C_k R_k}{\beta_i - R_k} e^{-R_k u}.
\end{aligned}$$

Thus, using (10.4.10), (10.4.11) becomes

$$\bar{q}_i(u) = \frac{q_i^* \sum_{k=1}^{n} \frac{C_k R_k}{\beta_i - R_k} e^{-R_k u}}{\theta \sum_{k=1}^{n} C_k e^{-R_k u}}; \quad i = 1, 2, \cdots, n. \tag{10.4.14}$$

□

We now consider the general density (10.4.1), showing that the severity density is still of the same form, but with different weights.

Theorem 10.4.1 If the claim amount distribution is given by (10.4.1) with $q_{ik} \geq 0$, then

$$g_u(y) = \sum_{i=1}^{n} \sum_{k=1}^{r} \overline{q}_{ik}(u) \frac{\beta_i^k y^{k-1} e^{-\beta_i y}}{(k-1)!}, \quad y \geq 0, \tag{10.4.15}$$

where $\overline{q}_{ik}(u) \geq 0$, $\sum_{i=1}^{n} \sum_{k=1}^{r} \overline{q}_{ik}(u) = 1$, and

$$\overline{q}_{ik}(u) = \frac{1}{\theta \psi(u)} \int_0^u \alpha_{ik}(u-x)\{-\psi'(x)\}dx + \frac{1}{1+\theta} \frac{\alpha_{ik}(u)}{\psi(u)}, \tag{10.4.16}$$

with

$$\alpha_{ik}(x) = e^{-\beta_i x} \sum_{j=k}^{r} w_{ij} \frac{(\beta_i x)^{j-k}}{(j-k)!}, \tag{10.4.17}$$

and

$$w_{ij} = \frac{\frac{1}{\beta_i} \sum_{k=j}^{r} q_{ik}}{\sum_{m=1}^{n} \frac{1}{\beta_m} \sum_{k=1}^{r} k q_{mk}}. \tag{10.4.18}$$

Proof: To prove that (10.4.15) holds with $\overline{q}_{ik}(u) \geq 0$ and $\sum_{i=1}^{n} \sum_{k=1}^{r} \overline{q}_{ik}(u) = 1$, it suffices to demonstrate that

$$h_{1,x}(y) = \sum_{i=1}^{n} \sum_{k=1}^{r} q_{ik}^*(x) \frac{\beta_i^k y^{k-1} e^{-\beta_i y}}{(k-1)!} \tag{10.4.19}$$

where $q_{ik}^*(x) \geq 0$, since (10.4.19) implies that

$$1 = \int_0^\infty h_{1,x}(y)dy = \sum_{i=1}^{n} \sum_{k=1}^{r} q_{ik}^*(x) \int_0^\infty \frac{\beta_i^k y^{k-1} e^{-\beta_i y}}{(k-1)!} dy = \sum_{i=1}^{n} \sum_{k=1}^{r} q_{ik}^*(x),$$

and thus from (10.3.9), (10.4.15)

$$\overline{q}_{ik}(u) = \frac{\int_0^u q_{ik}^*(u-x)\overline{H}_1(u-x)\{-\psi'(x)\}dx + \frac{\theta}{1+\theta}q_{ik}^*(u)\overline{H}_1(u)}{\int_0^u \overline{H}_1(u-x)\{-\psi'(x)\}dx + \frac{\theta}{1+\theta}\overline{H}_1(u)}. \tag{10.4.20}$$

Thus, it must be demonstrated that (10.4.19) holds, and that (10.4.20) is equal to (10.4.16). For notational convenience, let $q_i = \sum_{k=1}^{r} q_{ik}$, $\gamma_{ik} = q_{ik}/q_i$, and define $H^i(y) = 1 - \overline{H}^i(y)$ by $\frac{d}{dy}H^i(y) = \sum_{k=1}^{r}\gamma_{ik}\frac{\beta_i^k y^{k-1}e^{-\beta_i y}}{(k-1)!}$. Then (10.4.1) may be expressed as $\frac{d}{dy}H(y) = \sum_{i=1}^{n} q_i \frac{d}{dy}H^i(y)$, and integrating yields

$$\overline{H}(y) = \sum_{i=1}^{n} q_i \overline{H}^i(y). \tag{10.4.21}$$

Now, $\int_0^\infty \overline{H}^i(y)dy = \frac{1}{\beta_i}\sum_{k=1}^{r} k\gamma_{ik}$, and define $H_1^i(y) = 1 - \overline{H}_1^i(y)$ to be the equilibrium df of $H^i(y)$. Then $E(Y) = \int_0^\infty \overline{H}(y)dy = \sum_{i=1}^{n}\frac{q_i}{\beta_i}\sum_{k=1}^{r} k\gamma_{ik} = \sum_{i=1}^{n}\frac{1}{\beta_i}\sum_{k=1}^{r} kq_{ik}$. The equilibrium density $H_1'(y)$ may then be expressed as

$$H_1'(y) = \frac{\overline{H}(y)}{E(Y)} = \frac{\sum_{i=1}^{n} q_i \overline{H}^i(y)}{\sum_{i=1}^{n}\frac{1}{\beta_i}\sum_{k=1}^{r} kq_{ik}} = \frac{\sum_{i=1}^{n} q_i\left(\frac{1}{\beta_i}\sum_{k=1}^{r} k\gamma_{ik}\right)\frac{d}{dy}H_1^i(y)}{\sum_{i=1}^{n}\frac{1}{\beta_i}\sum_{k=1}^{r} kq_{ik}},$$

i.e. $H_1'(y) = \sum_{i=1}^{n}\eta_i \frac{d}{dy}H_1^i(y)$ where

$$\eta_i = \frac{\frac{1}{\beta_i}\sum_{m=1}^{r} mq_{im}}{\sum_{m=1}^{n}\frac{1}{\beta_m}\sum_{k=1}^{r} kq_{mk}}. \tag{10.4.22}$$

Thus

$$\overline{H}_1(y) = \sum_{i=1}^{n}\eta_i \overline{H}_1^i(y). \tag{10.4.23}$$

Let $h^i_{1,x}(y) = \frac{d}{dy}H^i_1(x+y)/\overline{H}^i_1(x)$ be the residual lifetime density associated with the equilibrium df $H^i_1(x)$. Then

$$h_{1,x}(y) = \frac{\frac{d}{dy}H_1(x+y)}{\overline{H}_1(x)} = \frac{\sum_{i=1}^{n}\eta_i \frac{d}{dy}H^i_1(x+y)}{\overline{H}_1(x)} = \frac{\sum_{i=1}^{n}\eta_i \overline{H}^i_1(x) h^i_{1,x}(y)}{\overline{H}_1(x)}.$$

That is,

$$h_{1,x}(y) = \sum_{i=1}^{n} \rho_i(x) h^i_{1,x}(y) \tag{10.4.24}$$

where

$$\rho_i(x) = \frac{\eta_i \overline{H}^i_1(x)}{\overline{H}_1(x)}. \tag{10.4.25}$$

Now, $h^i_{1,x}(y)$ is the residual lifetime pdf associated with the equilibrium df $H^i_1(y)$. By example (2.2.1),

$$\frac{d}{dy}H^i_1(y) = \sum_{k=1}^{r} \gamma^*_{ik} \frac{\beta_i^k y^{k-1} e^{-\beta_i y}}{(k-1)!},$$

where

$$\gamma^*_{ik} = \frac{\sum_{m=k}^{r} \gamma_{im}}{\sum_{m=1}^{r} m\gamma_{im}} = \frac{\sum_{m=k}^{r} q_{im}}{\sum_{m=1}^{r} m q_{im}}. \tag{10.4.26}$$

Thus, by example 2.3.1, it follows that

$$h^i_{1,x}(y) = \sum_{k=1}^{r} \gamma^*_{ik}(x) \frac{\beta_i^k y^{k-1} e^{-\beta_i y}}{(k-1)!} \tag{10.4.27}$$

where, from (2.3.10) and (2.1.13),

$$\gamma^*_{ik}(x) = \frac{e^{-\beta_i x} \sum_{j=k}^{r} \gamma^*_{ij} \frac{(\beta_i x)^{j-k}}{(j-k)!}}{\overline{H}^i_1(x)}. \tag{10.4.28}$$

Therefore, combining (10.4.24) and (10.4.27),

$$\begin{aligned} h_{1,x}(y) &= \sum_{i=1}^{n} \rho_i(x) \sum_{k=1}^{r} \gamma^*_{ik}(x) \frac{\beta_i^k y^{k-1} e^{-\beta_i y}}{(k-1)!} \\ &= \sum_{i=1}^{n} \sum_{k=1}^{r} \rho_i(x) \gamma^*_{ik}(x) \frac{\beta_i^k y^{k-1} e^{-\beta_i y}}{(k-1)!}, \end{aligned}$$

i.e. (10.4.19) holds with

$$q^*_{ik}(x) = \rho_i(x)\gamma^*_{ik}(x). \tag{10.4.29}$$

Now, from (10.4.25) and (10.4.28),

$$\begin{aligned} q^*_{ik}(x)\overline{H}_1(x) &= \rho_i(x)\overline{H}_1(x)\gamma^*_{ik}(x) \\ &= \eta_i\overline{H}^i_1(x)\gamma^*_{ik}(x) \\ &= e^{-\beta_i x}\sum_{j=k}^{r}(\eta_i\gamma^*_{ij})\frac{(\beta_i x)^{j-k}}{(j-k)!} \end{aligned}$$

and from (10.4.22) and (10.4.26),

$$\eta_i\gamma^*_{ij} = \frac{\frac{1}{\beta_i}\sum_{m=j}^{r} q_{im}}{\sum_{m=1}^{n}\frac{1}{\beta_m}\sum_{k=1}^{r} kq_{mk}} = w_{ij}$$

from (10.4.18). Thus, from (10.4.17), $q^*_{ik}(x)\overline{H}_1(x) = \alpha_{ik}(x)$. Since (10.2.4) holds, (10.4.16) follows from (10.4.20). □

We remark that (10.4.16) may be expressed as

$$\overline{q}_{ik}(u) = \frac{\alpha_{ik}(u)}{(1+\theta)\psi(u)} + \frac{1}{\theta\psi(u)}\int_0^u \alpha_{ik}(u-x)\{-\psi'(x)\}dx. \tag{10.4.30}$$

As mentioned, evaluation of $\overline{q}_{ik}(u)$ can be difficult unless a relatively simple analytic form exists for $\psi(x)$, as when $n = 1$ (mixture of Erlangs with the same scale parameter) or $r = 1$ (mixture of exponentials).

If $\psi(u)$ is complicated (i.e. when n and r are large), one approach to approximation would be to approximate the weights $\overline{q}_{ik}(u)$ by replacement of the ruin probability $\psi(x)$ by the Tijms approximation (see example 9.3.2)

$$\psi_T(x) = C_1 e^{-\kappa_1 x} + Ce^{-\kappa x}, \quad x \geq 0, \tag{10.4.31}$$

where $C_1 = (\frac{1}{1+\theta} - C)$. That is, approximate $\overline{q}_{ik}(u)$ by $\overline{q}_{ik,T}(u)$, where, using (10.4.20),

$$\overline{q}_{ik,T}(u) = \frac{\int_0^u \alpha_{ik}(u-x)\{-\psi'_T(x)\}dx + \frac{\theta}{1+\theta}\alpha_{ik}(u)}{\int_0^u \overline{H}_1(u-x)\{-\psi'_T(x)\}dx + \frac{\theta}{1+\theta}\overline{H}_1(u)}. \tag{10.4.32}$$

or

$$\overline{q}_{ik,T} = \frac{\int_0^u \alpha_{ik}(u-x)\{C_1\kappa_1 e^{-\kappa_1 x} + C\kappa e^{-\kappa x}\}dx + \frac{\theta}{1+\theta}\alpha_{ik}(u)}{\int_0^u \overline{H}_1(u-x)\{C_1\kappa_1 e^{-\kappa_1 x} + C\kappa e^{-\kappa x}\}dx + \frac{\theta}{1+\theta}\overline{H}_1(u)}. \tag{10.4.33}$$

The integrals in (10.4.33) are straightforward to evaluate using (10.4.17) and (10.4.23) but tedious, and we omit the details.

10.5 Further results

In this section, we briefly discuss approximations for the conditional distribution of the deficit $G_u(y)$, and the moments of the deficit $|U_T|$.

Equation (10.1.12) shows that the unconditional distribution of the deficit $G(u, y)$ may be expressed in terms of the probability of ruin $\psi(x)$. Thus an approximation to $G(u, y)$ and hence $G_u(y)$ may be obtained by replacement of $\psi(x)$ by the Tijms approximation $\psi_T(x)$ given in (10.4.31). Noting that $\psi(x)$ appears several times in (10.1.12), one may replace all or some of $\psi(x)$'s, which results in different approximations. As has been demonstrated in early chapters, $\psi(x)$ is a compound geometric tail and may be solved explicitly or may be approximated by other distributional tails. Thus, a proper way to approximate $G(u, y)$ and hence $G_u(y)$ is to replace $\psi(x)$ in the integral term of (10.1.12) by $\psi_T(x)$. In this case, the integral term becomes

$$\begin{aligned}
&\int_0^y \psi_T(u+y-x)dH_1(x) \\
= &\frac{C_1}{\kappa_1 E(Y)} e^{-\kappa_1(u+y)} \left\{ \int_0^y e^{\kappa_1 x} dH(x) + e^{\kappa_1 y}\overline{H}(y) - 1 \right\} \\
+ &\frac{C}{\kappa E(Y)} e^{-\kappa(u+y)} \left\{ \int_0^y e^{\kappa x} dH(x) + e^{\kappa y}\overline{H}(y) - 1 \right\} \qquad (10.5.1)
\end{aligned}$$

using integration by parts. See Willmot and Lin (1998) for further discussion of this idea.

Brown (1990) has shown that compound geometric df's are NWU. Thus, $\psi(x+y) \geq \psi(x)\psi(y)$ for $x \geq 0$ and $y \geq 0$. An upper bound for $G_u(y)$ is thus derived as a direct consequence of this observation.

Theorem 10.5.1

$$\overline{G}_u(y) \leq \frac{1+\theta}{\theta} \left\{ \frac{\psi(u+y)}{\psi(u)} - \psi(y) \right\}, \quad y \geq 0. \qquad (10.5.2)$$

Proof: It follows from the NWU property and (10.1.7) that

$$\begin{aligned}
\int_0^y \psi(u+y-x)dH_1(x) &\geq \psi(u)\int_0^y \psi(y-x)dH_1(x) \\
&= \psi(u)\{(1+\theta)\psi(y) - \overline{H}_1(y)\}.
\end{aligned}$$

Thus, from (10.1.12),

$$G(u, y) \geq H_1(y)\psi(u) + \frac{1+\theta}{\theta}\{\overline{H}_1(y)\psi(u) - \psi(u+y)\}$$

$$\begin{aligned}
&+ \frac{\psi(u)}{\theta}\{(1+\theta)\psi(y) - \overline{H}_1(y)\} \\
&= \{H_1(y) + \frac{1+\theta}{\theta}\overline{H}_1(y) - \frac{1}{\theta}\overline{H}_1(y)\}\psi(u) \\
&- \frac{1+\theta}{\theta}\{\psi(u+y) - \psi(u)\psi(y)\} \\
&= \psi(u) - \frac{1+\theta}{\theta}\{\psi(u+y) - \psi(u)\psi(y)\}.
\end{aligned}$$

Thus,

$$\overline{G}_u(y) = \frac{\psi(u) - G(u,y)}{\psi(u)} \leq \frac{1+\theta}{\theta}\left\{\frac{\psi(u+y)}{\psi(u)} - \psi(y)\right\}.$$

□

We now turn to the moments of the deficit at the time of ruin. Recall from section 10.1 that the expected discounted penalty $m(u) = E\{w(|U_T|)\, I(T < \infty)\}$ given in (10.1.1) satisfies the defective renewal equation (10.1.3). It is easy to see that if $w(x) = x$ one obtains the mean of the deficit $|U_T|$, and if $w(x) = x^k$ one obtains the k-th moment of the deficit. Thus, the mean of $|U_T|$ is also not difficult to obtain using the results obtained earlier, as in the following theorem.

Theorem 10.5.2

$$E\{|U_T| \mid T < \infty\} = \int_u^\infty \frac{\psi(x)}{\psi(u)}dx - \frac{E(Y^2)}{2\theta E(Y)}. \tag{10.5.3}$$

Proof: Let $w(x) = x$ in (10.1.1) to obtain from (10.1.3)

$$v(u) = \frac{\lambda}{c}\int_u^\infty \int_x^\infty (t-x)dH(t)dx,$$

and using (2.2.5),

$$v(u) = \frac{\lambda E(Y)}{c}\int_u^\infty \overline{H}_1(x)dx.$$

That is,

$$v(u) = \frac{1}{1+\theta}\int_u^\infty \overline{H}_1(x)dx. \tag{10.5.4}$$

Therefore, from (10.1.2), $m(u) = E\{|U_T|\ I(T < \infty)\}$ satisfies the defective renewal equation

$$m(u) = \frac{1}{1+\theta}\int_0^u m(u-x)dH_1(x) + \frac{1}{1+\theta}\int_u^\infty \overline{H}_1(x)dx. \tag{10.5.5}$$

Thus, from theorem 9.1.2 with $\phi = 1/(1+\theta)$ and $\overline{G}(x) = \psi(x)$, it follows that

$$m(u) = \frac{1}{\theta}\int_u^\infty \overline{H}_1(x)dx + \frac{1}{\theta}\int_0^u \psi(u-x)\overline{H}_1(x)dx - \frac{1}{\theta}\left\{\int_0^\infty \overline{H}_1(x)dx\right\}\psi(u).$$

But (9.2.42) holds, and from (2.2.3), it follows that

$$m(u) = \int_u^\infty \psi(x)dx - \frac{E(Y^2)}{2\theta E(Y)}\psi(u), \tag{10.5.6}$$

from which (10.5.3) follows. □

The result of theorem 10.5.2 suggests that (10.5.2) is likely to be a very good bound if θ is large. To see this, note that the compound geometric mean is

$$\int_0^\infty \psi(y)dy = \frac{E(Y^2)}{2\theta E(Y)}. \tag{10.5.7}$$

Thus, (10.5.3) may be expressed as

$$E\{|U_T| \mid T < \infty\} = \int_0^\infty \left\{\frac{\psi(y+u)}{\psi(u)} - \psi(y)\right\} dy. \tag{10.5.8}$$

But $E\{|U_T| \mid T < \infty\} = \int_0^\infty \overline{G}_u(y)dy$, and thus the inequality (10.5.2) implies that (by integrating)

$$E\{|U_T| \mid T < \infty\} \le (1 + \frac{1}{\theta})E\{|U_T| \mid T < \infty\},$$

which is equivalent to $1 \le 1 + \theta^{-1}$, and is obviously a tight inequality if θ is large.

As with the mean time to ruin in section 9.2, it is not difficult to evaluate the mean of $|U_T|$ when the claim amount distribution is a mixture of exponentials or a mixture of Erlang distributions. Furthermore, a Tijms approximation to the mean is easily established as in example 9.2.2. Substitution of $\psi_T(x) = C_1 e^{-\kappa_1 x} + Ce^{-\kappa x}$ into (10.5.3) yields

$$E\{|U_T| \mid T < \infty\} \approx \frac{\frac{C_1}{\kappa_1}e^{-\kappa_1 u} + \frac{C}{\kappa}e^{-\kappa u}}{C_1 e^{-\kappa_1 u} + Ce^{-\kappa u}} - \frac{E(Y^2)}{2\theta E(Y)}. \tag{10.5.9}$$

Higher moments of $|U_T|$ have been obtained by Lin and Willmot (2000). For notational convenience, define for $n = 2, 3, 4, \cdots$,

$$\begin{aligned} \delta_n(u) &= n\theta E(Y)\int_u^\infty (x-u)^{n-1}\psi(x)dx \\ &\quad - \sum_{j=0}^{n-2}\binom{n}{j}E(Y^{n-j})\int_u^\infty (x-u)^j\psi(x)dx. \end{aligned} \tag{10.5.10}$$

Then

$$E\{|U_T|^n \mid T < \infty\} = \frac{\delta_n(u)}{\theta E(Y)\psi(u)} - \frac{E(Y^{n+1})}{\theta(n+1)E(Y)}, \quad n = 2, 3, 4, \cdots. \tag{10.5.11}$$

It is not hard to see that (10.5.11) is straightforward to evaluate when the claim amount distribution is a mixture of exponentials or a mixture of Erlang distributions. Again, a Tijms approximation is easily obtained by substitution of $\psi_T(u)$ for $\psi(u)$ in (10.5.10).

11
Renewal risk processes

In this chapter we consider a generalization of the classical ruin theoretic model. The generalization involves replacement of the Poisson process of claims by a more general renewal process. The model may be formulated as a particular random walk, and as such also allows for interpretation in terms of the equilibrium waiting time distribution in the G/G/1 queue. These ideas are discussed in more detail in section 11.1, where various properties of the model are presented, including in particular Lundberg-type upper and lower bounds and a close relationship with the compound geometric distribution. Analytic complexities obviate the need for such bounds and approximations, and this approach has been followed by various authors such as Abate, Choudhury, and Whitt (1995). A more detailed analysis of this model may be found in Prabhu (1998), Cohen (1982), Resnick (1992), Ross (1996), or Feller (1971). See Grandell (1991, chapter 3), or Rolski et al (1999, section 6.5), for insurance applications.

The analysis of the model in terms of the compound geometric distribution alluded to above is complicated for an arbitrary distribution of interclaim/interarrival times, essentially because the compound geometric components are difficult to identify. In section 11.2 we demonstrate how they may be identified in the special case when the interclaim/interarrival distribution is from the so-called Coxian-2 (or Phase type-2) class of distributions. This identification allows for the derivation of various properties of the ruin/waiting time probabilities, as well as comparison with the classical model. Stronger results are obtained in section 11.3 in the further special case when the interclaim/interarrival distribution is that of the sum of two exponentially distributed random variables. In section 11.4, the delayed or modified renewal process is considered where the time to the first claim/arrival is allowed to have a different distribution. The ruin/waiting time probability is expressed as a function of that of the ordinary process of section 11.1, thus allowing for application of earlier results to the delayed model. The important special case of the delayed model, namely the equilibrium or stationary renewal process, is then considered.

11.1 General properties of the model

We shall begin with a stochastic formulation of the model, and then describe the ruin theoretic and queueing interpretations. Let $\{X_1, X_2, \cdots\}$ be a sequence of independent and identically distributed positive random variables with common df $A(x) = 1 - \overline{A}(x) = Pr(X \leq x)$, mean $E(X) = \int_0^\infty x dA(x)$, and Laplace-Stieltjes transform $\tilde{a}(s) = \int_0^\infty e^{-sx} dA(x)$. Next, let $\{Y_1, Y_2, \cdots\}$ be an independent and identically distributed sequence of positive random variables independent of $\{X_1, X_2, \cdots\}$ with common df $H(y) = 1 - \overline{H}(y) = Pr(Y \leq y)$, mean $E(Y) = \int_0^\infty y dH(y) < E(X)$, and Laplace-Stieltjes transform $\tilde{h}(s) = \int_0^\infty e^{-sy} dH(y)$. We also define $H_1(y) = 1 - \overline{H}_1(y) = \int_0^y \overline{H}(t)dt/E(Y)$ to be the equilibrium df of $H(y)$, and $\tilde{h}_1(s) = \int_0^\infty e^{-sy} dH_1(y)$.

Define $\{S_n = \sum_{k=1}^{n} (Y_k - X_k);\ n = 1, 2, \cdots\}$ to be a random walk, and the quantity of interest is

$$\psi(x) = Pr\left(\bigcup_{n=1}^{\infty} \{S_n > x\}\right), \quad x \geq 0. \tag{11.1.1}$$

Intuitively, $\psi(x) = Pr(S_n > x$ for some $n)$. We remark that $\psi(x) = 1,\ x \geq 0$, if $E(Y) \geq E(X)$. See Grimmett and Stirzaker (1992, section 11.5) or Rolski et al (1999, section 6.3.2), for example.

In connection with ruin theory, $\psi(x)$ is the probability of ultimate ruin beginning with initial surplus x, Y_k is the size of the k-th claim, and $X_k = cT_k$ where c is the premium rate per unit time and T_k is the time between the $(k-1)$-th and k-th claims. Then S_n is the net decrease in surplus up to and including the n-th claim. See Ross (1996, section 7.4) for details. Ruin probabilities in the classical model of section 7.3 are obtained when T_k (and thus X_k) is exponentially distributed. In keeping with this interpretation and by analogy with the classical model of section 7.3, we may refer to the inequality $E(Y) < E(X)$ as the 'positive loading condition'.

In a queueing context, $\psi(x)$ may be interpreted as the probability that the equilibrium actual waiting time in the G/G/1 queue exceeds x. In this case Y_k is the service time of the k-th customer and X_k the interarrival time between the $(k-1)$-th and k-th arrivals to the queue.

Explicit analysis of $\psi(x)$ is complicated in general, and we begin with an inequality obtained by Willmot (1996).

Theorem 11.1.1 Suppose that $B(x)$ is an absolutely continuous DFR df with failure rate $\mu_B(x) = -\frac{d}{dx}\ln \overline{B}(x)$ which satisfies $\mu_B(\infty) > 0$. If $B(x)$ also satisfies the generalized Lundberg condition

$$\tilde{a}\{\mu_B(\infty)\}E\{1/\overline{B}(Y)\} = 1, \tag{11.1.2}$$

then

$$\psi(x) \leq \sigma\overline{B}(x), \quad x \geq 0, \tag{11.1.3}$$

where

$$\frac{1}{\sigma} = \inf_{x\geq 0} \frac{\overline{B}(x)}{\overline{H}(x)} \int_x^\infty \{\overline{B}(y)\}^{-1} dH(y). \tag{11.1.4}$$

Proof: Define $\psi_0(x) = 0$ and $\psi_k(x) = Pr\left(\bigcup_{n=1}^{k}\{S_n > x\}\right)$ for $k = 1, 2, 3, \cdots$.

Then by conditioning on X_1 and Y_1, it follows by the law of total probability that

$$\psi_{k+1}(x) = \int_0^\infty \{\overline{H}(x+t) + \int_0^{x+t} \psi_k(x+t-y)dH(y)\}dA(t). \tag{11.1.5}$$

We shall now show inductively that $\psi_k(x) \leq \sigma\overline{B}(x)$ for $k = 1, 2, 3, \cdots$. Clearly, $\psi_0(x) = 0 \leq \sigma\overline{B}(x)$, and by the inductive hypothesis, $\psi_k(x) \leq \sigma\overline{B}(x)$. By (11.1.4), $\overline{H}(x) \leq \sigma\overline{B}(x)\int_x^\infty\{\overline{B}(y)\}^{-1}dH(y)$, and hence from (11.1.5),

$$\begin{aligned}
&\psi_{k+1}(x) \\
\leq\ & \int_0^\infty \left\{\sigma\overline{B}(x+t)\int_{x+t}^\infty \frac{dH(y)}{\overline{B}(y)} + \sigma\int_0^{x+t} \overline{B}(x+t-y)dH(y)\right\} dA(t) \\
\leq\ & \sigma\int_0^\infty \left\{\overline{B}(x+t)\int_{x+t}^\infty \frac{dH(y)}{\overline{B}(y)} + \overline{B}(x+t)\int_0^{x+t} \frac{dH(y)}{\overline{B}(y)}\right\} dA(t) \\
=\ & \sigma\int_0^\infty \overline{B}(x+t)\left\{\int_0^\infty \frac{dH(y)}{\overline{B}(y)}\right\} dA(t)
\end{aligned}$$

since $B(x)$ is DFR implies that $B(x)$ is NWU and hence $\overline{B}(x+t-y) \leq \overline{B}(x+t)/\overline{B}(y)$. But (11.1.2) implies that $\int_0^\infty dH(y)/\overline{B}(y) = 1/\tilde{a}\{\mu_B(\infty)\}$ and hence

$$\psi_{k+1}(x) \leq \frac{\sigma}{\tilde{a}\{\mu_B(\infty)\}}\int_0^\infty \overline{B}(x+t)dA(t).$$

Now, since $B(x)$ is DFR with $\mu_B(\infty) < \infty$, it follows that $B(x)$ is UWA which implies that $\overline{B}(x+t) \le \overline{B}(x)e^{-\mu_B(\infty)t}$, and substitution yields

$$\psi_{k+1}(x) \le \frac{\sigma}{\tilde{a}\{\mu_B(\infty)\}} \int_0^\infty \overline{B}(x)e^{-\mu_B(\infty)t} dA(t) = \sigma\overline{B}(x).$$

Finally, $\psi(x) = \lim_{k\to\infty} \psi_k(x) \le \sigma\overline{B}(x)$. □

In the special case $\overline{B}(x) = e^{-\kappa x}$, the generalized Lundberg condition (11.1.2) becomes the (ordinary) Lundberg condition, namely

$$\tilde{a}(\kappa)\tilde{h}(-\kappa) = 1, \tag{11.1.6}$$

and the plausibility of a solution $\kappa > 0$ follows from the convexity of $E(e^{tS_1}) = \tilde{a}(t)\tilde{h}(-t)$ for $t \ge 0$ together with $\frac{d}{dt}E(e^{tS_1})\big|_{t=0} = E(S_1) < 0$. See figure 11.1 for a graph of $E(e^{tS_1})$.

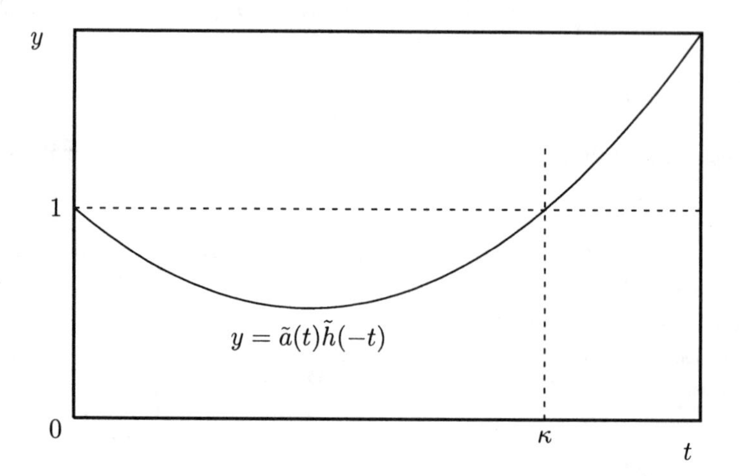

FIGURE 11.1. The graph of $y = \tilde{a}(t)\tilde{h}(-t)$

The following corollary to theorem 11.1.1 is a well-known result, given by Ross (1974). See also Rolski et al (1999, section 6.5.2) or Stoyan (1983, section 5.3).

Corollary 11.1.1 Suppose that $\kappa > 0$ satisfies the Lundberg condition (11.1.6). Then

$$\psi(x) \le \sigma_1 e^{-\kappa x}, \quad x \ge 0, \tag{11.1.7}$$

where σ_1 satisfies

$$\frac{1}{\sigma_1} = \inf_{x\ge 0} \frac{\int_x^\infty e^{\kappa y} dH(y)}{e^{\kappa x}\overline{H}(x)}. \tag{11.1.8}$$

Proof: Theorem 11.1.1 holds with $\overline{B}(x) = e^{-\kappa x}$. □

Clearly, $\sigma_1 \leq 1$, implying that $\psi(x) \leq e^{-\kappa x}$, $x \geq 0$.

The following lower bound is a dual to corollary 11.1.1, and is also given by Ross (1974), Rolski et al (1999, section 6.5.2) or Stoyan (1983, section 5.3). The present proof is essentially due to Wu (1996).

Theorem 11.1.2 Suppose that $\kappa > 0$ satisfies (11.1.6). Then

$$\psi(x) \geq \sigma_2 e^{-\kappa x}, \quad x \geq 0, \tag{11.1.9}$$

where

$$\frac{1}{\sigma_2} = \sup_{x \geq 0} \frac{\int_x^\infty e^{\kappa y} dH(y)}{e^{\kappa x}\overline{H}(x)}. \tag{11.1.10}$$

Proof: First consider a related renewal risk problem. Define the Esscher transformed df $H_\kappa^*(y)$ by

$$dH_\kappa^*(y) = \frac{e^{\kappa y} dH(y)}{\int_0^\infty e^{\kappa x} dH(x)}, \quad y \geq 0. \tag{11.1.11}$$

Similarly, let the df $A_\kappa(t)$ be defined by

$$dA_\kappa(t) = \frac{e^{-\kappa t} dA(t)}{\int_0^\infty e^{-\kappa x} dA(x)}, \quad t \geq 0. \tag{11.1.12}$$

Now, define $\psi_0^*(x) = 0$, and for $n = 0, 1, 2, \cdots$,

$$\psi_{n+1}^*(x) = \int_0^\infty \left\{ \overline{H}_\kappa^*(x+t) + \int_0^{x+t} \psi_n^*(x+t-y) dH_\kappa^*(y) \right\} dA_\kappa(t). \tag{11.1.13}$$

Next, consider $M_{S_1}(t) = \tilde{a}(t)\tilde{h}(-t)$. Since $M_{S_1}(0) = M_{S_1}(\kappa) = 1$ using (11.1.6), $M'_{S_1}(0) = E(Y_1) - E(X_1) < 0$, and $M''_{S_1}(t) > 0$ for $t \geq 0$, elementary analysis yields $M'_{S_1}(\kappa) > 0$. See figure 11.1 for example. But

$$M'_{S_1}(t) = \tilde{a}'(t)\tilde{h}(-t) - \tilde{a}(t)\tilde{h}'(-t)$$

and using (11.1.6),

$$\begin{aligned} M'_{S_1}(\kappa) &= \frac{\tilde{a}'(\kappa)\tilde{h}(-\kappa) - \tilde{a}(\kappa)\tilde{h}'(-\kappa)}{\tilde{a}(\kappa)\tilde{h}(-\kappa)} \\ &= \frac{\tilde{a}'(\kappa)}{\tilde{a}(\kappa)} - \frac{\tilde{h}'(-\kappa)}{\tilde{h}(-\kappa)} \\ &= -\int_0^\infty y\, dA_\kappa(y) + \int_0^\infty y\, dH_\kappa^*(y). \end{aligned}$$

Thus,

$$\int_0^\infty y dH_\kappa^*(y) > \int_0^\infty y dA_\kappa(y),$$

and by comparison of (11.1.13) with (11.1.5), it is clear from a renewal risk theoretical interpretation of the sequence $\{\psi_n^*(x); n = 0, 1, 2, \cdots\}$ that the positive loading condition is violated, and from the comment following (11.1.1) it follows that

$$\lim_{n\to\infty} \psi_n^*(x) = 1, \quad x \geq 0. \tag{11.1.14}$$

Returning to the original sequence defined recursively by (11.1.5), we will now prove by induction on n that

$$\psi_n(x) \geq \sigma_2 e^{-\kappa x} \psi_n^*(x); \quad n = 1, 2, \cdots. \tag{11.1.15}$$

It follows from (11.1.10) that

$$\overline{H}(x+t) \geq \sigma_2 e^{-\kappa(x+t)} \int_{x+t}^\infty e^{\kappa y} dH(y), \tag{11.1.16}$$

and so with $n = 1$,

$$\begin{aligned}
\psi_1(x) &= \int_0^\infty \overline{H}(x+t) dA(t) \\
&\geq \sigma_2 e^{-\kappa x} \int_0^\infty e^{-\kappa t} \int_{x+t}^\infty e^{\kappa y} dH(y) dA(t) \\
&= \sigma_2 e^{-\kappa x} \int_0^\infty \overline{H}_\kappa^*(x+t) dA_\kappa(t) \\
&= \sigma_2 e^{-\kappa x} \psi_1^*(x)
\end{aligned}$$

since (11.1.6) holds. Thus, (11.1.15) holds with $n = 1$, and assume true for n. Then from (11.1.5),

$$\begin{aligned}
\psi_{n+1}(x) &\geq \int_0^\infty \left\{ \sigma_2 e^{-\kappa(x+t)} \int_{x+t}^\infty e^{\kappa y} dH(y) \right. \\
&+ \left. \sigma_2 \int_0^{x+t} \psi_n^*(x+t-y) e^{-\kappa(x+t-y)} dH(y) \right\} dA(t) \\
&= \sigma_2 e^{-\kappa x} \left\{ \int_0^\infty e^{\kappa y} dH(y) \right\} \\
&\times \int_0^\infty e^{-\kappa t} \left\{ \overline{H}_\kappa^*(x+t) + \int_0^{x+t} \psi_n^*(x+t-y) dH_\kappa^*(y) \right\} dA(t) \\
&= \sigma_2 e^{-\kappa x} \int_0^\infty \left\{ \overline{H}_\kappa^*(x+t) + \int_0^{x+t} \psi_n^*(x+t-y) dH_\kappa^*(y) \right\} dA_\kappa(t) \\
&= \sigma_2 e^{-\kappa x} \psi_{n+1}^*(x),
\end{aligned}$$

where (11.1.16), (11.1.6) and (11.1.13) have also been used. Thus, (11.1.15) holds. Finally, combining (11.1.14) and (11.1.15) yields

$$\psi(x) = \lim_{n\to\infty} \psi_n(x) \geq \sigma_2 e^{-\kappa x} \lim_{n\to\infty} \psi_n^*(x) = \sigma_2 e^{-\kappa x},$$

and the theorem is proved. □

It follows from corollary 11.1.1 and theorem 11.1.2 that

$$\sigma_2 e^{-\kappa x} \leq \psi(x) \leq \sigma_1 e^{-\kappa x}, \quad x \geq 0, \tag{11.1.17}$$

if $\kappa > 0$ satisfies (11.1.6), with σ_1 and σ_2 given by (11.1.8) and (11.1.10) respectively. The following well-known result (e.g. Ross, 1996, section 7.4, or Rolski et al, 1999, p. 251) may now be proved.

Corollary 11.1.2 If $H(y) = 1 - e^{-y/E(Y)}$, $y \geq 0$, then

$$\psi(x) = \{1 - \kappa E(Y)\}e^{-\kappa x}, \; x \geq 0, \tag{11.1.18}$$

where $\kappa > 0$ satisfies

$$\tilde{a}(\kappa) = 1 - \kappa E(Y). \tag{11.1.19}$$

Proof: One finds easily that $\tilde{h}(s) = \{1 + sE(Y)\}^{-1}$ and that $\sigma_1 = \sigma_2 = 1 - \kappa E(Y)$ using (11.1.8) and (11.1.10). The result follows immediately from (11.1.17).

□

We remark that from proposition 6.1.1, if $H(y)$ is NWUC (NBUC), then (11.1.7) ((11.1.9)) becomes

$$\psi(x) \leq (\geq)\; e^{-\kappa x}/E\left\{e^{\kappa Y}\right\}, \quad x \geq 0. \tag{11.1.20}$$

If $A(x) = 1 - e^{-x/E(X)}$, the classical ruin model applies, and by the compound geometric nature of the tail $\psi(x)$ we obtain from corollary (6.1.5) that $\psi(x) \leq (\geq)\; \psi(0)e^{-\kappa x}$, $x \geq 0$ if $H_1(y)$ is NWUC (NBUC), implied if $H(y)$ is 2-NWU (2-NBU). Since this is an equality at 0, this improves (11.1.20) in this special case.

As far as explicit analysis of $\psi(x)$ is concerned in the general case, it can be shown even for the present model that $\psi(x)$ is still a compound geometric tail and may be expressed as

$$\psi(x) = \sum_{n=1}^{\infty} (1-\phi)\phi^n \overline{F}^{*n}(x), \quad x \geq 0. \tag{11.1.21}$$

See Embrechts et al. (1997, pp. 26-7), Rolski et al. (1999, section 6.5), and references therein for details. The difficulty with the use of (11.1.21) lies in evaluation of ϕ and $F(x)$, which would allow for application of the results of chapter 7 to the analysis of $\psi(x)$. This identification seems to be possible only in certain cases, however. For example, if $H(y)$ is exponential (e.g. Rolski et al, 1999, p. 248) then $F(x) = H(x)$ and $\phi = 1 - \kappa E(Y)$, and (11.1.18) holds. In the classical model with $A(x) = 1 - e^{-x/E(X)}$ one has $F(x) = H_1(x)$ and $\phi = E(Y)/E(X)$. If $H(x)$ is DFR, then each of $F(x)$ and $1-\psi(x)$ are also DFR, as shown by Szekli (1986) and Shanthikumar (1988), respectively. Asmussen (1992) has identified $F(x)$ and ϕ when either $H(y)$ or $A(x)$ is of phase type. For the even more general model where either $H(y)$ or $A(x)$ has rational Laplace transform, Cohen (1982) has obtained the Laplace transform of $\psi(x)$. See also De Smit (1995). Since the Laplace transform of $\psi(x)$ is available, the Tijms approximation to $\psi(x)$ may be used in this case, as in section 8.3.

It can be shown (e.g. Rolski et al, 1999, pp. 255-9) that $F(x)$ is non-arithmetic and $\kappa > 0$ satisfying the Lundberg condition (11.1.6) also satisfies $\int_0^\infty e^{\kappa y} dF(y) = 1/\phi$, and from the compound geometric representation (11.1.21),

$$\psi(x) \sim \frac{1-\phi}{\phi\kappa \int_0^\infty y e^{\kappa y} dF(y)} e^{-\kappa x}, \quad x \to \infty. \tag{11.1.22}$$

See also Grandell (1991, p. 65). Clearly, use of (11.1.22) requires identification of ϕ and $F(x)$.

In the next section we obtain ϕ and $F(x)$ in (11.1.21) when $A(x)$ has a Coxian-2 density (e.g. Tijms, 1994, pp. 360-4). Various analytic results follow from the identification of ϕ and $F(x)$.

11.2 The Coxian-2 case

In this section we show that analysis of $\psi(x)$ is relatively straightforward in the special case when $A(y)$ is a Coxian-2 df (Tijms, pp. 360-4) with Laplace-Stieltjes transform

$$\tilde{a}(s) = (1-p)\frac{\lambda_1}{\lambda_1 + s} + p\frac{\lambda_1\lambda_2}{(\lambda_1+s)(\lambda_2+s)}, \tag{11.2.1}$$

where $\lambda_1 > 0$, $\lambda_2 > 0$, and $p \leq 1$ (with $p < 0$ not excluded). In addition, we assume in what follows that $p \neq 0$ and $\lambda_2 \neq \lambda_1(1-p)$, since it is easily seen from (11.2.1) that in these cases $A(y)$ is the df of an exponential random variable with mean $1/\lambda_1$ and $1/\lambda_2$ respectively. We wish to exclude the exponential case, since this is simply the classical model. It is not difficult to show that $A(y)$ is IFR if $\lambda_1 p(1-p) < \lambda_2 p$, whereas if $\lambda_1 p(1-p) > \lambda_2 p$ then $A(y)$ is DFR. Thus, the behavior of $A(y)$ is very different in these two cases (see section 2.1 for details), and are key conditions in the subsequent analysis.

The distribution of the sum of two exponentials with possibly different means is obtained with $p = 1$ and the $E_{1,2}$ density (a mixture of an exponential and an Erlang-2 distribution with the same scale parameter, as in example 2.1.3) is obtained with $p > 0$ and $\lambda_1 = \lambda_2$. In addition, the H_2 density of the mixture of two exponentials with Laplace-Stieltjes transform

$$\tilde{a}(s) = (1 - p_1)\frac{\lambda_1}{\lambda_1 + s} + p_1\frac{\lambda_2}{\lambda_2 + s} \tag{11.2.2}$$

where $\lambda_1 < \lambda_2$ and $0 < p_1 < 1$ may be put in the form (11.2.1) with $p = p_1(\lambda_1 - \lambda_2)/\lambda_1$, as is easily verified. The mean is easily seen to be

$$E(X) = \frac{\lambda_2 + \lambda_1 p}{\lambda_1\lambda_2}. \tag{11.2.3}$$

In what follows, a technical result is needed and is now stated.

Lemma 11.2.1 The equation

$$\tilde{a}(-s)\tilde{h}(s) = 1 \tag{11.2.4}$$

may be expressed as

$$\lambda_1\lambda_2 E(Y)\tilde{h}_1(s) + \lambda_1(1 - p)\tilde{h}(s) = \lambda_1 + \lambda_2 - s, \tag{11.2.5}$$

and has exactly one positive root s_1 satisfying

$$\lambda_1\lambda_2\{E(X) - E(Y)\} < s_1 < \lambda_1 + \lambda_2.$$

Moreover, if $\lambda_1 p(1 - p) < (>)\ \lambda_2 p$, then $s_1 < (>)\ \lambda_2 + \lambda_1 p$.

Proof: It follows from (11.2.1) that (11.2.4) may be expressed as

$$\{\lambda_1(1 - p)(\lambda_2 - s) + \lambda_1\lambda_2 p\}\tilde{h}(s) = (\lambda_1 - s)(\lambda_2 - s),$$

that is

$$\{\lambda_1\lambda_2 - \lambda_1(1 - p)s\}\tilde{h}(s) = \lambda_1\lambda_2 - (\lambda_1 + \lambda_2)s + s^2,$$

or

$$(\lambda_1 + \lambda_2)s - s^2 = \lambda_1\lambda_2\{1 - \tilde{h}(s)\} + \lambda_1(1 - p)s\tilde{h}(s).$$

Since $1 - \tilde{h}(s) = sE(Y)\tilde{h}_1(s)$, (11.2.5) follows by division by s. Now let $y_1(s) = \lambda_1 + \lambda_2 - s$. Then $y_1(s)$ is a straight line which decreases from $y_1(0) = \lambda_1 + \lambda_2$ to $y_1(s) = 0$ at $s = \lambda_1 + \lambda_2$. Next let $y_2(s) = \lambda_1\lambda_2 E(Y)\tilde{h}_1(s)$ $+\lambda_1(1 - p)\tilde{h}(s)$. Then

$$y_2(0) = \lambda_1\lambda_2 E(Y) + \lambda_1(1 - p) < \lambda_1\lambda_2 E(X) + \lambda_1(1 - p) = \lambda_1 + \lambda_2 = y_1(0)$$

using (11.2.3). Also, $y_2'(s) < 0$ and $y_2''(s) > 0$ for $s > 0$. See figure 11.2 for a graph of $y_1(s)$ and $y_2(s)$.

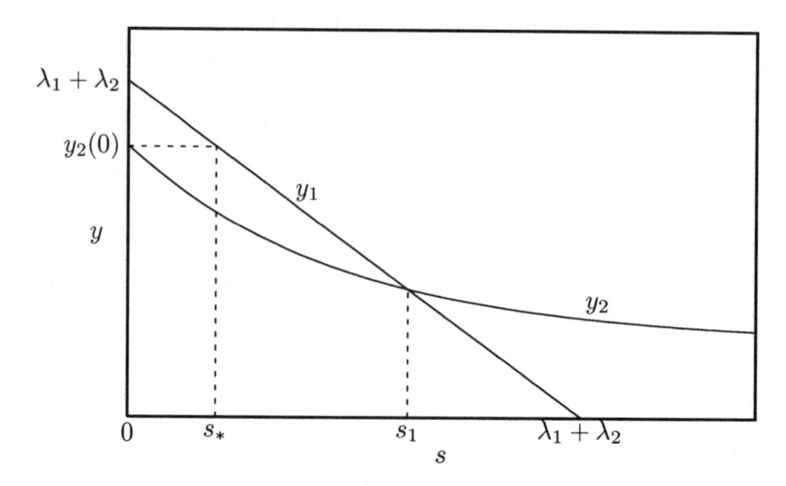

FIGURE 11.2. The graph of $y_1(s)$ and $y_2(s)$

Clearly, there is a positive root s_1 to the equation $y_1(s) = y_2(s)$ and since $y_1(s) \leq 0$ for $s \geq \lambda_1 + \lambda_2$, it follows that $s_1 < \lambda_1 + \lambda_2$. Also, since $y_2(s)$ is decreasing from $y_2(0)$, the intersection point of $y_1(s)$ and $y_2(s)$ must occur at a point to the right of s_* where $y_1(s_*) = y_2(0)$. That is, $\lambda_1 + \lambda_2 - s_* = \lambda_1\lambda_2 E(Y) + \lambda_1(1-p)$ which implies that

$$s_* = \lambda_2 + \lambda_1 p - \lambda_1\lambda_2 E(Y) = \lambda_1\lambda_2\{E(X) - E(Y)\}.$$

Now, if $p = 1$, it is clear that $\lambda_1 p(1-p) < \lambda_2 p$, and from the above discussion, $s_1 < \lambda_2 + \lambda_1 p$. Next, suppose that $p < 1$, and it is clear that for $s > 0$

$$\begin{aligned} y_2(s) &= \lambda_1\lambda_2\left\{\frac{1-\tilde{h}(s)}{s}\right\} + \lambda_1(1-p)\tilde{h}(s) \\ &= \frac{\lambda_1\lambda_2}{s} + \frac{\lambda_1}{s}\{s(1-p) - \lambda_2\}\tilde{h}(s). \end{aligned}$$

Therefore $y_2(\frac{\lambda_2}{1-p}) = \lambda_1(1-p)$. Also,

$$y_1(\frac{\lambda_2}{1-p}) = \lambda_1 + \lambda_2 - \frac{\lambda_2}{1-p} = \lambda_1 - \lambda_2(\frac{p}{1-p}).$$

Then $y_2(\frac{\lambda_2}{1-p}) > (<)\ y_1(\frac{\lambda_2}{1-p})$ is equivalent to $\lambda_1 p(1-p) < (>)\ \lambda_2 p$. However, it is clear from the above discussion that $y_2(\frac{\lambda_2}{1-p}) > (<)\ y_1(\frac{\lambda_2}{1-p})$ implies that the root s_1 must lie to the left (right) of $\lambda_2/(1-p)$, i.e., $s_1 < (>)\ \lambda_2/(1-p)$. Therefore, $\lambda_1 p(1-p) < (>)\ \lambda_2 p$ implies that $s_1 < (>)\ \lambda_2/(1-p)$. But s_1 satisfies (11.2.5), which may be expressed as

$$\lambda_2 + \lambda_1 p - s_1 = \lambda_1\lambda_2 E(Y)\tilde{h}_1(s_1) - \lambda_1(1-p)\{1 - \tilde{h}(s_1)\}$$

$$= \lambda_1(1-p)E(Y)\tilde{h}_1(s_1)\{\frac{\lambda_2}{1-p} - s_1\},$$

from which it is clear that $s_1 < (>) \lambda_2/(1-p)$ implies that $s_1 < (>) \lambda_2 + \lambda_1 p$. □

We remark that if $p < 1$, the important inequality $\lambda_2 + \lambda_1 p < (>) \lambda_2/(1-p)$ is simply a restatement of $\lambda_1 p(1-p) < (>) \lambda_2 p$. Thus $s_1 < (>) \lambda_2 + \lambda_1 p$ implies that $s_1 < (>) \lambda_2/(1-p)$, which is also evident from the proof of lemma 11.2.1. Each of the above inequalities will be used in what follows.

Furthermore, it is clear that if $p > 0$, then $\lambda_2 > (<) \lambda_1(1-p)$ implies that $s_1 < (>) \lambda_2 + \lambda_1 p$. In particular, for the E_{12} mixture (example 2.1.3) with $\lambda_1 = \lambda_2$, it is clear that $s_1 < \lambda_2 + \lambda_1 p$. On the other hand, for the H_2 density of the mixture of two exponentials with $\tilde{a}(s)$ given by (11.2.2), it is clear that $s_1 > \lambda_2 + \lambda_1 p$ since the inequality $p_1 < 1$ may be restated as $\lambda_1 p(1-p) > \lambda_2 p$.

Also, we remark that from (11.2.4) it follows that $-s_1$ is a negative root of the Lundberg equation (11.1.6).

Next, let $K(x) = 1 - \overline{K}(x)$ be defined by

$$\overline{K}(x) = \frac{\int_0^\infty e^{-s_1 t}\overline{H}(x+t)dt}{\int_0^\infty e^{-s_1 t}\overline{H}(t)dt} \tag{11.2.6}$$

where s_1 satisfies the conditions of lemma 11.2.1. Evidently, (11.2.6) is of the form (9.2.5) with $\rho = s_1$. Thus, by analogy with (9.2.20), the equilibrium df of $K(x)$ is $K_1(x) = 1 - \overline{K}_1(x) = \int_0^x \overline{K}(t)dt / \int_0^\infty \overline{K}(t)dt$ where

$$\overline{K}_1(x) = \frac{\int_0^\infty e^{-s_1 t}\overline{H}_1(x+t)dt}{\int_0^\infty e^{-s_1 t}\overline{H}_1(t)dt}. \tag{11.2.7}$$

The main result of this section may now be stated, and is a special case of a result in Willmot (1999).

Theorem 11.2.1 Suppose that $\tilde{a}(s)$ is the Coxian-2 Laplace-Stieltjes transform given by (11.2.1). Let s_1 be the unique positive root of equation (11.2.5). Then $\psi(x)$ satisfies (11.1.21) with

$$\phi = 1 - \frac{\lambda_1\lambda_2\{E(X) - E(Y)\}}{s_1}, \tag{11.2.8}$$

and

$$F(x) = (1-\theta)K(x) + \theta K_1(x) \tag{11.2.9}$$

where

$$\theta = \frac{\lambda_2\{1-\tilde{h}_1(s_1)\}}{\lambda_2\{1-\tilde{h}_1(s_1)\}+(1-p)s_1\tilde{h}_1(s_1)}. \tag{11.2.10}$$

Proof: It can be shown using Wiener-Hopf factorization that $\psi(x) = Pr(L > x)$ where L has Laplace-Stieltjes transform

$$E(e^{-sL}) = \frac{\lambda_1\lambda_2\{E(X)-E(Y)\}(s-s_1)s}{s_1\{(\lambda_1-s)(\lambda_2-s)-\lambda_1\lambda_2\tilde{h}(s)+\lambda_1(1-p)s\tilde{h}(s)\}}. \tag{11.2.11}$$

See De Smit (1995, p. 300) or Cohen (1982, p. 330). It is worth noting that an elementary but tedious derivation of (11.2.11) is possible using the approach of Dickson (1998).

It follows from lemma 11.2.1 that $0 < \phi < 1$ where ϕ is given by (11.2.8). Then (11.2.11) may be expressed as

$$E(e^{-sL}) = \frac{(1-\phi)(s-s_1)s}{s^2-(\lambda_1+\lambda_2)s+\lambda_1\lambda_2\{1-\tilde{h}(s)\}+\lambda_1(1-p)s\tilde{h}(s)}$$

and since $1-\tilde{h}(s) = sE(Y)\tilde{h}_1(s)$,

$$E(e^{-sL}) = \frac{(1-\phi)(s-s_1)}{s-(\lambda_1+\lambda_2)+\lambda_1\lambda_2E(Y)\tilde{h}_1(s)+\lambda_1(1-p)\tilde{h}(s)}.$$

This may be rewritten as

$$\begin{aligned} & E(e^{-sL}) \\ = & \frac{(1-\phi)(s-s_1)}{(s-s_1)-\{(\lambda_1+\lambda_2)-s_1-\lambda_1\lambda_2E(Y)\tilde{h}_1(s)-\lambda_1(1-p)\tilde{h}(s)\}} \\ = & (1-\phi)/\{1-\phi\tilde{f}(s)\} \end{aligned}$$

where

$$\tilde{f}(s) = \frac{(\lambda_1+\lambda_2)-s_1-\lambda_1\lambda_2E(Y)\tilde{h}_1(s)-\lambda_1(1-p)\tilde{h}(s)}{\phi(s-s_1)}.$$

The theorem will be proved if we can show that $\tilde{f}(s) = \int_0^\infty e^{-sx}dF(x)$ where $F(x)$ is given by (11.2.9). Now, since s_1 satisfies (11.2.5), it follows that

$$\tilde{f}(s) = \frac{\lambda_1\lambda_2E(Y)\{\tilde{h}_1(s_1)-\tilde{h}_1(s)\}+\lambda_1(1-p)\{\tilde{h}(s_1)-\tilde{h}(s)\}}{\phi(s-s_1)}.$$

But, combining (11.2.8), (11.2.3), and (11.2.5), it follows that

$$\begin{aligned} \phi &= \frac{s_1-\lambda_1\lambda_2E(X)+\lambda_1\lambda_2E(Y)}{s_1} \\ &= \frac{s_1-(\lambda_1+\lambda_2)+\lambda_1\lambda_2E(Y)+\lambda_1(1-p)}{s_1} \\ &= \frac{\lambda_1\lambda_2E(Y)\{1-\tilde{h}_1(s_1)\}+\lambda_1(1-p)\{1-\tilde{h}(s_1\}}{s_1}. \end{aligned}$$

Thus,

$$\tilde{f}(s) = \frac{\lambda_1\lambda_2 E(Y)s_1\{\tilde{h}_1(s_1) - \tilde{h}_1(s)\} + \lambda_1(1-p)s_1\{\tilde{h}(s_1) - \tilde{h}(s)\}}{(s-s_1)(\lambda_1\lambda_2 E(Y)\{1-\tilde{h}_1(s_1)\} + \lambda_1(1-p)\{1-\tilde{h}(s_1)\})}.$$

Now, from the analogy between (11.2.6) and (9.2.5), it follows from (9.2.15) that

$$\tilde{k}(s) = \int_0^\infty e^{-sx}dK(x) = \frac{s_1}{s_1 - s}\left\{\frac{\tilde{h}(s) - \tilde{h}(s_1)}{1-\tilde{h}(s_1)}\right\}, \tag{11.2.12}$$

and with H replaced by H_1, from (11.2.7) that

$$\tilde{k}_1(s) = \int_0^\infty e^{-sx}dK_1(x) = \frac{s_1}{s_1 - s}\left\{\frac{\tilde{h}_1(s) - \tilde{h}_1(s_1)}{1-\tilde{h}_1(s_1)}\right\}. \tag{11.2.13}$$

Therefore

$$\begin{aligned}\tilde{f}(s) &= \frac{\lambda_1\lambda_2 E(Y)\{1-\tilde{h}_1(s_1)\}\tilde{k}_1(s) + \lambda_1(1-p)\{1-\tilde{h}(s_1)\}\tilde{k}(s)}{\lambda_1\lambda_2 E(Y)\{1-\tilde{h}_1(s_1)\} + \lambda_1(1-p)\{1-\tilde{h}(s_1)\}} \\ &= \theta\tilde{k}_1(s) + (1-\theta)\tilde{k}(s)\end{aligned}$$

with θ given by (11.2.10), since $1 - \tilde{h}(s_1) = s_1 E(Y)\tilde{h}_1(s_1)$. The result follows by inversion of the transform. □

The identification of ϕ and $F(x)$ in theorem 11.2.1 yields (in principle) an explicit solution for $\psi(x)$ from the compound geometric tail (11.1.2). The following example is an immediate consequence of this identification.

Example 11.2.1 Mixture of Erlangs

If

$$H'(x) = \sum_{k=1}^{r} q_k \frac{\beta(\beta x)^{k-1}e^{-\beta x}}{(k-1)!},$$

then from example 9.2.1,

$$K'(x) = \sum_{k=1}^{r} q_k(s_1)\frac{\beta(\beta x)^{k-1}e^{-\beta x}}{(k-1)!}$$

where

$$q_k(s_1) = \frac{\sum_{j=k}^{r} q_j(\frac{\beta}{\beta+s_1})^{j-k}}{\sum_{j=1}^{r} q_j \sum_{i=0}^{j-1}(\frac{\beta}{\beta+s})^i}, \quad k = 1, 2, \cdots, r,$$

and since $K_1(x)$ is the equilibrium df of $K(x)$, it follows from example 2.2.1 that

$$K_1'(x) = \sum_{k=1}^{r} q_k^*(s_1) \frac{\beta(\beta x)^{k-1} e^{-\beta x}}{(k-1)!}$$

where

$$q_k^*(s_1) = \frac{\sum_{j=k}^{r} q_j(s_1)}{\sum_{j=1}^{r} j q_j(s_1)}, \quad k = 1, 2, \cdots, r.$$

Then from (11.2.9) and (11.2.10),

$$F'(x) = \sum_{k=1}^{r} q_k^* \frac{\beta(\beta x)^{k-1} e^{-\beta x}}{(k-1)!}$$

where

$$q_k^* = (1-\theta) q_k(s_1) + \theta q_k^*(s_1), \quad k = 1, 2, \cdots, r.$$

Finally, as in example 7.1.1,

$$\psi(x) = e^{-\beta x} \sum_{k=0}^{\infty} \overline{C}_k \frac{(\beta x)^k}{k!}$$

where

$$\sum_{k=0}^{\infty} \overline{C}_k z^k = \frac{1 - \frac{1-\phi}{1-\phi \sum_{k=1}^{r} q_k^* z^k}}{1-z}.$$

□

The following corollary may be proved as a result of the identification of $F(x)$.

Corollary 11.2.1 The following class implications hold.

a) If $H(x)$ is DFR then $F(x)$ is DFR.

b) If $H(x)$ is IMRL then $F(x)$ is IMRL.

c) If $H(x)$ is 3-DFR then $F(x)$ is 3-DFR.

Proof: Part a) follows from the preservation of the DFR property under mixing and (11.2.9), and is actually a special case of the result of Szekli (1986). If $H(x)$ is IMRL it follows from theorem 9.2.2.b) that $K(x)$ is IMRL and hence $K_1(x)$ is IMRL. Since the IMRL property is preserved under mixing (section 2.3), it follows from (11.2.9) that $F(x)$ is IMRL, proving b). To prove c), let $K_2(x) = \int_0^x \overline{K}_1(y)dy / \int_0^\infty \overline{K}_1(y)dy$ be the equilibrium df of $K_1(x)$. Then from (11.2.9) and (2.2.16), the equilibrium df $F_1(x)$ of $F(x)$ satisfies

$$F_1(x) = (1-\theta_*)K_1(x) + \theta_* K_2(x),$$

where

$$\theta_* = \frac{\theta \int_0^\infty \overline{K}_1(y)dy}{(1-\theta)\int_0^\infty \overline{K}(y)dy + \theta \int_0^\infty \overline{K}_1(y)dy}.$$

If $H(x)$ is 3-DFR then $H_1(x)$ is IMRL, and from (11.2.7) and theorem 9.2.2.b) it follows that $K_1(x)$ is IMRL and hence $K_2(x)$ is IMRL. Thus, $F_1(x)$ is IMRL since it is a mixture of $K_1(x)$ and $K_2(x)$ as shown above, and so $F(x)$ is 3-DFR. □

We remark that upper bounds on the ruin probability $\psi(x)$ follow as in section 7.1 from the compound geometric representation (11.1.21) and the class properties of corollary 11.2.1.

There is a close relationship between $F(x)$ given by (11.2.9) and $H_1(x)$, given in the following corollary.

Corollary 11.2.2 Suppose that $\tilde{a}(s)$ is given by (11.2.1) and $H(x)$ is 2-NWU (2-NBU). Then if $\lambda_1 p(1-p) < \lambda_2 p$, it follows that

$$\overline{F}(x) \geq (\leq)\, \overline{H}_1(x), \quad x \geq 0, \tag{11.2.14}$$

whereas if $\lambda_1 p(1-p) > \lambda_2 p$, (11.2.14) is reversed, i.e.

$$\overline{F}(x) \leq (\geq)\, \overline{H}_1(x), \quad x \geq 0. \tag{11.2.15}$$

Proof: In an identical manner to the derivation of (9.2.28), it follows that

$$\overline{H}_1(x) = \tilde{h}_1(s_1)\overline{K}(x) + \{1-\tilde{h}_1(s_1)\}\overline{K}_1(x), \quad x \geq 0. \tag{11.2.16}$$

Therefore, combining (11.2.16) with (11.2.9) yields

$$\overline{F}(x) = \overline{H}_1(x) + \{\tilde{h}_1(s_1) - (1-\theta)\}\{\overline{K}_1(x) - \overline{K}(x)\}. \tag{11.2.17}$$

But from (11.2.10), it follows that

$$1-\theta = \frac{s_1(1-p)\tilde{h}_1(s_1)}{\lambda_2\{1-\tilde{h}_1(s_1)\} + s_1(1-p)\tilde{h}_1(s_1)} = \frac{s_1(1-p)\tilde{h}_1(s_1)}{\lambda_2\{1-\tilde{h}_1(s_1)\}}\theta,$$

i.e.

$$(1-\theta)\{1-\tilde{h}_1(s_1)\} = \frac{\theta(1-p)}{\lambda_2}s_1\tilde{h}_1(s_1).$$

That is,

$$1-\theta-\tilde{h}_1(s_1)+\theta\tilde{h}_1(s_1) = \frac{\theta(1-p)}{\lambda_2}s_1\tilde{h}_1(s_1),$$

or

$$\tilde{h}_1(s_1) - (1-\theta) = \frac{\theta}{\lambda_2}\{\lambda_2 - (1-p)s_1\}\tilde{h}_1(s_1). \tag{11.2.18}$$

Now, if $p = 1$ then $\theta = 1$ and $\tilde{h}_1(s_1) - (1-\theta) > 0$. If $p < 1$ then $\lambda_1 p(1-p) < (>) \lambda_2 p$ implies that $s_1 < (>) \lambda_2 + \lambda_1 p$ from lemma 11.2.1. But $\lambda_1 p(1-p) < (>) \lambda_2 p$ is a restatement of $\lambda_2 + \lambda_1 p < (>) \lambda_2/(1-p)$. Thus, $\lambda_1 p(1-p) < (>) \lambda_2 p$ implies that $(1-p)s_1 < (>) \lambda_2$, which in turn implies from (11.2.18) that $\tilde{h}_1(s_1) - (1-\theta) > (<) 0$. Also, if $H(x)$ is 2-NWU (2-NBU), then theorem 9.2.2.f) implies that $K(x)$ is NWUE (NBUE), i.e. $\overline{K}_1(x) \geq (\leq) \overline{K}(x)$. The result follows from (11.2.17). □

Bounds for $F(x)$ in terms of other distributions besides $H_1(x)$ may also be obtained, as in the following corollary.

Corollary 11.2.3 Suppose that $\overline{W}(x)$ satisfies $\overline{H}(x+t) \geq (\leq) \overline{W}(x)\overline{H}(t)$. Then $\overline{F}(x) \geq (\leq) \overline{W}(x)$. In particular,

a) if $H(x)$ is NWU (NBU), then $\overline{F}(x) \geq (\leq) \overline{H}(x)$;

b) if $H(x)$ is UBA (UWA) then $\overline{F}(x) \geq (\leq) e^{-x/r_H(\infty)}$.

Proof: As shown in the proof of theorem 9.2.3 c), $\overline{H}_1(x+t) \geq (\leq) \overline{W}(x)\overline{H}_1(t)$. Thus, from (11.2.6) and (11.2.7), $\overline{K}(x) \geq (\leq) \overline{W}(x)$ and $\overline{K}_1(x) \geq (\leq) \overline{W}(x)$. Therefore, from (11.2.9),

$$\overline{F}(x) \geq (\leq) (1-\theta)\overline{W}(x) + \theta\overline{W}(x) = \overline{W}(x).$$

Then a) and b) are the special cases $\overline{W}(x) = \overline{H}(x)$ and $\overline{W}(x) = e^{-x/r_H(\infty)}$ respectively. □

Since $H(x)$ is NWU (NBU) implies that $H(x)$ is NWUE (NBUE), i.e. $\overline{H}_1(x) \geq (\leq) \overline{H}(x)$, corollary 11.2.3.a) is weaker than (11.2.14), but is useful in conjunction with (11.2.15) when $\lambda_1 p(1-p) > \lambda_2 p$.

11.3 The sum of two exponentials

Stronger results may be obtained when $p = 1$ in (11.2.1), which becomes

$$\tilde{a}(s) = \frac{\lambda_1\lambda_2}{(\lambda_1 + s)(\lambda_2 + s)}, \tag{11.3.1}$$

the Laplace transform of the density of the sum of two exponentials with means $1/\lambda_1$ and $1/\lambda_2$ respectively, and an E_2 density if $\lambda_1 = \lambda_2$. Then $\theta = 1$ in (11.2.10), and (11.2.9) becomes

$$F(x) = K_1(x) \tag{11.3.2}$$

with $K_1(x) = 1 - \overline{K}_1(x)$ given by (11.2.7). In the special case with $\lambda_1 = \lambda_2$, this result was obtained by Dickson and Hipp (1998).

Corollary 11.2.1 may be strengthened in this case.

Corollary 11.3.1 Suppose that $\tilde{a}(s)$ is given by (11.3.1). Then the following class implications hold.

a) If $H(x)$ is IMRL (DMRL) then $F(x)$ is DFR (IFR).

b) If $H(x)$ is 3-DFR (3-IFR) then $F(x)$ is IMRL (DMRL).

c) If $H(x)$ is UWAE (UBAE) then $F(x)$ is UWA (UBA).

d) If $H_1(x)$ is 2-NWU (2-NBU) then $F(x)$ is NWUE (NBUE).

Proof: If $H(x)$ is IMRL (DMRL) then $H_1(x)$ is DFR (IFR) and from (11.2.7) and theorem 9.2.2.a), $F(x)$ is DFR (IFR), proving a). If $H(x)$ is 3-DFR (3-IFR) then $H_1(x)$ is IMRL (DMRL) and from (11.2.7) and theorem 9.2.2.b), $F(x)$ is IMRL (DMRL), proving b). To prove c), if $H(x)$ is UWAE (UBAE) then from the comment following theorem 2.4.3, $H_1(x)$ is UWA (UBA), and from (11.2.7) and theorem 9.2.2.d), $F(x)$ is UWA(UBA). Similarly, d) follows from (11.2.7) and theorem 9.2.2.f). □

Bounds on the ruin probability $\psi(x)$ are obtainable from section 7.1 using corollary 11.3.1 and the compound geometric representation (11.1.21). Also, as mentioned following corollary 10.2.3, $H_1(x)$ is 2-NWU (2-NBU) is equivalent to $H(x)$ is 3-NWU (3-NBU). See Fagiuoli and Pellerey (1994).

A class property of $1 - \psi(x)$ may also be derived.

Corollary 11.3.2 Suppose that $\tilde{a}(s)$ satisfies (11.3.1) and $H(x)$ is IMRL. Then $1 - \psi(x)$ is DFR.

Proof: Since $H(x)$ is IMRL then from theorem 9.2.2 b) it follows that $K(x)$ is IMRL, i.e. $F(x) = K_1(x)$ is DFR. Thus $1 - \psi(x)$ is DFR from Shanthikumar (1988). □

More can be said about stochastic ordering than when $p < 1$.

Corollary 11.3.3 Suppose that $\tilde{a}(s)$ is given by (11.3.1). Then the following stochastic implications hold.

a) $\overline{F}(x) \leq \dfrac{\lambda_1\lambda_2 E(Y)}{s_1 - \lambda_1\lambda_2\{E(X) - E(Y)\}}\overline{H}_1(x)$.

b) If $\overline{W}(x)$ satisfies $\overline{H}_1(x+t) \geq (\leq)\ \overline{H}_1(t)\overline{W}(x)$, then $\overline{F}(x) \geq (\leq)\ \overline{W}(x)$. In particular

   i) if $H(y)$ is 2-NWU (2-NBU) then $\overline{F}(x) \geq (\leq)\ \overline{H}_1(x)$.

ii) if $H(y)$ is NWUC (NBUC) then $\overline{F}(x) \geq (\leq)\ \overline{H}(x)$.

iii) if $H(y)$ is NWUE (NBUE) then $\overline{F}(x) \geq (\leq)\ e^{-x/E(Y)}$.

iv) if $H(y)$ is UBAE (UWAE) then $\overline{F}(x) \geq (\leq)\ e^{-x/r_H(\infty)}$.

Proof: It follows from (11.2.16) and (11.3.2) that $\overline{H}_1(x) \geq \{1-\tilde{h}_1(s_1)\}\overline{F}(x)$. But from (11.2.5) and (11.2.3) with $p = 1$,

$$\tilde{h}_1(s_1) = \frac{\lambda_1 + \lambda_2 - s_1}{\lambda_1\lambda_2 E(Y)} = \frac{\lambda_1\lambda_2 E(X) - s_1}{\lambda_1\lambda_2 E(Y)}.$$

Thus,

$$1 - \tilde{h}_1(s_1) = \frac{\lambda_1\lambda_2 E(Y) - \{\lambda_1\lambda_2 E(X) - s_1\}}{\lambda_1\lambda_2 E(Y)} = \frac{s_1 - \lambda_1\lambda_2\{E(X) - E(Y)\}}{\lambda_1\lambda_2 E(Y)}$$

which proves a). To prove b), the result follows from (11.2.7) and the proof of corollary 10.2.2. □

11.4 Delayed and equilibrium renewal risk processes

In this section we consider the same model as in section 11.1, but with one minor variation. Namely, the random variable X_1 is assumed to be independent of $\{X_2, X_3, \cdots\}$ and $\{Y_1, Y_2, \cdots\}$ as before but may have a different distribution than that of X_k for $k = 2, 3, \cdots$. For a delayed or modified renewal process, define $A^d(x) = 1 - \overline{A}^d(x) = Pr(X_1 \leq x)$, and let $\tilde{a}^d(s) = \int_0^\infty e^{-sx} dA^d(x)$.

The reason for this assumption is that the variables $\{X_1, X_2, \cdots\}$ normally represent times between events (as described following (11.1.1)), and when the process begins at time 0 it may well be the case that an event did not occur at that time, but rather at some time in the past. Hence the distribution of the time until the first event after time 0 may be different than the others. It can be shown (e.g. Cox, 1962, section 5.2) that the limiting df of the time until the next event in the ordinary renewal process is the equilibrium df $A_1(x) = 1 - \overline{A}_1(x) = \int_0^\infty \overline{A}(t)dt/E(X)$. Therefore, a useful choice is $A^d(x) = A_1(x)$ with $\tilde{a}^d(s) = \tilde{a}_1(s) = \int_0^\infty e^{-sx} dA_1(x)$. In this case we shall refer to the delayed renewal risk process as the equilibrium (or stationary) renewal risk process. See Cox (1962, section 2.2), Thorin (1975), Grandell (1991, section 3.2), Ross (1996, section 3.5) and Rolski et al (1999, pp. 212-3), for further discussion and analysis.

We shall continue to use the notation and assumptions of section 11.1 with two exceptions. First, the notation for the sequence $\{X_1, X_2, \cdots\}$ now applies to the sequence $\{X_2, X_3, \cdots\}$, i.e. excluding X_1. Second, we shall continue to use $\psi(x)$ in connection with the ordinary (not delayed) model of section 11.1, and we now define for the delayed process

$$\psi^d(x) = Pr\left(\bigcup_{n=1}^{\infty}\left\{\sum_{k=1}^{n}(Y_k - X_k) > x\right\}\right), \quad x \geq 0. \tag{11.4.1}$$

As before, the positive loading condition $E(Y) < E(X)$ is assumed, where X refers to an arbitrary X_k for $k = 2, 3, \cdots$. For the equilibrium process with $A^d(x) = A_1(x)$, we shall write $\psi^d(x) = \psi^e(x)$.

To begin the analysis, we note that the process behaves exactly like the ordinary process of section 11.1 once the first event X_1 has occurred. Therefore, by conditioning on X_1 and Y_1, it follows by the law of total probability that

$$\psi^d(x) = \int_0^\infty \left\{\overline{H}(x+t) + \int_0^{x+t} \psi(x+t-y)dH(y)\right\} dA^d(t). \tag{11.4.2}$$

This fundamental relationship expresses the delayed probability $\psi^d(x)$ in terms of the ordinary probability $\psi(x)$, and in what follows results for $\psi(x)$ can then be used to obtain results for $\psi^d(x)$.

We now consider bounds of the type considered in section 11.1.

Theorem 11.4.1 Suppose that $B(x)$ is an absolutely continuous DFR df with failure rate $\mu_B(x) = -\frac{d}{dx}\ln \overline{B}(x)$ which satisfies $\mu_B(\infty) > 0$. If $B(x)$ also satisfies

$$\tilde{a}\{\mu_B(\infty)\}E\{1/\overline{B}(Y)\} = 1, \tag{11.4.3}$$

then

$$\psi^d(x) \leq \sigma \frac{\tilde{a}^d\{\mu_B(\infty)\}}{\tilde{a}\{\mu_B(\infty)\}}\overline{B}(x), \; x \geq 0, \tag{11.4.4}$$

where σ satisfies (11.1.4).

Proof: It follows from (11.1.4) that

$$\overline{H}(x+t) \leq \sigma \overline{B}(x+t) \int_{x+t}^\infty \{\overline{B}(y)\}^{-1} dH(y),$$

and hence (11.4.2) and theorem 11.1.1 imply that

$$\begin{aligned}
&\psi^d(x)\\
\leq\; & \int_0^\infty \left\{\sigma\overline{B}(x+t)\int_{x+t}^\infty \frac{dH(y)}{\overline{B}(y)} + \sigma\int_0^{x+t}\overline{B}(x+t-y)dH(y)\right\} dA^d(t)\\
\leq\; & \sigma\int_0^\infty \left\{\overline{B}(x+t)\int_{x+t}^\infty \frac{dH(y)}{\overline{B}(y)} + \overline{B}(x+t)\int_0^{x+t}\frac{dH(y)}{\overline{B}(y)}\right\} dA^d(t)\\
=\; & \frac{\sigma}{\tilde{a}\{\mu_B(\infty)\}}\int_0^\infty \overline{B}(x+t)dA^d(t),
\end{aligned}$$

where the last line follows from (11.4.3). Since $B(x)$ is DFR with $\mu_B(\infty) < \infty$, $B(x)$ is also UWA and therefore $\overline{B}(x+t) \le \overline{B}(x)e^{-\mu_B(\infty)t}$, and so

$$\psi^d(x) \le \frac{\sigma \overline{B}(x)}{\tilde{a}\{\mu_B(\infty)\}} \int_0^\infty e^{-\mu_B(\infty)t} dA^d(t),$$

which is (11.4.4). □

The following is the most important special case of this result.

Corollary 11.4.1 Suppose that $\kappa > 0$ satisfies

$$\tilde{a}(\kappa)\tilde{h}(-\kappa) = 1. \tag{11.4.5}$$

Then,

$$\psi^d(x) \le \sigma_1 \frac{\tilde{a}^d(\kappa)}{\tilde{a}(\kappa)} e^{-\kappa x}, \; x \ge 0, \tag{11.4.6}$$

where σ_1 satisfies (11.1.8). In particular,

$$\psi^e(x) \le \sigma_1 \frac{E(Y)}{E(X)} \tilde{h}_1(-\kappa) e^{-\kappa x}, \; x \ge 0. \tag{11.4.7}$$

Proof: Clearly, (11.4.6) is the special case of (11.4.4) with $\overline{B}(x) = e^{-\kappa x}$. In the equilibrium case with $\tilde{a}^d(\kappa) = \tilde{a}_1(\kappa)$, one has using (11.4.5)

$$\frac{\tilde{a}_1(\kappa)}{\tilde{a}(\kappa)} = \frac{1-\tilde{a}(\kappa)}{\kappa E(X)\tilde{a}(\kappa)} = \frac{\tilde{h}(-\kappa)-1}{\kappa E(X)} = \frac{E(Y)}{E(X)} \tilde{h}_1(-\kappa), \tag{11.4.8}$$

and (11.4.7) follows. □

It follows from Ross (1996, p.405) that if $\overline{A}(x) \le \overline{A}^d(x)$ then $1 - \tilde{a}(\kappa) \le 1 - \tilde{a}^d(\kappa)$, i.e. $\tilde{a}^d(\kappa)/\tilde{a}(\kappa) \le 1$, and from (11.4.6) one obtains the simpler but weaker inequality $\psi^d(x) \le e^{-\kappa x}$ since $\sigma_1 \le 1$. In particular, $\psi^e(x) \le e^{-\kappa x}$ if $A(x)$ is NWUE.

We now present the following lower bound.

Theorem 11.4.2 Suppose that $\kappa > 0$ satisfies (11.4.5). Then

$$\psi^d(x) \ge \sigma_2 \frac{\tilde{a}^d(\kappa)}{\tilde{a}(\kappa)} e^{-\kappa x}, \; x \ge 0, \tag{11.4.9}$$

where σ_2 is given by (11.1.10). In particular,

$$\psi^e(x) \ge \sigma_2 \frac{E(Y)}{E(X)} \tilde{h}_1(-\kappa) e^{-\kappa x}, \; x \ge 0. \tag{11.4.10}$$

Proof: It follows from theorem 11.1.2 and (11.4.2) that

$$\begin{aligned}
&\psi^d(x)\\
\geq\ & \int_0^\infty \sigma_2 \left\{ e^{-\kappa(x+t)} \int_{x+t}^\infty e^{\kappa y} dH(y) + \int_0^{x+t} e^{-\kappa(x+t-y)} dH(y) \right\} dA^d(t)\\
\leq\ & \sigma_2 e^{-\kappa x} \int_0^\infty e^{-\kappa t} \left\{ \int_{x+t}^\infty e^{\kappa y} dH(y) + \int_0^{x+t} e^{\kappa y} dH(y) \right\} dA^d(t)\\
=\ & \sigma_2 e^{-\kappa x} \tilde{h}(-\kappa) \int_0^\infty e^{-\kappa t} dA^d(t),
\end{aligned}$$

and (11.4.9) follows using (11.4.5). Thus (11.4.10) follows from (11.4.8).

□

One may combine corollary 11.4.1 and theorem 11.4.2 to obtain for $x \geq 0$,

$$\sigma_2 \frac{\tilde{a}^d(\kappa)}{\tilde{a}(\kappa)} e^{-\kappa x} \leq \psi^d(x) \leq \sigma_1 \frac{\tilde{a}^d(\kappa)}{\tilde{a}(\kappa)} e^{-\kappa x}, \tag{11.4.11}$$

and

$$\sigma_2 \frac{E(Y)}{E(X)} \tilde{h}_1(-\kappa) e^{-\kappa x} \leq \psi^e(x) \leq \sigma_1 \frac{E(Y)}{E(X)} \tilde{h}_1(-\kappa) e^{-\kappa x}, \tag{11.4.12}$$

where $\kappa > 0$ satisfies (11.4.5), and σ_1 and σ_2 are given by (11.1.8) and (11.1.10) respectively. We remark that as mentioned by Rolski et al (1999, p. 500), Lundberg bounds for delayed renewal risk processes of the type in (11.4.11) and (11.4.12) are standard and well-known, as are Cramer-Lundberg asymptotic results of the following type. See Grandell (1991, p. 69) for example.

Theorem 11.4.3 Suppose that $\kappa > 0$ satisfies (11.4.5). Then

$$\psi^d(x) \sim \frac{(1-\phi)\tilde{a}^d(\kappa)}{\phi\kappa\tilde{a}(\kappa)\int_0^\infty y e^{\kappa y} dF(y)} e^{-\kappa x},\ x \to \infty, \tag{11.4.13}$$

and

$$\psi^e(x) \sim \frac{(1-\phi)E(Y)\tilde{h}_1(-\kappa)}{\phi\kappa E(X)\int_0^\infty y e^{\kappa y} dF(y)} e^{-\kappa x},\ x \to \infty. \tag{11.4.14}$$

Proof: Consider (11.4.2). Since (11.4.6) implies that $\tilde{h}(-\kappa) < \infty$, it follows that $\lim_{x\to\infty} e^{\kappa x}\overline{H}(x) = 0$ as $e^{\kappa x}\overline{H}(x) \leq \int_x^\infty e^{\kappa y} dH(y)$. It follows from (11.1.7) that $e^{\kappa(x-y)}\psi(x-y) \leq 1$, for $y \leq x$ and bounded covergence yields

$$\begin{aligned}
&\lim_{x\to\infty} e^{\kappa x} \left\{ \overline{H}(x) + \int_0^x \psi(x-y) dH(y) \right\}\\
=\ & \lim_{x\to\infty} \int_0^x \{e^{\kappa(x-y)}\psi(x-y)\} e^{\kappa y} dH(y)
\end{aligned}$$

$$= \int_0^x \{\lim_{x\to\infty} e^{\kappa(x-y)}\psi(x-y)\}e^{\kappa y}dH(y)$$

$$= \frac{(1-\phi)\tilde{h}(-\kappa)}{\phi\kappa\int_0^\infty ye^{\kappa y}dF(y)},$$

using (11.1.22). An immediate implication of the above is that $e^{\kappa x}\left\{\overline{H}(x)+\int_0^x \psi(x-y)dH(y)\right\}$ is bounded for all $x \geq 0$.

Returning to (11.4.2), one has with the help of (11.4.6) and bounded convergence again,

$$\lim_{x\to\infty} e^{\kappa x}\psi^d(x)$$

$$= \lim_{x\to\infty}\int_0^\infty e^{\kappa(x+t)}\left\{\overline{H}(x+t)+\int_0^{x+t}\psi(x+t-y)dH(y)\right\}e^{-\kappa t}dA^d(t)$$

$$= \int_0^\infty \lim_{x\to\infty} e^{\kappa(x+t)}\left\{\overline{H}(x+t)+\int_0^{x+t}\psi(x+t-y)dH(y)\right\}e^{-\kappa t}dA^d(t)$$

$$= \frac{(1-\phi)\tilde{a}^d(\kappa)}{\phi\kappa\tilde{a}(\kappa)\int_0^\infty ye^{\kappa y}dF(y)}.$$

Then (11.4.14) follows from (11.4.8). □

The asymptotic results in theorem 11.4.3 are of use as long as the compound geometric values ϕ and $F(y)$ are available. In the exponential case with $H(y)=1-e^{-y/E(Y)}$, (11.4.13) and (11.4.14) are equalities for $x \geq 0$, as may be verified using $F(y)=H(y)$ (e.g. Rolski et al, 1999, p. 248) and the following result (e.g. Grandell, 1991, p. 69).

Corollary 11.4.2 If $H(y)=1-e^{-y/E(Y)}$, $y>0$, then

$$\psi^d(x)=\tilde{a}^d(\kappa)e^{-\kappa x},\ x\geq 0, \tag{11.4.15}$$

and

$$\psi^e(x)=\frac{E(Y)}{E(X)}e^{-\kappa x},\ x\geq 0, \tag{11.4.16}$$

where $\kappa>0$ satisfies $\tilde{a}(\kappa)=1-\kappa E(Y)$.

Proof: One finds easily that $\sigma_1=\sigma_2=1-\kappa E(Y)$ using (11.1.8) and (11.1.10), and since $\tilde{h}(s)=\{1+sE(Y)\}^{-1}$, (11.4.5) implies that $\tilde{a}(\kappa)=1-\kappa E(Y)=\sigma_1=\sigma_2$. Thus (11.4.15) follows from (11.4.11). In the equilibrium case, $\tilde{h}_1(-\kappa)=\tilde{h}(-\kappa)=1/\tilde{a}(\kappa)=1/\sigma_1=1/\sigma_2$, and (11.4.16) follows from (11.4.12). □

The following ordering between the delayed ruin probability $\psi^d(x)$ and that of the ordinary model $\psi(x)$ is both simple and intuitive.

Proposition 11.4.1 If $\overline{A}^d(x) \geq (\leq)\ \overline{A}(x),\ x \geq 0$, then

$$\psi^d(x) \leq (\geq)\ \psi(x),\ x \geq 0. \tag{11.4.17}$$

Proof: Let $\psi(x) = Pr(L > x)$ as in section 11.1, and let $\overline{H}(y) = Pr(Y > y)$ where L and Y are independent. Thus (11.4.2) may be expressed as

$$\psi^d(x) = \int_0^\infty Pr(L + Y > x + t) dA^d(t).$$

Since $Pr(L + Y > x + t)$ is nonincreasing in t, it follows from Ross (1996, p. 405) that $\psi^d(x) \leq (\geq)\ \psi(x)$ since (11.4.2) holds for the ordinary renewal risk process without the superscripts. □

In the equilibrium situation, further analysis is possible which yields an alternative expression for $\psi^e(x)$. The following result is given by Grandell (1991, pp. 67-9). See also Rolski et al (1999, p. 499) for a more general result.

Theorem 11.4.4 For the equilibrium renewal risk process,

$$\psi^e(x) = \frac{E(Y)}{E(X)} \left\{ \int_0^x \psi(x-y) dH_1(y) + \overline{H}_1(x) \right\},\ x \geq 0. \tag{11.4.18}$$

Proof: It follows from (11.4.2) that

$$1 - \psi^d(u) = \int_0^\infty \int_0^{u+t} \{1 - \psi(u + t - y)\} dH(y) dA^d(t), \tag{11.4.19}$$

which in the equilibrium case may be expressed as

$$\begin{aligned} 1 - \psi^e(u) &= \int_0^\infty \frac{\overline{A}(t)}{E(X)} \int_0^{u+t} \{1 - \psi(u + t - y)\} dH(y) dt \\ &= \frac{1}{E(X)} \int_u^\infty \overline{A}(t - u) \int_0^t \{1 - \psi(t - y)\} dH(y) dt. \end{aligned}$$

Since $\overline{A}(0) = 1$, this expression may be differentiated (Grandell, 1991, p. 68) using Leibniz's rule to obtain

$$\begin{aligned} -E(X)\frac{d}{du}\psi^e(u) &= -\int_0^u \{1 - \psi(u - y)\} dH(y) \\ &+ \int_0^\infty \int_0^{u+t} \{1 - \psi(u + t - y)\} dH(y) dA(t). \end{aligned}$$

In the ordinary model, (11.4.19) holds with no superscripts, and thus

$$E(X)\frac{d}{du}\psi^e(u) = \int_0^u \{1 - \psi(u - y)\} dH(y) - \{1 - \psi(u)\}. \tag{11.4.20}$$

Integration of (11.4.20) over $(0,\ x)$ yields

$$\begin{aligned} & E(X)\left\{\psi^e(x)-\psi^e(0)\right\} \\ = & \int_0^x\int_0^u\{1-\psi(u-y)\}\,dH(y)du-\int_0^x\{1-\psi(u)\}\,du \\ = & \int_0^x\int_0^{x-y}\{1-\psi(u)\}\,dudH(y)-\int_0^x\{1-\psi(u)\}\,du \end{aligned}$$

by reversing the order of integration. But integration by parts yields

$$\begin{aligned} & \int_0^x\int_0^{x-y}\{1-\psi(u)\}\,dudH(y) \\ = & \int_0^x\{1-\psi(u)\}\,du-\int_0^x\{1-\psi(x-y)\}\,\overline{H}(y)dy, \end{aligned}$$

and so

$$\psi^e(x)=\psi^e(0)-\frac{1}{E(X)}\int_0^x\{1-\psi(x-y)\}\,\overline{H}(y)dy. \qquad (11.4.21)$$

Now, $\psi^e(\infty)=0$, $\psi(\infty)=0$ (e.g. Grandell, 1991, p. 5), $\int_0^\infty\overline{H}(y)dy=E(Y)$, and from (11.4.21) with $x\to\infty$ monotone convergence yields

$$0=\psi^e(0)-\frac{1}{E(X)}\int_0^\infty\{1-0\}\,\overline{H}(y)dy,$$

i.e.,

$$\psi^e(0)=\frac{E(Y)}{E(X)}. \qquad (11.4.22)$$

Finally, substitution of (11.4.22) into (11.4.21) results in

$$\begin{aligned} \psi^e(x) &= \frac{E(Y)}{E(X)}\left\{1-\int_0^x\{1-\psi(x-y)\}\,dH_1(y)\right\} \\ &= \frac{E(Y)}{E(X)}\left\{\overline{H}_1(x)+\int_0^x\psi(x-y)dH_1(y)\right\}, \end{aligned}$$

which is (11.4.18). □

A useful feature of (11.4.18) is the fact that a relatively simple Tijms approximation for $\psi^e(x)$ is obtainable by substituting a Tijms approximation for the compound geometric tail $\psi(x)$ into the integral term on the right-hand side of (11.4.18). This approximation is exact for certain choices of $H(x)$. More generally, if $\psi(x)$ is known as in example 11.2.1, then (11.4.18) simply involves a convolution.

In the classical case with $A(x) = 1 - e^{-x/E(X)}$, $\phi = E(Y)/E(X)$, and $F(x) = H_1(x)$, $\psi(x)$ becomes

$$\psi^c(x) = \sum_{n=1}^{\infty} \left\{1 - \frac{E(Y)}{E(X)}\right\} \left\{\frac{E(Y)}{E(X)}\right\}^n \overline{H}_1^{*n}(x), \quad x \geq 0. \tag{11.4.23}$$

Also in the classical case, since $A(x) = A_1(x)$, it follows that the equilibrium probability $\psi^e(x)$ also equals $\psi^c(x)$, and therefore (11.4.18) implies that

$$\psi^c(x) = \frac{E(Y)}{E(X)} \left\{ \int_0^x \psi^c(x-y) dH_1(y) + \overline{H}_1(x) \right\}, \ x \geq 0. \tag{11.4.24}$$

The following ordering of the ruin probabilities is essentially due to Rolski et al (1999, pp.511-2), who actually obtain more general results.

Theorem 11.4.5 If $A(x)$ is NWUE (NBUE), then

$$\psi^c(x) \leq (\geq)\ \psi^e(x) \leq (\geq)\ \psi(x), \ x \geq 0. \tag{11.4.25}$$

Proof: Since $A(x)$ is NWUE (NBUE), one has $\overline{A}_1(x) \geq (\leq)\ \overline{A}(x)$. It follows from proposition 11.4.1 that $\psi^e(x) \leq (\geq)\ \psi(x)$, proving the right-hand side of (11.4.25). Therefore, it follows from (11.4.18) that $\psi^e(x)$ satisfies the defective renewal inequality

$$\psi^e(x) \geq (\leq)\ \frac{E(Y)}{E(X)} \left\{ \int_0^x \psi^e(x-y) dH_1(y) + \overline{H}_1(x) \right\}, \ x \geq 0. \tag{11.4.26}$$

The left-hand side of (11.4.25) is a consequence of (11.4.24) and (11.4.26), as we now demonstrate (see Rolski et al, 1999, p.511, for an alternative proof). Let $\tau_n(x)$ be the solution of the equation

$$\tau_n(x) = \frac{E(Y)}{E(X)} \left\{ \int_0^x \tau_n(x-y) dH_1(y) + \overline{H}_1(x) \right\} - (+)\ \frac{1}{n}.$$

Then (11.4.26) yields $\psi^e(0) > (<)\ \tau_n(0)$ for $n > 0$. We shall prove by contradiction that $\psi^e(x) > (<)\ \tau_n(x)$ for all $x \geq 0$. Suppose that $\psi^e(x) \leq (\geq)\tau_n(x)$ for some $x > 0$. By continuity, there must be some smallest value $x_0 > 0$ such that $\psi^e(x_0) = \tau_n(x_0)$ and $\psi^e(x) > (<)\ \tau_n(x)$ for all $0 \leq x < x_0$. Thus, (11.4.26) yields

$$\begin{aligned} 0 &= \psi^e(x_0) - \tau_n(x_0) \\ &\geq (\leq)\ \frac{E(Y)}{E(X)} \int_0^{x_0} \{\psi^e(x_0 - y) - \tau_n(x_0 - y)\} dH_1(y) + (-)\frac{1}{n} > (<)\ 0, \end{aligned}$$

a contradiction. Therefore, no such x_0 exists, and $\psi^e(x) > (<)\ \tau_n(x)$ for all $x \geq 0$. From (11.4.24),

$$\psi^e(x) \geq (\leq)\ \lim_{n\to\infty} \tau_n(x) = \psi^c(x),$$

and the left-hand side of (11.4.25) is proved. □

It is worth noting that if $A(x)$ is NWUE (NBUE), then (11.4.22) and (11.4.25) with $x = 0$ yields

$$\phi \geq (\leq) \frac{E(Y)}{E(X)}, \tag{11.4.27}$$

where $\phi = \psi(0)$ is the compound geometric parameter in (11.1.21).

Bibliography

Abate, J., Choudhury, G., and W. Whitt (1995). "Exponential approximations for tail probabilities in queues, 1: waiting times," *Operations Research,* 43, 885-901.

Abate, J., and W. Whitt (1999). "Explicit M/G/1 waiting-time distributions for a class of long-tail service-time distributions," *Operations Research Letters,* 25, 25-31.

Abramowitz, M. and I. Stegun (1965). *Handbook of Mathematical Functions,* Dover, New York.

Alzaid, A. (1994). "Aging concepts for items of unknown age," *Communications in Statistics - Stochastic Models*, 10, 649-659.

Asmussen, S. (1987). *Applied Probability and Queues,* John Wiley, Chichester.

Asmussen, S. (1992). "Phase-type representations in random walk and queueing problems," *The Annals of Probability*, 20, 772-789.

Asmussen, S. (2000). *Ruin Probability*, World Scientific, Singapore, in print.

Barlow, R. and F. Proschan (1965). *Mathematical Theory of Reliability,* John Wiley, New York.

Barlow, R. and F. Proschan (1975). *Statistical Theory of Reliability and Life Testing: Probability Models,* Holt, Rinehart and Winston, New York.

Block, H., and T. Savits (1980). "Laplace transforms for classes of life distributions," *Annals of Probability*, 8, 465-474.

Bondesson, L. (1983). "On preservation of classes of life distributions under reliability operations: some complementary results," *Naval Research Logistics Quarterly*, 30, 443-447.

Bowers, N., Gerber, H., Hickman, J., Jones, D., and C. Nesbitt (1997). *Actuarial Mathematics*, 2nd Edition, Society of Actuaries, Schaumburg.

Brown, M. (1990). "Error bounds for exponential approximations of geometric convolutions," *Annals of Probability*, 18, 1388-1402.

Brown, M. and S. Ross (1969). "Some results for infinite server Poisson queues," *Journal of Applied Probability,* 6, 604-611.

Cai, J. (1998). "A unified study of bounds and asymptotic estimates for renewal equations and compound distributions with applications to insurance risk analysis," *PhD Thesis*, Department of Mathematics and Statistics, Concordia University, Montreal.

Cai, J. and J. Garrido (1998). "Aging properties and bounds for ruin probabilities and stop-loss premiums," *Insurance: Mathematics and Economics*, 23, 33-43.

Cai, J. and J. Garrido (1999). "A unified approach to the study of tail probabilities of compound distributions," *Journal of Applied Probability,* 36, 1058-1073.

Cai, J., and V. Kalashnikov (2000). "NWU Property of a class of Random Sums," *Journal of Applied Probability*, to appear.

Cai, J. and Y. Wu (1997). "Some improvements on the Lundberg bound for the ruin probability," *Statistics and Probability Letters,* 33, 395-403.

Cao, J. and Y. Wang (1991). " The NBUC and NWUC classes of life distributions," *Journal of Applied Probability,* 28, 473-479; Correction(1992), 29, 753.

Chaudhry, M and J. Templeton (1983). *A First Course in Bulk Queues,* John Wiley, New York.

Cohen, J. (1982). *The Single Server Queue*, (rev.). North Holland, Amsterdam.

Cox, D. (1962). *Renewal Theory,* Methuen, London.

Deligonul, Z. (1985). "An approximate solution of the integral equation of renewal theory," *Journal of Applied Probability*, 22, 926-931.

De Smit, J. (1995). "Explicit Weiner-Hopf Factorizations for the Analysis of Multidimensional Queues," In *Advances in Queueing: Theory, Methods, and Open Problems* (J.H. Dshalalow, ed.), CRC Press, Boca Baton, 293-309.

Dickson, D. (1992). "On the distribution of surplus prior to ruin," *Insurance: Mathematics and Economics*, 11, 191-207.

Dickson, D. (1998). "On a class of renewal risk processes," *North American Actuarial Journal*, 2, 60-73.

Dickson, D., and A. Egídio dos Reis (1996). "On the distribution of the duration of negative surplus," *Scandinavian Actuarial Journal*, 148-164.

Dickson, D., and C. Hipp (1998). "Ruin probabilities for Erlang (2) risk processes," *Insurance: Mathematics and Economics*, 22, 251-262.

Dufresne, F. and H. Gerber (1991). "Risk theory for the compound Poisson process that is perturbed by diffusion," *Insurance: Mathematics and Economics*, 10, 51-59.

Ebrahimi, N. (1986). "Classes of discrete decreasing and increasing mean-residual-life distributions," *IEEE Transactions on Reliability*, R-35, 403-405.

Egídio dos Reis, A. (1993). "How long is the surplus below zero?" *Insurance: Mathematics and Economics*, 12, 23-38.

Embrechts, P. and C. Goldie (1982). "On convolution tails," *Stochastic Processes and their Applications*, 13, 263-278.

Embrechts, P., C. Goldie and N. Veraverbeke (1979). "Subexponentiality and infinite divisibility," *Zeitschrift Fur Wahrscheinlichkeitstheorie und Verwandte Gebiete*, 49, 335-347.

Embrechts, P., C. Klüppelberg, and T. Mikosch (1997). *Modelling Extremal Events*, Springer-Verlag, Berlin.

Embrechts, P., M. Maejima and J. Teugels (1985). "Asymptotic behaviour of compound distributions," *ASTIN Bulletin*, 15, 45-48.

Esary, J., A. Marshall and F. Proschan (1973). "Shock models and wear processes," *Annals of Probability*, 1, 627-649.

Fagiuoli, E. and F. Pellerey (1993). "New partial orderings and applications," *Naval Research Logistics*, 40, 829-842.

Fagiuoli, E. and F. Pellerey (1994). "Preservation of certain classes of life distributions under Poisson shock models," *Journal of Applied Probability*, 31, 458-465.

Feller, W. (1968). *An Introduction to Probability Theory and Its Applications*, Vol 1, 3rd edition, John Wiley, New York.

Feller, W. (1971). *An Introduction to Probability Theory and Its Applications*, Vol 2, 2nd edition, John Wiley, New York.

Gerber, H. (1973). "Martingales in risk theory," *Vereinigung Schweizerischer Versicherungs Mathematiker Mitteilungen,* 73, 205-216.

Gerber, H. (1979). *An Introduction to Mathematical Risk Theory,* S.S. Huebner Foundation, University of Pennsylvania, Philadelphia.

Gerber, H. (1988). "Mathematical fun with the compound binomial process," *ASTIN Bulletin*, 18, 161-168.

Gerber, H. (1994). "Martingales and tail probabilities," *ASTIN Bulletin*, 24, 145-146.

Gerber, H., Goovaerts, M., and R. Kaas (1987). "On the probability and severity of ruin," *ASTIN Bulletin*, 17, 151-163.

Gerber, H. and E. S.W. Shiu (1998). "On the time value of ruin," *North American Actuarial Journal*, 2, 48-78.

Gertsbakh, I. (1984). "Asymptotic methods in reliability theory: a review," *Advances in Applied Probability*, 16, 147-175.

Grandell, J. (1991). *Aspects of Risk Theory*, Springer-Verlag, New York.

Grandell. J. (1997). *Mixed Poisson Processes,* Chapman and Hall, London.

Grimmett, G., and D. Stirzaker (1992). *Probability and Random Processes,* 2nd edition, Oxford University Press, Oxford.

Hansen, B., and J. Frenk (1991). "Some monotonicity properties of the delayed renewal function," *Journal of Applied Probability*, 28, 811-821.

Hesselager, O., Wang, S., and G. Willmot (1998). "Exponential and scale mixtures and equilibrium distributions," *Scandinavian Actuarial Journal*, 125-142.

Jorgensen, B. (1982). *Statistical Properties of the Generalized Inverse Gaussian Distribution*, Lecture Notes in Statistics 9. Springer Verlag, New York.

Kalashnikov, V. (1994a). *Mathematical Methods in Queueing Theory*, Kluwer Academic Publishers, Dordrecht.

Kalashnikov, V. (1994b). *Topics on Regenerative Processes*, CRS Press, Boca Raton.

Kalashnikov, V. (1996). "Two-sided bounds of ruin probabilities," *Scandinavian Actuarial Journal*, No.1, 1-18.

Kalashnikov, V. (1997a). *Geometric Sums: Bounds for Rare Events with Applications,* Kluwer Academic Publishers, Dordrecht.

Kalashnikov, V. (1997b). "A simple proof of the Cramer formula," *University of Copenhagen - Laboratory of Actuarial Mathematics*, Working Paper No. 149.

Kalashnikov, V. (1999). "Bounds of ruin probability in the presence of large claims and their comparison," *North American Actuarial Journal,* 116-129.

Karlin, S., and H. Taylor (1975). *A First Course in Stochastic Processes,* (2nd Edition), Academic Press, New York.

Klefsjo, B. (1981). "HNBUE survival under some shock models," *Scandinavian Journal of Statistics*, 8, 39-47.

Kingman, J.F.C. (1964). "A martingale inequality in the theory of queues." *Proc. Camb. Phil. Soc.* 60, 359-361.

Kingman, J.F.C. (1970). "Inequatlties in the theory of queues." *JRSS* B32, 102-110.

Klugman, S., Panjer, H., and G. Willmot (1998). *Loss Models - From Data to Decisions*, John Wiley, New York.

Lemaire, J. (1995). *Bonus-Malus Systems in Automobile Insurance*, Kluwer, Boston.

Lin, X. (1996). "Tail of compound distributions and excess time," *Journal of Applied Probability,* 33, 184-195.

Lin, X. and G.E. Willmot (1999). "Analysis of a defective renewal equation arising in ruin theory," *Insurance: Mathematics and Economics*, 25, 63-84.

Lin, X. and G.E. Willmot (2000). "The moments of the time of ruin, the surplus before ruin, and the deficit at ruin," *Insurance: Mathematics and Economics*, to appear.

Michel, R. (1989). "Representation of a time-discrete probability of eventual ruin," *Insurance: Mathematics and Economics*, 8, 149-152.

Neuts, M. (1986). "Generalizations of the Pollaczek-Khinchin integral equation in the theory of queue," *Advences in Applied Probability.* 18, 952-990.

Panjer, H.H. and G.E. Willmot (1992). *Insurance Risk Models,* Society of Actuaries, Schaumburg.

Prabhu, N. (1998). *Stochastic Storage Processes*, 2nd edition, Springer-Verlag, New York.

Resnick, S. (1992). *Adventures in Stochastic Processes*, Birkhauser, Boston.

Rolski, T., H. Schmidli, V. Schmidt, and J. Teugels (1999). *Stochastic Processes for Insurance and Finance*, John Wiley, Chichester.

Ross, S. (1974). "Bounds on the delay distribution in GI/G/1 queues," *Journal of Applied Probability*, 11, 417-421.

Ross, S. (1996). *Stochastic Processes*, 2nd edition, John Wiley, New York.

Ruohonen, M. (1988). "A model for the claim number process," *ASTIN Bulletin*, 18, 57-68.

Schmidli, H. (1994). "Diffusion approximations for a risk process with the possibility of borrowing and investment," *Communications in Statistics, Stochastic Models*, 10, 365-388.

Schmidli, H. (1999). "On the distribution of the surplus prior to and at ruin," *ASTIN Bulletin*, 29, 227-244.

Shaked, M. and J. Shanthikumar (1994). *Stochastic Orders and their Applications*, Academic Press, San Diego.

Shanthikumar, J. (1988). "DFR property of first passage times and its preservation under geometric compounding," *Annals of Probability*, 16, 397-406.

Shiu, E. (1988). "Calculation of the probability of eventual ruin by Beekman's convolution series," *Insurance: Mathematics and Economics*, 7, 41-47.

Shiu, E. (1989). "The probability of eventual ruin in the compound binomial model," *ASTIN Bulletin*, 19, 179-190.

Sparre-Andersen, E. (1957). "On the collective theory of risk in the case of contagion between the claims," *Transactions XV International Congress of Actuaries*, New York, 2, 219-227.

Steutel, (1970). *Preservation of Infinite Divisibility under Mixing and Related Topics*, Math. Centre Tracts 33, Math. Centre, Amsterdam.

Stoyan, D. (1983). *Comparison Methods for Queues and Other Stochastic Models*, John Wiley, Chichester.

Szekli, R. (1986). "On the concavity of the waiting time distribution in some GI/G/1 queues," *Journal of Applied Probability*, 23, 555-561.

Taylor, G. (1976). "Use of differential and integral inequalities to bound ruin and queueing probabilities," *Scandinavian Actuarial Journal*, 197-208.

Taylor, H. and S. Karlin (1998). *An Introduction to Stochastic Modeling,* (Third edition), Academic Press, San Diego.

Thorin, O. (1975). "Stationarity aspects of the Sparre Andersen risk process and the corresponding ruin probabilities," *Scandinavian Actuarial Journal,* 87-98.

Tijms, H. (1986). *Stochastic Modelling and Analysis: A Computational Approach,* John Wiley, Chichester.

Tijms, H. (1994). *Stochastic Models: An Algorithmic Approach,* John Wiley, Chichester.

Van Harn, K. (1978). *Classifying Infinitely Divisible Distributions by Functional Equations,* Math. Centre Tracts 103, Math. Centre, Amsterdam.

Van Hoorn, M. (1984). *Algorithms and Approximations for Queueing Systems,* CWI Tract 8, Amsterdam.

Vinogradov, O.P. (1973). "The definition of distribution functions with increasing hazard rate in terms of the Laplace Transform," *Theor. Prob. Appl.,* 18, 811-814.

Widder, D. (1946). *The Laplace Transform,* Princeton University Press.

Willmot, G.E. (1988a). "Sundt and Jewell's family of discrete distributions," *Astin Bulletin,* 18, 17-29.

Willmot, G.E. (1988b). "Further use of Shiu's approach to the evaluation of ultimate ruin probabilities," *Insurance: Mathematics and Economics,* 7, 275-281.

Willmot, G.E. (1989a). "Limiting tail behaviour of some discrete compound distributions," *Insurance: Mathematics and Economics,* 8, 175-185.

Willmot, G.E. (1989b). "The total claims distribution under inflationary conditions," *Scandinavian Actuarial Journal,* 1-12.

Willmot, G.E. (1990). "Asymptotic tail behaviour of Poisson mixtures with applications," *Advances in Applied Probability,* 22, 147-159.

Willmot, G.E. (1993). "Ruin probabilities in the compound binomial model," *Insurance: Mathematics and Economics,* 12, 133-142.

Willmot, G.E. (1994). "Refinements and distributional generalizations of Lundberg's inequality," *Insurance: Mathematics and Economics,* 15, 49-63.

Willmot, G.E. (1996). "A non-exponential generalization of an inequality arising in queueing and insurance risk," *Journal of Applied Probability,* 33, 176-183.

Willmot, G.E. (1997a). "On the relationship between bounds on the tails of compound distributions," *Insurance: Mathematics and Economics*, 19, 95-103.

Willmot, G.E. (1997b). "Bounds for compound distributions based on mean residual lifetimes and equilibrium distributions," *Insurance: Mathematics and Economics*, 21, 25-42.

Willmot, G.E. (1998). "On a class of approximations for ruin and waiting time probabilities," *Operations Research Letters*, 22, 27-32.

Willmot, G.E. (1999). "A Laplace transform representation in a class of renewal queueing and risk processes," *Journal of Applied Probability*, 36, 570-584.

Willmot, G.E. (2000). "On evaluation of the conditional distribution of the deficit at the time of ruin," *Scandinavian Actuarial Journal*, to appear.

Willmot, G.E. and J. Cai (1999). "Aging and other distributional properties of discrete compound geometric distributions and related random sums," preprint.

Willmot, G.E. and J. Cai (2000). "On classes of lifetime distributions with unknown age," *Probability in the Engineering and Informational Sciences*, to appear.

Willmot, G.E. and S. Drekic (2000). "On the transient analysis of the $M^X/M/\infty$ queue," preprint.

Willmot, G.E. and X. Lin (1994). "Lundberg bounds on the tails of compound distributions," *Journal of Applied Probability*, 31, 743-756.

Willmot, G.E. and X. Lin (1996). "Bounds on the tails of convolutions of compound distributions," *Insurance: Mathematics and Economics*, 18, 29-33. Erratum, 18, 219.

Willmot, G.E. and X. Lin (1997a). "Simplified bounds on the tails of compound distributions," *Journal of Applied Probability*, 34, 127-133.

Willmot, G.E. and X. Lin (1997b). "Upper bounds for the tail of the compound negative binomial distribution," *Scandinavian Actuarial Journal*, 138-148.

Willmot, G.E. and X. Lin (1998). "Exact and approximate properties of the distribution of surplus before and after ruin," *Insurance: Mathematics and Economics*, 23, 91-110.

Wolff, R. (1989). *Stochastic Modeling and the Theory of Queues*, Prentice Hall, New Jersey.

Wu, Y. (1996). Personal communication.

Symbol Index

$A(z)$: generating function of a_n, 34
$B(y)$: NWU or NBU df associated with bounds, 66
$E(Y)$: mean of Y, 9
$E_{r_1,r_2,\ldots}$: Erlang mixture, 13
$F(y)$: distribution function, individual claim amount distribution, 8, 52
$F^{*n}(y)$: n-fold convolution of $F(y)$, 52
$F_1(y)$: equilibrium df of $F(y)$, 14
$G(u,y)$: unconditional df of deficit at ruin, 184
$G(x)$: aggregate claims distribution function, compound distribution function, 52
$G_u(y)$: conditional df of deficit, 186
$H(y)$: service time distribution in queueing, claim amount distribution in risk model, 126, 129
$H_x(y)$: dominating residual lifetime df, 82, 83, 94
$I(A)$: indicator function, 160
$K(x)$: mixing kernel of Poisson mixture, 38
$K_\lambda(x)$: modified Bessel function, 101
L: maximal aggregate loss, 130
N: counting random variable, number of claims, 28, 52
N^*: discrete equilibrium random variable of N, 32
N_t: Poisson process, 129
$P(z)$: probability generating function, 34
S_k: aggregate claims process in discrete time, 124
S_t: aggregate claims process in continuous time, 129
T: time of ruin, 129
T_y: residual lifetime random variable, 18
$U(a,b,x)$: Kummer's confluent hypergeometric function, 17
U_t: surplus process, 129
$V(x)$: dominating distribution function, 66

W_t: Wiener process, 177
X: random sum, aggregate claims, 52
Y: individual claim amount random variable, 52
$\Gamma(\alpha, x)$: incomplete gamma function, 100
$\alpha_1(x)$: function associated exponential upper bound, 70
$\alpha_2(x)$: function associated exponential lower bound, 76
δ: force of interest, 59, 160
κ: adjustment coefficient, 54, 94, 97, 108, 130, 142, 155
λ: Poisson parameter, 38, 129
$\mu(y)$: failure rate, 8
$\overline{F}(y)$: survival function, 8
$\overline{F}^{*n}(y)$; tail of n-fold convolution of $F(y)$, 52
$\overline{G}_T(x)$: Tijms approximation to $\overline{G}(x)$, 142
ϕ: bound on ratio of discrete tail probabilities, 34
$\phi(x)$: exponential tail associated with $\overline{K}(y)$, 39
ϕ_1: upper bound on ratio of discrete tail probabilities, 66
ϕ_2: lower bound on ratio of discrete tail probabilities, 74
$\psi(u)$: probability of ruin, 129
ψ_x: discrete-time ruin probability, 124
$\tau(x, z)$: reduced form of $c(x, z)$, 82
θ: relative security loading, 129
$\tilde{a}(s)$: Laplace-Stieltjes transform of $A(x)$, 210
a_n: tail probability of N, 28
c: premium rate, 129
$c(x, z)$: function associated with general upper and lower bounds, 66
$c_1(x)$: general upper bound, 66
$c_2(x)$: general lower bound, 74
$f(y)$: probability density function, 8
$f_1(y)$: equilibrium probability density function, 14
f_n: probability distribution of a positive counting random variable, 31
h_n: discrete failure rate, 28
$k(x)$: density of mixing df $K(x)$, 46
$m(x)$: solution of defective renewal equation, 152
$m_T(x)$: Tijms approximation to $m(x)$, 158
p_n: probability function of N, 28
$r(x)$: mean residual lifetime, 19
$r_*(x)$: minimum function of $r(x)$, 87
$r_K(x)$: mean residual lifetime of $K(y)$, 41
r_n: mean residual lifetime of N, 32
$w(x)$: penalty function, 160
z_0: radius of convergence of $P(z)$, 34

Author Index

Abate, 130, 209, 235
Abramowitz, 17, 235
Alzaid, 27, 235
Asmussen, 4, 51, 216, 235

Barlow, 7, 10, 11, 235
Block, 7, 32, 39, 43, 128, 236
Bondesson, 19, 20, 236
Bowers, 130, 236
Brown, 4, 63, 93, 107, 115, 236

Cai, 3, 27, 33, 34, 43, 69–71, 73, 74, 77, 78, 82, 112, 115–117, 120, 123, 128, 134, 236, 242
Cao, 7, 236
Cauchy, 10
Chaudhry, 65, 236
Choudhury, 209, 235
Cohen, 4, 148, 209, 220, 236
Cox, 226, 236

De Smit, 4, 148, 216, 220, 237
Deligonul, 158, 236
Dickson, 184, 189, 220, 225, 237
Drekic, 64, 242
Dufresne, 174, 177, 237

Ebrahimi, 32, 237
Egídio dos Reis, 184, 189, 197, 237
Embrechts, 1, 3, 53, 54, 105, 148, 215, 237
Esary, 13, 34, 237

Fagiuoli, 7, 24, 33, 34, 192, 225, 237
Feller, 3, 10, 56, 113, 116, 151, 152, 156, 157, 161, 178, 209, 238
Frenk, 116, 238

Garrido, 3, 70, 71, 73, 77, 78, 82, 112, 236
Gerber, 1, 4, 7, 61, 73, 107, 124, 130, 131, 160, 161, 164, 173, 174, 177, 198, 199, 236–238
Gertsbakh, 4, 107, 238
Goldie, 1, 3, 54, 105, 237
Goovaerts, 130, 198, 199, 238

Grandell, 7, 37–39, 43, 49, 53, 58, 59, 64, 209, 216, 226, 229–232, 238
Grimmett, 78, 210, 238

Hansen, 116, 238
Hesselager, 17, 238
Hickman, 236
Hipp, 225, 237

Jones, 236
Jorgensen, 101, 102, 238

Kaas, 130, 198, 199, 238
Kalashnikov, 3, 7, 33, 43, 107, 109, 115, 128, 131, 236, 238
Karlin, 62, 116, 152, 157, 239, 241
Kingman, 107, 239
Klüppelberg, 237
Klefsjo, 34, 239
Klugman, 30, 37, 53, 116, 239

Lemaire, 108, 239
Lin, 3, 7, 73, 78, 79, 82, 87, 94, 112, 134, 140, 158, 160, 163, 167, 168, 174, 176, 183, 205, 207, 239, 242

Maejima, 1, 53, 237
Marshall, 13, 237
Michel, 124, 239
Mikosch, 237

Nesbitt, 236
Neuts, 4, 5, 51, 174, 239

Panjer, 30, 37, 53, 107, 108, 116, 132, 239
Pellerey, 7, 24, 33, 34, 192, 225, 237
Prabhu, 3, 209, 240
Proschan, 7, 10, 11, 13, 235, 237

Resnick, 151, 152, 157, 159, 209, 240
Rolski, 3, 29, 73, 170, 173, 209, 210, 212, 213, 215, 216, 226, 229–231, 233, 240
Ross, 3, 5, 58, 61–63, 76, 107, 128, 143, 152, 159, 180, 209, 210, 212, 213, 215, 226, 228, 231, 236, 240
Ruohonen, 49, 240

Savits, 7, 32, 39, 43, 128, 236
Schmidli, 240
Schmidt, 240
Schwarz, 10
Shaked, 25, 27, 96, 167, 240
Shanthikumar, 25, 27, 96, 116, 123, 167, 216, 225, 240
Shiu, 4, 124, 130, 160, 161, 164, 238, 240
Sparre-Andersen, 240
Stegun, 17, 235
Steutel, 10, 240
Stirling, 48, 55
Stirzaker, 78, 210, 238
Stoyan, 4, 107, 127, 212, 213, 240
Szekli, 216, 240

Taylor,G., 107, 109, 240
Taylor,H., 62, 116, 152, 157, 239, 241
Templeton, 65, 236
Teugels, 1, 53, 237, 240
Thorin, 226, 241
Tijms, 141, 147, 152, 216, 241

Van Harn, 114, 116, 148, 241
Van Hoorn, 179, 241
Veraverbeke, 1, 237
Vinogradov, 39, 241

Wang,S., 238
Wang,Y., 7, 236
Whitt, 130, 209, 235
Widder, 42, 241
Willmot, 7, 27, 30, 31, 34, 37, 48, 49, 53, 54, 59, 64, 73, 82,

94, 107, 108, 116, 117, 120, 123–125, 132, 134, 140, 147, 158, 160, 163, 167, 168, 174, 176, 183, 188, 205, 207, 211, 219, 238, 239, 242
Wolff, 24, 28, 242
Wu, 69, 73, 74, 111, 112, 213, 236, 242

Subject Index

(a, b) class, 30
2-NBU, 24
2-NWU, 24
2-UBA, 28
2-UWA, 28
3-DFR, 24
3-IFR, 24

accident insurance, 108
adjustment coefficient, 108, 130, 142, 155
age-dependent branching process, 158
aggregate claims, 51, 52
aggregate claims process, 129
asymptotic geometric, 141
automobile insurance, 108, 132

batch arrival infinite server queue, 63
binomial distribution, 30
binomial-exponential, 55
bulk queue, 236
Burr distribution, 85

Cauchy-Schwarz inequality, 10
classical risk model, 107, 129
combination of exponentials, 198
compound distribution, 1, 52
compound geometric distribution, 107
compound geometric-exponential, 55
compound mixed Poisson distribution, 57
compound modified geometric distribution, 108
compound negative binomial distribution, 132
compound Pascal-exponential distribution, 54, 136
conditional distribution of deficit, 186
counting random variable, 28
Coxian-2 density, 147, 216
Cramer-Lundberg asymptotic ruin formula, 130

D-DFR, 29
D-DMRL, 32
D-IFR, 29

D-IMRL, 32
D-NBU, 33
D-NWU, 33
defective renewal equation, 151, 152
deficit at ruin, 183
delayed renewal process, 210, 226
DFR, 9
discrete compound geometric distribution, 114
discrete equilibrium random variable, 32
discrete failure rate, 28
discrete ruin model, 124
distribution function (df), 8
DMRL, 19
dominance assumption, 34, 66, 74, 112
dominating distribution function, 66
DS-DFR, 29
DS-IFR, 29
DS-NWU, 33

equilibrium distribution function, 14
equilibrium renewal process, 210, 226
Esscher transform, 213
expected discounted penalty, 160
exponential bound, 76, 94
exponential distibution, 9
extended truncated negative binomial (ETNB), 31

failure rate, 8
first order properties, 81
force of mortality, 8

G/G/1 queue, 93, 107, 211
gamma distribution, 11, 16, 22
general Erlang mixture, 198
generalized adjustment equation, 66, 74
generalized inverse Gaussian distribution, 100
geometric distribution, 30

hazard rate, 8

IFR, 9
IMRL, 19
incomplete gamma function, 100
incurred but not reported claims (IBNR), 62
individual claim amount distribution, 52
inflation model, 59
insurance loss, 8
insurance portfolio, 51, 129
interclaim/interarrival distribution, 209
interclaim/interarrival time, 209
inverse gamma distribution, 101
inverse Gaussian distribution, 101

Jensen's inequality, 88

Kummer's confluent hypergeometric function, 17

Laplace-Stieltjes transform, 210
log-convex (log-concave), 9, 11, 29
logarithmic distribution, 30, 108
Lundberg adjustment equation, 54, 94, 108, 142, 155
Lundberg asymptotics, 1, 3
Lundberg inequality, 131

$M^X/G/\infty$ queue, 63
$M^X/M/\infty$ queue, 63
$M/G/\infty$ queue, 62
M/G/1 equilibrium queue length distribution, 126
M/G/c queue, 179
Markov's inequality, 23
maximal aggregate loss, 107, 130
mean of deficit, 206
mean residual lifetime (MRL), 19
mixed Poisson distribution, 38
mixed Poisson thinning, 62
mixing distribution, 38

mixture of Erlangs, 12, 15, 20, 52, 113, 153, 163, 192, 221
mixture of exponentials, 10, 18
modified Bessel function, 101
modified discrete probability distribution, 31
modified geometric-combinations of exponentials, 55
moment of deficit, 207

NBU, 24
NBUC, 25
NBUE, 24
negative binomial distribution, 30, 48, 132
nonexponential bound, 77, 112
number of claims distribution, 52, 129
NWU, 24
NWUC, 25
NWUE, 24

Pareto bound, 97
Pareto tail, 3, 87, 97
penalty function, 160
Phase type-2 distribution, 209
Poisson distribution, 30
Poisson process, 129
premium rate, 129
probability density function (pdf), 8
probability generating function (pgf), 34
probability of ruin, 129
product based bound, 100

random sum, 1, 52
relative security loading, 129
renewal risk process, 209
residual lifetime random variable, 18, 82
risk premium, 51
ruin with diffusion, 177

secondary distribution, 107
service time distribution, 126
shifted mixing distribution, 48
Stirling's formula, 48
stop-loss premium, 51
subexponential distribution, 54
surplus process, 107, 129
survival function (sf), 8

tail behaviour, 46
Tijms approximation, 142
time of ruin, 129

UBA, 27
UBAE, 27
UWA, 27
UWAE, 27

Wald's identity, 73
Wiener-Hopf factorization, 220

zero-modified discrete probability distribution, see modified discrete probability distribution, 31

Lecture Notes in Statistics

For information about Volumes 1 to 82, please contact Springer-Verlag

Vol. 83: C. Gatsonis, J. Hodges, R. Kass, N. Singpurwalla (Editors), Case Studies in Bayesian Statistics. xii, 437 pages, 1993.

Vol. 84: S. Yamada, Pivotal Measures in Statistical Experiments and Sufficiency. vii, 129 pages, 1994.

Vol. 85: P. Doukhan, Mixing: Properties and Examples. xi, 142 pages, 1994.

Vol. 86: W. Vach, Logistic Regression with Missing Values in the Covariates. xi, 139 pages, 1994.

Vol. 87: J. Müller, Lectures on Random Voronoi Tessellations.vii, 134 pages, 1994.

Vol. 88: J. E. Kolassa, Series Approximation Methods in Statistics. Second Edition, ix, 183 pages, 1997.

Vol. 89: P. Cheeseman, R.W. Oldford (Editors), Selecting Models From Data: AI and Statistics IV. xii, 487 pages, 1994.

Vol. 90: A. Csenki, Dependability for Systems with a Partitioned State Space: Markov and Semi-Markov Theory and Computational Implementation. x, 241 pages, 1994.

Vol. 91: J.D. Malley, Statistical Applications of Jordan Algebras. viii, 101 pages, 1994.

Vol. 92: M. Eerola, Probabilistic Causality in Longitudinal Studies. vii, 133 pages, 1994.

Vol. 93: Bernard Van Cutsem (Editor), Classification and Dissimilarity Analysis. xiv, 238 pages, 1994.

Vol. 94: Jane F. Gentleman and G.A. Whitmore (Editors), Case Studies in Data Analysis. viii, 262 pages, 1994.

Vol. 95: Shelemyahu Zacks, Stochastic Visibility in Random Fields. x, 175 pages, 1994.

Vol. 96: Ibrahim Rahimov, Random Sums and Branching Stochastic Processes. viii, 195 pages, 1995.

Vol. 97: R. Szekli, Stochastic Ordering and Dependence in Applied Probability. viii, 194 pages, 1995.

Vol. 98: Philippe Barbe and Patrice Bertail, The Weighted Bootstrap. viii, 230 pages, 1995.

Vol. 99: C.C. Heyde (Editor), Branching Processes: Proceedings of the First World Congress. viii, 185 pages, 1995.

Vol. 100: Wlodzimierz Bryc, The Normal Distribution: Characterizations with Applications. viii, 139 pages, 1995.

Vol. 101: H.H. Andersen, M.Højbjerre, D. Sørensen, P.S.Eriksen, Linear and Graphical Models: for the Multivariate Complex Normal Distribution. x, 184 pages, 1995.

Vol. 102: A.M. Mathai, Serge B. Provost, Takesi Hayakawa, Bilinear Forms and Zonal Polynomials. x, 378 pages, 1995.

Vol. 103: Anestis Antoniadis and Georges Oppenheim (Editors), Wavelets and Statistics. vi, 411 pages, 1995.

Vol. 104: Gilg U.H. Seeber, Brian J. Francis, Reinhold Hatzinger, Gabriele Steckel-Berger (Editors), Statistical Modelling: 10th International Workshop, Innsbruck, July 10-14th 1995. x, 327 pages, 1995.

Vol. 105: Constantine Gatsonis, James S. Hodges, Robert E. Kass, Nozer D. Singpurwalla(Editors), Case Studies in Bayesian Statistics, Volume II. x, 354 pages, 1995.

Vol. 106: Harald Niederreiter, Peter Jau-Shyong Shiue (Editors), Monte Carlo and Quasi-Monte Carlo Methods in Scientific Computing. xiv, 372 pages, 1995.

Vol. 107: Masafumi Akahira, Kei Takeuchi, Non-Regular Statistical Estimation. vii, 183 pages, 1995.

Vol. 108: Wesley L. Schaible (Editor), Indirect Estimators in U.S. Federal Programs. viii, 195 pages, 1995.

Vol. 109: Helmut Rieder (Editor), Robust Statistics, Data Analysis, and Computer Intensive Methods. xiv, 427 pages, 1996.

Vol. 110: D. Bosq, Nonparametric Statistics for Stochastic Processes. xii, 169 pages, 1996.

Vol. 111: Leon Willenborg, Ton de Waal, Statistical Disclosure Control in Practice. xiv, 152 pages, 1996.

Vol. 112: Doug Fischer, Hans-J. Lenz (Editors), Learning from Data. xii, 450 pages, 1996.

Vol. 113: Rainer Schwabe, Optimum Designs for Multi-Factor Models. viii, 124 pages, 1996.

Vol. 114: C.C. Heyde, Yu. V. Prohorov, R. Pyke, and S. T. Rachev (Editors), Athens Conference on Applied Probability and Time Series Analysis Volume I: Applied Probability In Honor of J.M. Gani. viii, 424 pages, 1996.

Vol. 115: P.M. Robinson, M. Rosenblatt (Editors), Athens Conference on Applied Probability and Time Series Analysis Volume II: Time Series Analysis In Memory of E.J. Hannan. viii, 448 pages, 1996.

Vol. 116: Genshiro Kitagawa and Will Gersch, Smoothness Priors Analysis of Time Series. x, 261 pages, 1996.

Vol. 117: Paul Glasserman, Karl Sigman, David D. Yao (Editors), Stochastic Networks. xii, 298, 1996.

Vol. 118: Radford M. Neal, Bayesian Learning for Neural Networks. xv, 183, 1996.

Vol. 119: Masanao Aoki, Arthur M. Havenner, Applications of Computer Aided Time Series Modeling. ix, 329 pages, 1997.

Vol. 120: Maia Berkane, Latent Variable Modeling and Applications to Causality. vi, 288 pages, 1997.

Vol. 121: Constantine Gatsonis, James S. Hodges, Robert E. Kass, Robert McCulloch, Peter Rossi, Nozer D. Singpurwalla (Editors), Case Studies in Bayesian Statistics, Volume III. xvi, 487 pages, 1997.

Vol. 122: Timothy G. Gregoire, David R. Brillinger, Peter J. Diggle, Estelle Russek-Cohen, William G. Warren, Russell D. Wolfinger (Editors), Modeling Longitudinal and Spatially Correlated Data. x, 402 pages, 1997.

Vol. 123: D. Y. Lin and T. R. Fleming (Editors), Proceedings of the First Seattle Symposium in Biostatistics: Survival Analysis. xiii, 308 pages, 1997.

Vol. 124: Christine H. Müller, Robust Planning and Analysis of Experiments. x, 234 pages, 1997.

Vol. 125: Valerii V. Fedorov and Peter Hackl, Model-oriented Design of Experiments. viii, 117 pages, 1997.

Vol. 126: Geert Verbeke and Geert Molenberghs, Linear Mixed Models in Practice: A SAS-Oriented Approach. xiii, 306 pages, 1997.

Vol. 127: Harald Niederreiter, Peter Hellekalek, Gerhard Larcher, and Peter Zinterhof (Editors), Monte Carlo and Quasi-Monte Carlo Methods 1996, xii, 448 pages, 1997.

Vol. 128: L. Accardi and C.C. Heyde (Editors), Probability Towards 2000, x, 356 pages, 1998.

Vol. 129: Wolfgang Härdle, Gerard Kerkyacharian, Dominique Picard, and Alexander Tsybakov, Wavelets, Approximation, and Statistical Applications, xvi, 265 pages, 1998.

Vol. 130: Bo-Cheng Wei, Exponential Family Nonlinear Models, ix, 240 pages, 1998.

Vol. 131: Joel L. Horowitz, Semiparametric Methods in Econometrics, ix, 204 pages, 1998.

Vol. 132: Douglas Nychka, Walter W. Piegorsch, and Lawrence H. Cox (Editors), Case Studies in Environmental Statistics, viii, 200 pages, 1998.

Vol. 133: Dipak Dey, Peter Müller, and Debajyoti Sinha (Editors), Practical Nonparametric and Semiparametric Bayesian Statistics, xv, 408 pages, 1998.

Vol. 134: Yu. A. Kutoyants, Statistical Inference For Spatial Poisson Processes, vii, 284 pages, 1998.

Vol. 135: Christian P. Robert, Discretization and MCMC Convergence Assessment, x, 192 pages, 1998.

Vol. 136: Gregory C. Reinsel, Raja P. Velu, Multivariate Reduced-Rank Regression, xiii, 272 pages, 1998.

Vol. 137: V. Seshadri, The Inverse Gaussian Distribution: Statistical Theory and Applications, xi, 360 pages, 1998.

Vol. 138: Peter Hellekalek, Gerhard Larcher (Editors), Random and Quasi-Random Point Sets, xi, 352 pages, 1998.

Vol. 139: Roger B. Nelsen, An Introduction to Copulas, xi, 232 pages, 1999.

Vol. 140: Constantine Gatsonis, Robert E. Kass, Bradley Carlin, Alicia Carriquiry, Andrew Gelman, Isabella Verdinelli, Mike West (Editors), Case Studies in Bayesian Statistics, Volume IV, xvi, 456 pages, 1999.

Vol. 141: Peter Müller, Brani Vidakovic (Editors), Bayesian Inference in Wavelet Based Models, xi, 394 pages, 1999.

Vol. 142: György Terdik, Bilinear Stochastic Models and Related Problems of Nonlinear Time Series Analysis: A Frequency Domain Approach, xi, 258 pages, 1999.

Vol. 143: Russell Barton, Graphical Methods for the Design of Experiments, x, 208 pages, 1999.

Vol. 144: L. Mark Berliner, Douglas Nychka, and Timothy Hoar (Editors), Case Studies in Statistics and the Atmospheric Sciences, x, 208 pages, 2000.

Vol. 145: James H. Matis and Thomas R. Kiffe, Stochastic Population Models, viii, 220 pages, 2000.

Vol. 146: Wim Schoutens, Stochastic Processes and Orthogonal Polynomials, xiv, 163 pages, 2000.

Vol. 147: Jürgen Franke, Wolfgang Härdle, and Gerhard Stahl, Measuring Risk in Complex Stochastic Systems, xvi, 272 pages, 2000.

Vol. 148: S.E. Ahmed and Nancy Reid, Empirical Bayes and Likelihood Inference, x, 200 pages, 2000.

Vol. 149: D. Bosq, Linear Processes in Function Spaces: Theory and Applications, xv, 296 pages, 2000.

Vol. 150: Tadeusz Caliński and Sanpei Kageyama, Block Designs: A Randomization Approach, Volume I: Analysis, ix, 313 pages, 2000.

Vol. 151: Håkan Andersson and Tom Britton, Stochastic Epidemic Models and Their Statistical Analysis: ix, 152 pages, 2000.

Vol. 152: David Ríos Insua and Fabrizio Ruggeri, Robust Bayesian Analysis: xiii, 435 pages, 2000.

Vol. 153: Parimal Mukhopadhyay, Topics in Survey Sampling, x, 303 pages, 2000.

Vol. 154: Regina Kaiser and Agustín Maravall, Measuring Business Cycles in Economic Time Series, vi, 190 pages, 2000.

Vol. 155: Leon Willenborg and Ton de Waal, Elements of Statistical Disclosure Control, xvii, 289 pages, 2000.

Vol. 156: Gordon Willmot and X. Sheldon Lin, Lundberg Approximations for Compound Distributions with Insurance Applications, xi, 272 pages, 2000.